Teacher Edition

Eureka Math® Grade K Module 5

Special thanks go to the Gordon A. Cain Center and to the Department of Mathematics at Louisiana State University for their support in the development of *Eureka Math.*

For a free *Eureka Math* Teacher
Resource Pack, Parent Tip
Sheets, and more please visit
https://eurekamath.greatminds.org/teacher-resource-pack

Published by the Great Minds

Printed in the U.S.A.
This book may be purchased from the publisher at eureka-math.org
BAB 10 9 8 7 6 5 4 3
ISBN 978-1-63255-819-0

Eureka Math: A Story of Units® Contributors

Katrina Abdussalaam, Curriculum Writer
Tiah Alphonso, Program Manager—Curriculum Production
Kelly Alsup, Lead Writer / Editor, Grade 4
Catriona Anderson, Program Manager—Implementation Support
Debbie Andorka-Aceves, Curriculum Writer
Eric Angel, Curriculum Writer
Leslie Arceneaux, Lead Writer / Editor, Grade 5
Kate McGill Austin, Lead Writer / Editor, Grades PreK–K
Adam Baker, Lead Writer / Editor, Grade 5
Scott Baldridge, Lead Mathematician and Lead Curriculum Writer
Beth Barnes, Curriculum Writer
Bonnie Bergstresser, Math Auditor
Bill Davidson, Fluency Specialist
Jill Diniz, Program Director
Nancy Diorio, Curriculum Writer
Nancy Doorey, Assessment Advisor
Lacy Endo-Peery, Lead Writer / Editor, Grades PreK–K
Ana Estela, Curriculum Writer
Lessa Faltermann, Math Auditor
Janice Fan, Curriculum Writer
Ellen Fort, Math Auditor
Peggy Golden, Curriculum Writer
Maria Gomes, Pre-Kindergarten Practitioner
Pam Goodner, Curriculum Writer
Greg Gorman, Curriculum Writer
Melanie Gutierrez, Curriculum Writer
Bob Hollister, Math Auditor
Kelley Isinger, Curriculum Writer
Nuhad Jamal, Curriculum Writer
Mary Jones, Lead Writer / Editor, Grade 4
Halle Kananak, Curriculum Writer
Susan Lee, Lead Writer / Editor, Grade 3
Jennifer Loftin, Program Manager—Professional Development
Soo Jin Lu, Curriculum Writer
Nell McAnelly, Project Director

Ben McCarty, Lead Mathematician / Editor, PreK–5
Stacie McClintock, Document Production Manager
Cristina Metcalf, Lead Writer / Editor, Grade 3
Susan Midlarsky, Curriculum Writer
Pat Mohr, Curriculum Writer
Sarah Oyler, Document Coordinator
Victoria Peacock, Curriculum Writer
Jenny Petrosino, Curriculum Writer
Terrie Poehl, Math Auditor
Robin Ramos, Lead Curriculum Writer / Editor, PreK–5
Kristen Riedel, Math Audit Team Lead
Cecilia Rudzitis, Curriculum Writer
Tricia Salerno, Curriculum Writer
Chris Sarlo, Curriculum Writer
Ann Rose Sentoro, Curriculum Writer
Colleen Sheeron, Lead Writer / Editor, Grade 2
Gail Smith, Curriculum Writer
Shelley Snow, Curriculum Writer
Robyn Sorenson, Math Auditor
Kelly Spinks, Curriculum Writer
Marianne Strayton, Lead Writer / Editor, Grade 1
Theresa Streeter, Math Auditor
Lily Talcott, Curriculum Writer
Kevin Tougher, Curriculum Writer
Saffron VanGalder, Lead Writer / Editor, Grade 3
Lisa Watts-Lawton, Lead Writer / Editor, Grade 2
Erin Wheeler, Curriculum Writer
MaryJo Wieland, Curriculum Writer
Allison Witcraft, Math Auditor
Jessa Woods, Curriculum Writer
Hae Jung Yang, Lead Writer / Editor, Grade 1

Board of Trustees

Mathematics Curriculum

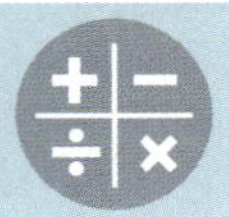

Table of Contents

GRADE K • MODULE 5

Numbers 10–20 and Counting to 100

GK-M5-TE-B5-1.3.1-01.2016

Grade K • Module 5

Numbers 10–20 and Counting to 100

OVERVIEW

Students have worked intensively within 10 and have often counted to 30 using the Rekenrek during Fluency Practice. This sets the stage for Module 5, where students clarify the meaning of the 10 ones and some ones within a teen number and extend that understanding to count to 100. In Topic A, students start at the concrete level, counting 10 straws.

T: Count straws with me into piles of ten.

S: 1, 2, 3, 4, 5, 6, 7, 8, 9, 10. 1, 2, 3, 4, 5, 6, 7, 8, 9, 10. 1, 2, 3, ..., 8, 9, 10. 1, 2, 3, ..., 8, 9, 10.

T: Let's count the piles!

S: 1 pile, 2 piles, 3 piles, 4 piles.

Thus, Kindergarten students learn to comfortably talk about 10 ones, setting the foundation for the critical Grade 1 step of understanding 1 ten. They next separate 10 objects from within concrete and pictorial counts up to 20, analyzing the total as 10 ones and no ones or 10 ones and some ones (**K.CC.1, K.NBT.1**). They see two distinct sets which are then counted the Say Ten way: ten 1, ten 2, ten 3, ten 4, ten 5, ten 6, ten 7, ten 8, ten 9, 2 tens. Students hear the separation of the 10 ones and some ones as they count, solidifying their understanding as they also return to regular counting: eleven, twelve, thirteen, ..., etc. (**K.CC.5**)

In Topic B, the two distinct sets of ones are composed, or brought together, through the use of the Hide Zero cards (pictured below) and number bonds. Students represent the whole number numerically while continuing to separate the count of 10 ones from the count of the remaining ones with drawings and materials (**K.NBT.1**). Emerging from Topic B, students should be able to model and write a teen number without forgetting that the 1 in 13 represents 10 ones (**K.CC.3**).

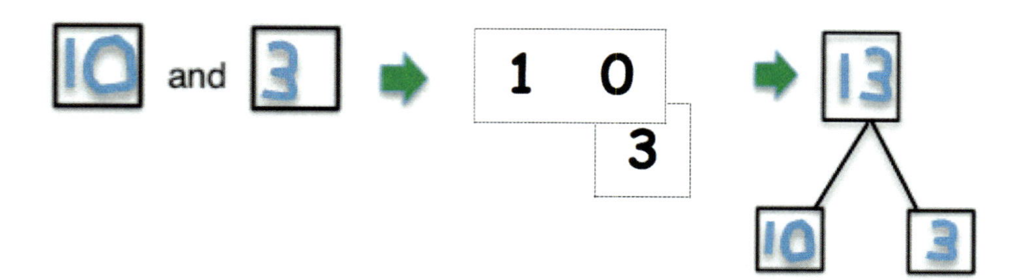

Topic C opens with students making a simple Rekenrek to 20 (pictured below) and modeling numbers thereon. The tens can be seen both as two lines with a color change at the five or two parallel unicolor fives.

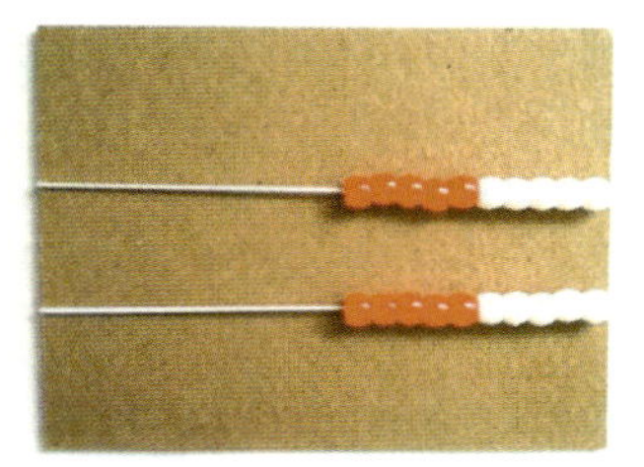

In Topic C, the focus is now on the decomposition of the total teen quantity so that one part is ten ones. This is what makes Topic C a step forward from Topics A and B. Previously, the ten and ones were always separated when modeled pictorially or with materials. Now, the entire teen number is a whole quantity represented both concretely and pictorially in different configurations: towers or linear configurations, arrays (including the 10-frame or 5-groups), and circles. Students decompose the total into 10 ones and some ones. Through their experiences with the different configurations, students have practice both separating 10 ones within teen numbers and counting or conservation as they count quantities arranged in different ways and, as always, use math talk to share their observations (**K.CC.5**). They also come to know each successive teen number as one larger than the previous number (**K.CC.4a**).

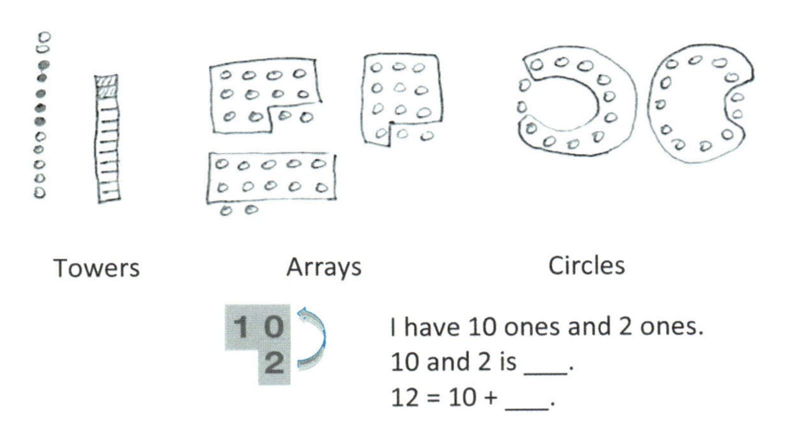
Towers
Arrays
Circles
10
2
I have 10 ones and 2 ones.
10 and 2 is ___.
12 = 10 + ___.

In Topic D, students extend their understanding of counting teen numbers to numbers from 21 to 100. They first count by tens both the Say Ten way—1 ten, 2 tens, 3 tens, 4 tens, etc.—and the regular way: twenty, thirty, forty, etc. They then count by ones to 100, first within a decade and finally across the decade (**K.CC.1, K.CC.2**). Topic D involves the Grade 1 standard **1.NBT.1** because students also write their numbers from 21 to 100.

The writing of larger numbers has been included because of the range of activities they make possible. The writing of these numbers is not assessed nor emphasized, however. Topic D closes with an optional exploration of numbers on the Rekenrek, bringing together counting with decomposition and finding embedded numbers within larger numbers. This lesson is optional because it does not directly address a particular Kindergarten standard.

In Topic E, students apply their skill with the decomposition and composition of teen numbers. In Lesson 20, they represent both compositions and decompositions as addition statements (**K.NBT.1**). In Lesson 21, they model teen quantities with materials in a number bond and hide one part. The hidden part is represented as an addition sentence with a hidden part (e.g., 10 + ___ = 13 or 13 = ___ + 3). The missing addend aligns Lesson 21 to the Grade 1 standard **1.OA.8**. In Lesson 22, students apply their skill with decomposition into 10 ones and some ones to compare the some ones of two numbers and thus to compare the teen numbers. They *stand* on the structure of the 10 ones and use what they know of numbers 1–9 (MP.7). Comparison of numbers 1–9 is a Kindergarten standard (**K.CC.6, K.CC.7**).

In Lesson 23, students reason about situations to determine whether they are decomposing a teen number (as 10 ones and some ones) or composing 10 ones and some ones to find a teen number. They analyze their number sentences that represent each situation to determine if they started with the total or the parts and if they composed or decomposed, for example, 13 = 10 + 3 or 10 + 3 = 13 (**K.NBT.1**). Throughout the lesson, students draw the number of objects presented in the situation (**K.CC.5**).

The module closes with a culminating task, wherein students integrate all the methods they have used up until now to show decomposition. For example, they are instructed, "Open your mystery bag. Show the number of objects in your bag in different ways using the materials you choose" (MP.5). This experience also serves as a part of the End-of-Module Assessment, allowing students to demonstrate skill and understanding using all they have learned throughout the module.

Notes on Pacing for Differentiation

If pacing is a challenge, consider the following modifications and omissions. Consider collaborating with a specialist teacher to have students build the Rekenrek from Lesson 10 (e.g., make a Rekenrek in art, practice counting in foreign language class), or plan an event to engage families in math activities such as these.

If writing numbers 21–100 overwhelms students, omit the Problem Sets in Lessons 15, 16, and 17. Instead, complete the verbal counting activities in the lessons that prepare them for numeral writing to 100 as required in Grade 1. This allows for the completion of these three lessons in just one or two days.

Lesson 19 is exploratory in nature and addresses some standards beyond the level of Kindergarten. It works well as an extension lesson if students are advancing quickly, but if pacing is a challenge, it could be omitted.

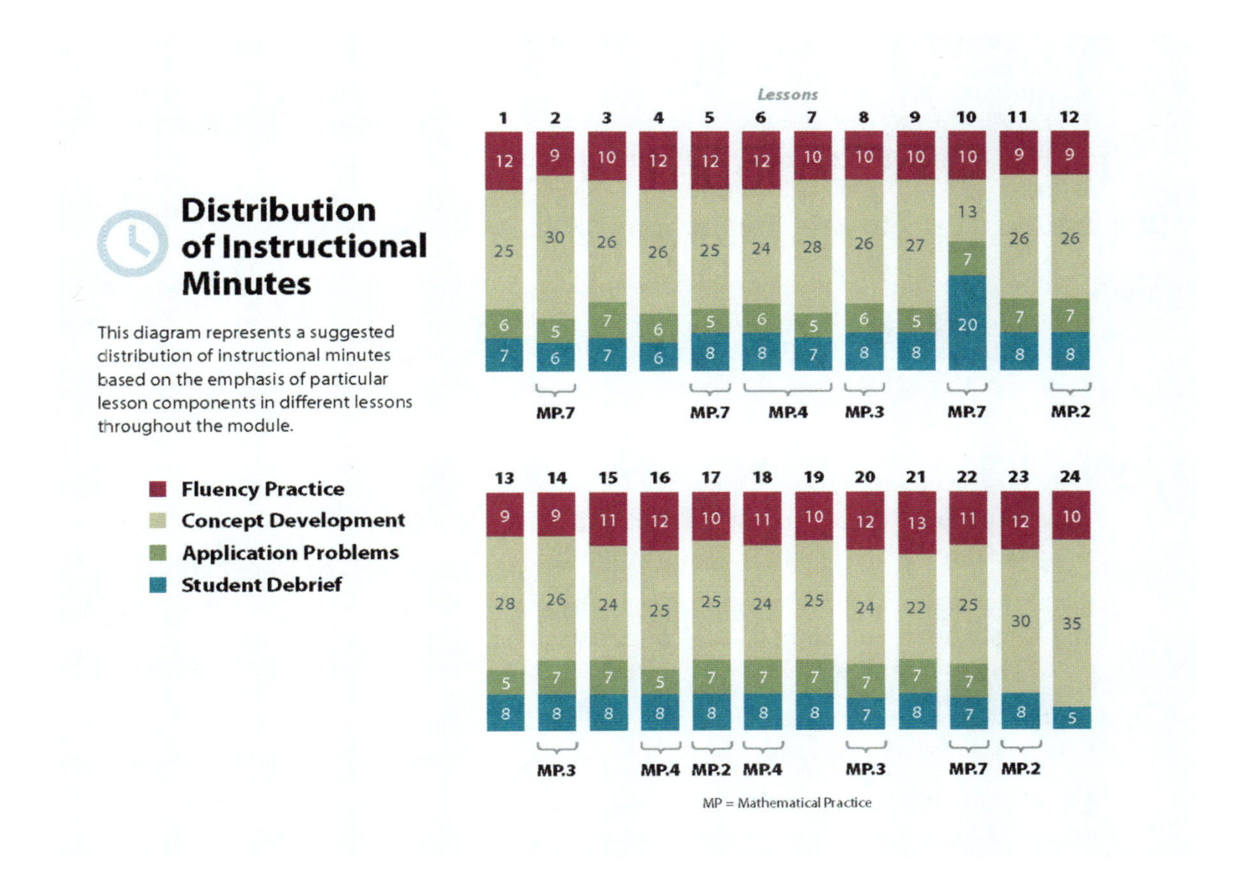

Focus Grade Level Standards

Know number names and the count sequence.

K.CC.1 Count to 100 by ones and by tens.

K.CC.2 Count forward beginning from a given number within the known sequence (instead of having to begin at 1).

K.CC.3 Write numbers from 0 to 20. Represent a number of objects with a written numeral 0–20 (with 0 representing a count of no objects).

Count to tell the number of objects.[1]

K.CC.4 Understand the relationship between numbers and quantities; connect counting to cardinality.

b. Understand that the last number name said tells the number of objects counted. The number of objects is the same regardless of their arrangement or the order in which they were counted.

[1]K.CC.4a is addressed in Module 1; K.CC.4d is addressed in Module 6.

c. Understand that each successive number name refers to a quantity that is one larger.

K.CC.5 Count to answer "how many?" questions about as many as 20 things arranged in a line, a rectangular array, or a circle, or as many as 10 things in a scattered configuration; given a number from 1–20, count out that many objects.

Work with numbers 11–19 to gain foundations for place value.

K.NBT.1 Compose and decompose numbers from 11 to 19 into ten ones and some further ones, e.g., by using objects or drawings, and record each composition or decomposition by a drawing or equation (e.g., 18 = 10 + 8); understand that these numbers are composed of ten ones and one, two, three, four, five, six, seven, eight, or nine ones.

Focus Standards for Mathematical Practice

MP.2 **Reason abstractly and quantitatively**. Students represent teen numerals with concrete objects separated as 10 ones and some ones.

MP.3 **Construct viable arguments and critique the reasoning of others**. Students explain their thinking about teen numbers as 10 ones and some ones and how to represent those numbers as addition sentences.

MP.4 **Model with mathematics**. Students model teen quantities with number bonds, place value cards, and teen numbers.

MP.7 **Look for and make use of structure**. Students use the structure of 10 ones to reason about teen numbers. They compare teen numbers using the structure of the 10 ones to compare the some ones.

Overview of Module Topics and Lesson Objectives

Standards		Topics and Objectives		Days
K.CC.1 **K.NBT.1** K.CC.2 K.CC.4a K.CC.4b K.CC.4c K.CC.5	A	**Count 10 Ones and Some Ones**		5
		Lesson 1:	Count straws into piles of ten; count the piles as 10 ones.	
		Lesson 2:	Count 10 objects within counts of 10 to 20 objects, and describe as 10 ones and ___ ones.	
		Lesson 3:	Count and circle 10 objects within images of 10 to 20 objects, and describe as 10 ones and ___ ones.	
		Lesson 4:	Count straws the Say Ten way to 19; make a pile for each ten.	
		Lesson 5:	Count straws the Say Ten way to 20; make a pile for each ten.	

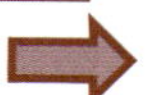

Standards		Topics and Objectives	Days
K.CC.3 **K.NBT.1** K.CC.1 K.CC.2 K.CC.4a K.CC.4b K.CC.4c K.CC.5	B	**Compose Numbers 11–20 from 10 Ones and Some Ones; Represent and Write Teen Numbers** Lesson 6: Model with objects and represent numbers 10 to 20 with place value or Hide Zero cards. Lesson 7: Model and write numbers 10 to 20 as number bonds. Lesson 8: Model teen numbers with materials from abstract to concrete. Lesson 9: Draw teen numbers from abstract to pictorial.	4
K.CC.4b **K.CC.4c** **K.CC.5** **K.NBT.1** K.CC.3 K.CC.4a	C	**Decompose Numbers 11–20, and Count to Answer "How Many?" Questions in Varied Configurations** Lesson 10: Build a Rekenrek to 20. Lesson 11: Show, count, and write numbers 11 to 20 in tower configurations increasing by 1—a pattern of *1 larger*. Lesson 12: Represent numbers 20 to 11 in tower configurations decreasing by 1—a pattern of *1 smaller*. Lesson 13: Show, count, and write to answer *how many* questions in linear and array configurations. Lesson 14: Show, count, and write to answer *how many* questions with up to 20 objects in circular configurations.	5
		Mid-Module Assessment: Topics A–C (Interview-style assessment)	3
K.CC.1 **K.CC.2** K.CC.3 K.CC.4c K.CC.5 K.NBT.1 1.NBT.1[2]	D	**Extend the Say Ten and Regular Count Sequence to 100** Lesson 15: Count up and down by tens to 100 with Say Ten and regular counting. Lesson 16: Count within tens by ones. Lesson 17: Count across tens when counting by ones through 40. Lesson 18: Count across tens by ones to 100 with and without objects. Lesson 19: Explore numbers on the Rekenrek. (Optional)	5

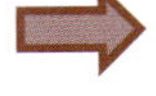

[2]Students write numbers 21–100, aligned to Grade 1 standard 1.NBT.1.

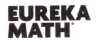

Standards		Topics and Objectives		Days
K.CC.5 **K.NBT.1** K.CC.1 K.CC.2 K.CC.3 K.CC.4c K.CC.6 1.OA.8[3] 1.NBT.3[4]	E	**Represent and Apply Compositions and Decompositions of Teen Numbers** Lesson 20: Lesson 21: Lesson 22: Lesson 23: Lesson 24:	 Represent teen number compositions and decompositions as addition sentences. Represent teen number decompositions as 10 ones and some ones, and find a hidden part. Decompose teen numbers as 10 ones and some ones; compare *some ones* to compare the teen numbers. Reason about and represent situations, decomposing teen numbers into 10 ones and some ones and composing 10 ones and some ones into a teen number. Culminating Task—Represent teen number decompositions in various ways.	5
		End-of-Module Assessment: Topics D–E (Interview-style assessment)		3
Total Number of Instructional Days				**30**

Terminology

New or Recently Introduced Terms

Hide Zero card (front)

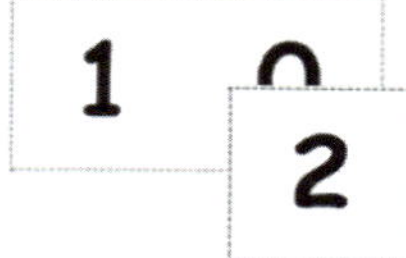

Hide Zero card (back)

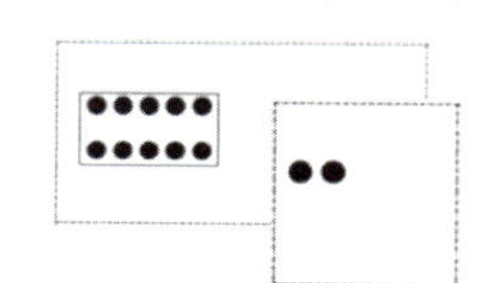

- 10 and __
- 10 ones and some ones
- 10 plus
- Hide Zero cards (called Place Value cards in later grades, pictured to the right)
- Regular counting by ones from 11 to 20 (eleven, twelve, thirteen, etc.)
- Regular counting by tens to 100 (e.g., ten, twenty, thirty, forty, fifty, sixty, seventy, eighty, ninety, one hundred)
- Say Ten counting by tens to 100 (e.g., 1 ten, 2 tens, 3 tens, 4 tens, 5 tens, 6 tens, 7 tens, 8 tens, 9 tens, 10 tens)
- Teen numbers

Familiar Terms and Symbols[5]

- 10-frame
- 5-group

[3]While using concrete materials, a hidden part is related to 10 + ___. Missing addends are aligned to 1.OA.8.
[4]Kindergarten standards K.CC.6 and K.CC.7 compare numbers to 10. Grade 1's standard 1.NBT.3 compares numbers greater than 10.
[5]These are terms and symbols students have used or seen previously.

- Circle 10 ones
- Circular count
- Count 10 ones
- Dot path, empty path, number path
- Linear count
- Number bond
- Number tower
- Part, whole, total
- Say Ten counting (e.g., 11–20 is spoken as "ten one, ten two, ten three, ten four, ten five, ten six, ten seven, ten eight, ten nine, two tens")
- Scatter count

Suggested Tools and Representations

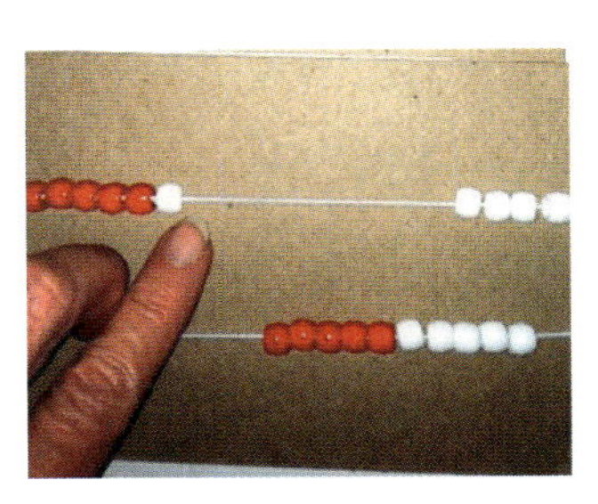

- 50 sticks or straws for each group of 2 students
- Student-made Rekenrek (pictured to the right): 10 red and 10 white pony beads, 1 cardboard strip, 2 elastics
- 1 egg carton per pair of students with 2 slots cut off to make a carton with 10 slots
- Hide Zero cards (called Place Value cards in later grades)
- Objects to put in the egg carton such as mandarin oranges, plastic eggs, or beans
- Single and double 10-frames
- Linking cubes: ideally 10 of two different colors per student
- Number bond template

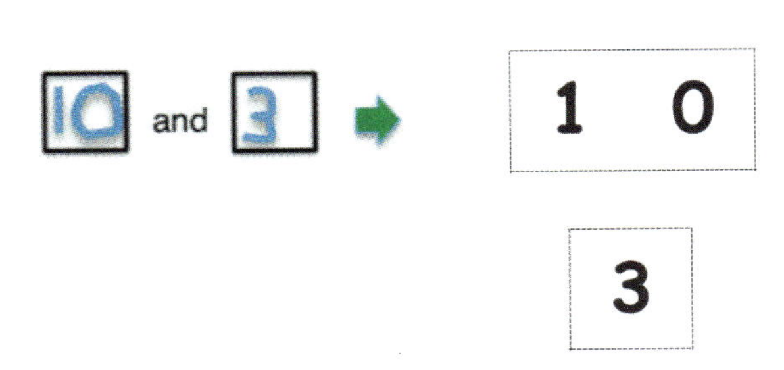

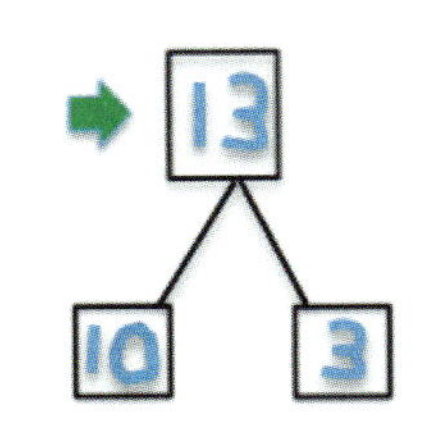

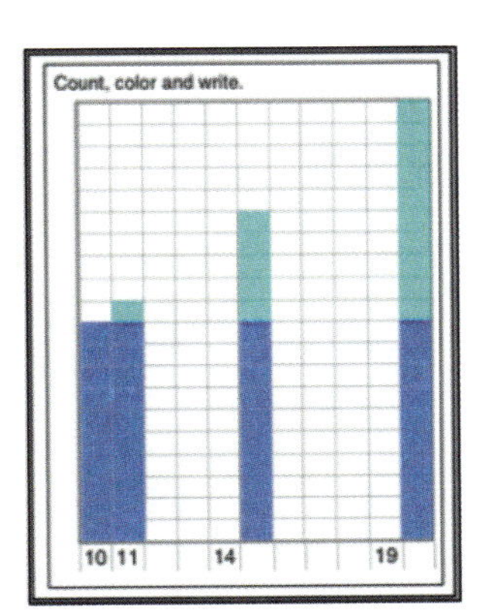

Homework

Homework at the K–1 level is not a convention in all schools. In this curriculum, homework is an opportunity for additional practice of the content from the day's lesson. The teacher is encouraged, with the support of parents, administrators, and colleagues, to discern the appropriate use of homework for his students. Fluency exercises can also be considered as an alternative homework assignment.

Scaffolds[6]

The scaffolds integrated into *A Story of Units*® give alternatives for how students access information as well as express and demonstrate their learning. Strategically placed margin notes are provided within each lesson elaborating on the use of specific scaffolds at applicable times. They address many needs presented by English language learners, students with disabilities, students performing above grade level, and students performing below grade level. Many of the suggestions are organized by Universal Design for Learning (UDL) principles and are applicable to more than one population. To read more about the approach to differentiated instruction in *A Story of Units*, please refer to "How to Implement *A Story of Units*."

Assessment Summary

Assessment Type	Administered	Format	Standards Addressed
Mid-Module Assessment Task	After Topic C	Interview with Rubric	K.CC.1 K.CC.3 K.CC.4bc K.CC.5 K.NBT.1
End-of-Module Assessment Task	After Topic E	Interview with Rubric	K.CC.1 K.CC.2 K.CC.3 K.CC.4 K.CC.5 K.NBT.1
Culminating Task	Last Instructional Day, Lesson 24	Cooperative Group Task	K.NBT.1

[6]Students with disabilities may require Braille, large print, audio, or special digital files. Please visit the website www.p12.nysed.gov/specialed/aim for specific information on how to obtain student materials that satisfy the National Instructional Materials Accessibility Standard (NIMAS) format.

Topic A

Count 10 Ones and Some Ones

K.CC.1, K.NBT.1, K.CC.2, K.CC.4a, K.CC.4b, K.CC.4c, K.CC.5

Focus Standards:		K.CC.1	Count to 100 by ones and by tens.
		K.NBT.1	Compose and decompose numbers from 11 to 19 into ten ones and some further ones, e.g., by using objects or drawings, and record each composition or decomposition by a drawing or equation (e.g., 18 = 10 + 8); understand that these numbers are composed of ten ones and one, two, three, four, five, six, seven, eight, or nine ones.
Instructional Days:		5	
Coherence	**-Links from:**	GPK–M5	Addition and Subtraction Stories and Counting to 20
	-Links to:	G1–M2	Introduction to Place Value Through Addition and Subtraction Within 20

In Topic A, students count two separate parts within teen numbers: 10 ones and some ones. They start by counting piles of 10 straws to understand 10 ones. In Lesson 2, students separate 10 ones and some ones from within teen quantities using an egg carton cut off to have 10 compartments. Continuing with decomposing, in Lesson 3, students circle 10 ones within teen quantities at the pictorial level. In Lessons 4 and 5, students count their 10 ones and some ones to 20 the Say Ten way (e.g., ten 1, ten 2, ten 3, ten 4, ten 5, ten 6, ten 7, ten 8, ten 9, 2 tens).[1]

[1] In the NBT Progression on page 5, this is referred to as the East Asian way of counting.

A Teaching Sequence Toward Mastery of Counting 10 Ones and Some Ones	
Objective 1:	**Count straws into piles of ten; count the piles as 10 ones.** **(Lesson 1)**
Objective 2:	**Count 10 objects within counts of 10 to 20 objects, and describe as 10 ones and ___ ones.** **(Lesson 2)**
Objective 3:	**Count and circle 10 objects within images of 10 to 20 objects, and describe as 10 ones and ___ ones.** **(Lesson 3)**
Objective 4:	**Count straws the Say Ten way to 19; make a pile for each ten.** **(Lesson 4)**
Objective 5:	**Count straws the Say Ten way to 20; make a pile for each ten.** **(Lesson 5)**

Lesson 1

Objective: Count straws into piles of ten; count the piles as 10 ones.

Suggested Lesson Structure

Fluency Practice	(12 minutes)
Application Problem	(6 minutes)
Concept Development	(25 minutes)
Student Debrief	(7 minutes)
Total Time	**(50 minutes)**

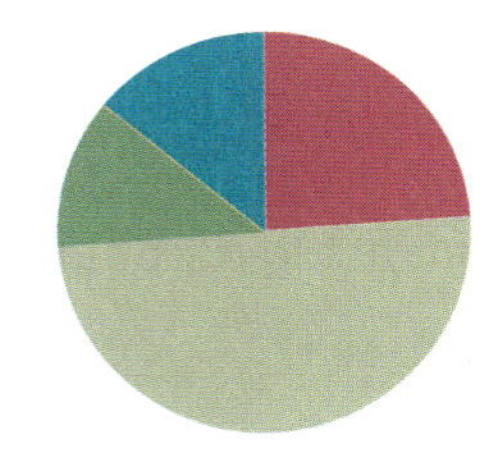

Fluency Practice (12 minutes)

- Finger Counting from Left to Right **K.CC.2, K.CC.4** (2 minutes)
- 5-Group Flashes: Partners to 5 **K.CC.1, K.CC.4** (4 minutes)
- 5-Group Flashes: Partners to 10 **K.CC.2** (6 minutes)

Finger Counting from Left to Right (2 minutes)

Note: This variation of counting the Math Way maintains students' abilities to model counting sequences within 10 on fingers.

Count by ones within 10 on fingers from left to right, from pinky on the left hand as 1 to pinky on the right hand as 10.

Hover the fingers as if playing the piano. Drop each finger as it is counted, and leave it down. Start and end at different numbers (e.g., count from 5 to 7). (The five fingers of the left hand have played. Students say, "6, 7," while playing the thumb and pointer finger of the right hand.)

5-Group Flashes: Partners to 5 (4 minutes)

Materials: (T) Large 5-group cards (Fluency Template 1)
(S) 5-group cards (Fluency Template 2)

Note: Reviewing compositions of 5 leads to proficiency with the core fluency for the grade, **K.OA.5**, add and subtract within 5.

T: (Show 4 dots.) How many dots do you see?
S: 4.
T: How many more to make 5?

NOTES ON MULTIPLE MEANS OF ENGAGEMENT:

To help English language learners develop oral language skills, alternate between choral response and written response. Provide personal white boards for students to write the answer during frame flashes.

S: 1.
T: Say the number sentence.
S: 4 and 1 makes 5.

Continue with the following possible sequence: 3, 2, 1, 4, 2, 3, 5, 0, 5. Have students play with a partner. Give pairs sets of 5-group cards.

5-Group Flashes: Partners to 10 (6 minutes)

Materials: (T) Large 5-group cards (Fluency Template 1) (S) 5-group cards (Fluency Template 2)

Note: Reviewing partners to 10 prepares students to decompose 10 in the Application Problem.

T: (Show 9 dots.) How many dots do you see?
S: 9.
T: How many more does nine need to be 10?
S: 1.

Repeat for possible sequence: 8, 5, 7, 6, 1, 4, 3, 5, 2, 9. Have students play with a partner. Give pairs sets of cards.

Application Problem (6 minutes)

Marta loves to share her peanuts at recess. She counted 10 peanuts into the hands of her friend Joey. Draw a picture of the peanuts in Joey's hands.

Note: There is more than one possible solution to this problem.

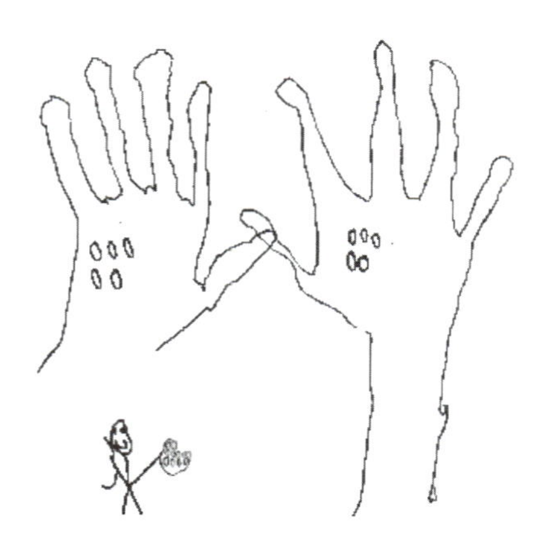

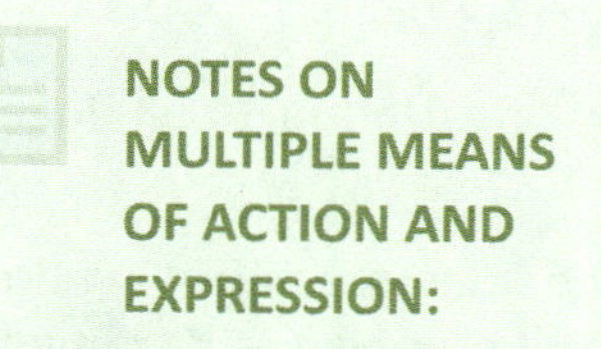

NOTES ON MULTIPLE MEANS OF ACTION AND EXPRESSION:

For students who are working above grade level, provide extensions to the Application Problem such as:

1. If Marta had 15 peanuts to start with, how many does she have left?
2. How many more peanuts does Marta need to have 10 in her hand?
3. Draw a picture to show Marta's peanuts.

Concept Development (25 minutes)

Materials: (S) 1 egg carton cut to have 10 compartments for each pair of students, 10 bags with different items in each (suggestions to the right), 40 straws

Bag Contents:
- 8 clothespins
- 8 pasta shells
- 8 beads
- 9 3-inch by 5-inch cards
- 9 pennies
- 9 crayons
- 10 erasers
- 10 linking cubes
- 10 walnuts in the shell
- 10 play dollars

T: Count to find out how many slots there are in your egg carton. Wait for the signal to tell me. (Pause. When all are ready, give the signal.)
S: 10.
T: Each team will explore 10 bags. Find out which bags have 10 things in them.

Have students in pairs investigate each bag by placing the materials into the egg carton to see if there are enough to count 10 ones. After counting the items in the bag, students will pass it to the next pair on a signal.

T: (Allow time for students to investigate all 10 bags.) Discuss with the pair next to you, which bags had 10 things?
S: The erasers, the linking cubes, the walnuts, and the play dollars!
T: How many times did we count 10 things?
S: 4 times!
T: Now, we are going to count these straws into 4 piles of 10 to match the erasers, linking cubes, walnuts, and play dollars.
T: Count with me to match the number of erasers.
S: 1, 2, 3, 4, 5, 6, 7, 8, 9, 10.
T: 1 pile! Let's count another pile to match the number of linking cubes.
S: 1, 2, 3, 4, 5, 6, 7, 8, 9, 10.
T: How many piles of 10 do we have now?
S: 2 piles!
T: Let's count another pile to match the number of walnuts.

Continue with the walnuts and play dollars.

T: Let's count how many piles of 10 we made.
S: 1 pile, 2 piles, 3 piles, 4 piles.
T: How many straws are in each pile?
S: 10 straws.
T: Let's count the bags of 10, too.
S: 1 bag, 2 bags, 3 bags, 4 bags.
T: How many things are in each bag?
S: 10 things.

T: Talk to your partner about what is the same and different about the bags of things and the piles of straws.

T: (Allow time.) How many times did we count **10 ones** when we were counting the straws?

S: 4.

T: How many times did we count 10 things when we were counting the things in the bags?

S: 4.

T: How many of the bags didn't have 10 things?

S: 6 bags!

Problem Set (5 minutes)

Students should do their personal best to complete the Problem Set within the allotted time.

Have students circle the pictures that show 10 things.

Note: Students have been counting linear, array, circular, and scatter configurations through 10 since the first module (**K.CC.5**). They have further developed skill in circling pictorial sets in Module 4 when learning to add and subtract.

Student Debrief (7 minutes)

Lesson Objective: Count straws into piles of ten; count the piles as 10 ones.

The Student Debrief is intended to invite reflection and active processing of the total lesson experience.

Invite students to review their solutions for the Problem Set. They should check work by comparing answers with a partner before going over answers as a class. Look for misconceptions or misunderstandings that can be addressed in the Debrief.

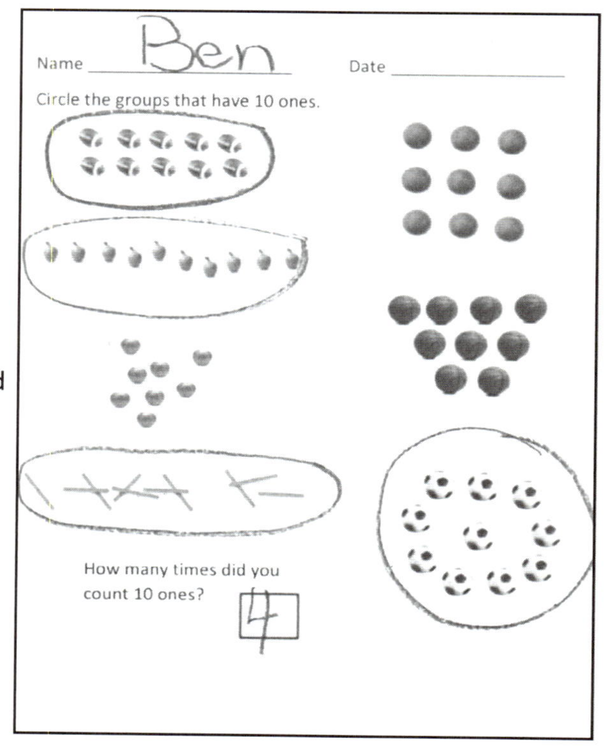

- Have students bring their Problem Set to the meeting area and discuss with a partner which things they circled and why. Suggested sentence frames:

  "I circled __________ because I counted 10 of them."
  "I didn't circle __________ because I counted _____ of them."

- Have them count the number of sets of 10 ones they counted.
- Help students to remember that there were also 4 piles of 10 straws and 4 bags with 10 things in them. Have them discuss how the Problem Set is the same as and different from their work with the bags and straws. Would you ever put apples or soccer balls in bags of 10?

- To review and apply **K.OA.4**, discuss how many objects the other groups are missing to make 10. Have students draw in the missing objects and circle all the sets of 10 ones. "Now, how many times did we count 10 ones?"

Exit Ticket (3 minutes)

After the Student Debrief, instruct students to complete the Exit Ticket. A review of their work will help with assessing students' understanding of the concepts that were presented in today's lesson and planning more effectively for future lessons. The questions may be read aloud to students.

Homework

Homework at the K–1 level is not a convention in all schools. In this curriculum, homework is an opportunity for additional practice of the content from the day's lesson. The teacher is encouraged, with the support of parents, administrators, and colleagues, to discern the appropriate use of homework for his or her students. Fluency exercises can also be considered as an alternative homework assignment.

Name ______________________________ Date ______________

Circle the groups that have 10 ones.

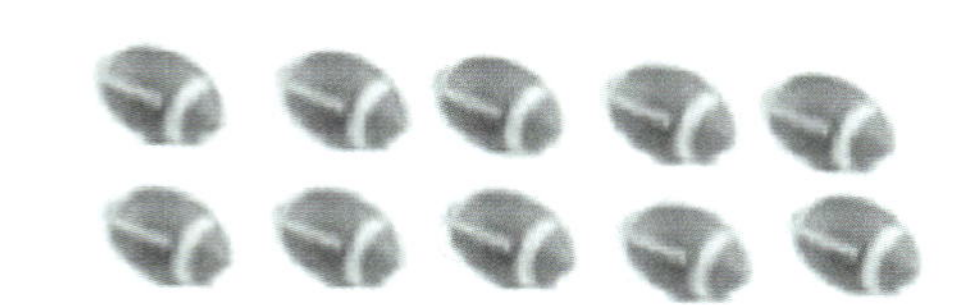

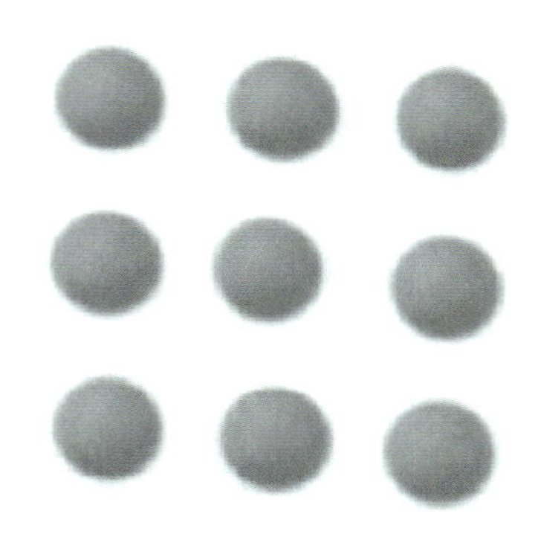

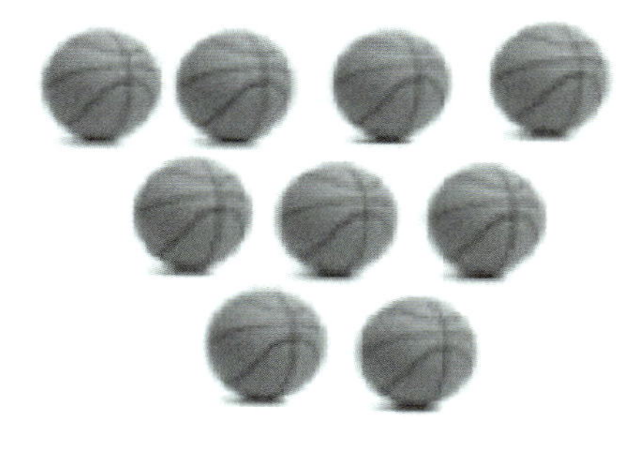

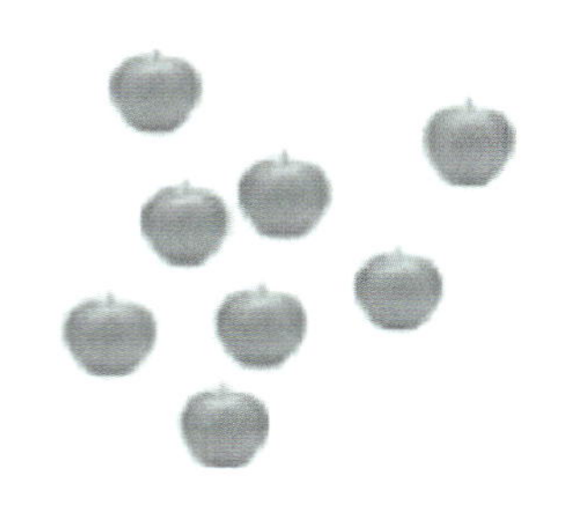

How many times did you count 10 ones?

Name ____________________ Date ____________

Circle the groups that have 10 things.

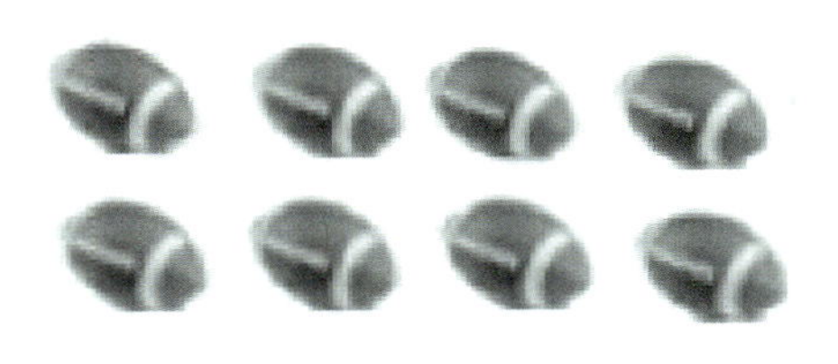

How many times did you count 10 things?

Name ______________________________ Date ______________

Circle 10.

Count the number of times you circled 10 ones. Tell a friend or an adult how many times you circled 10 ones.

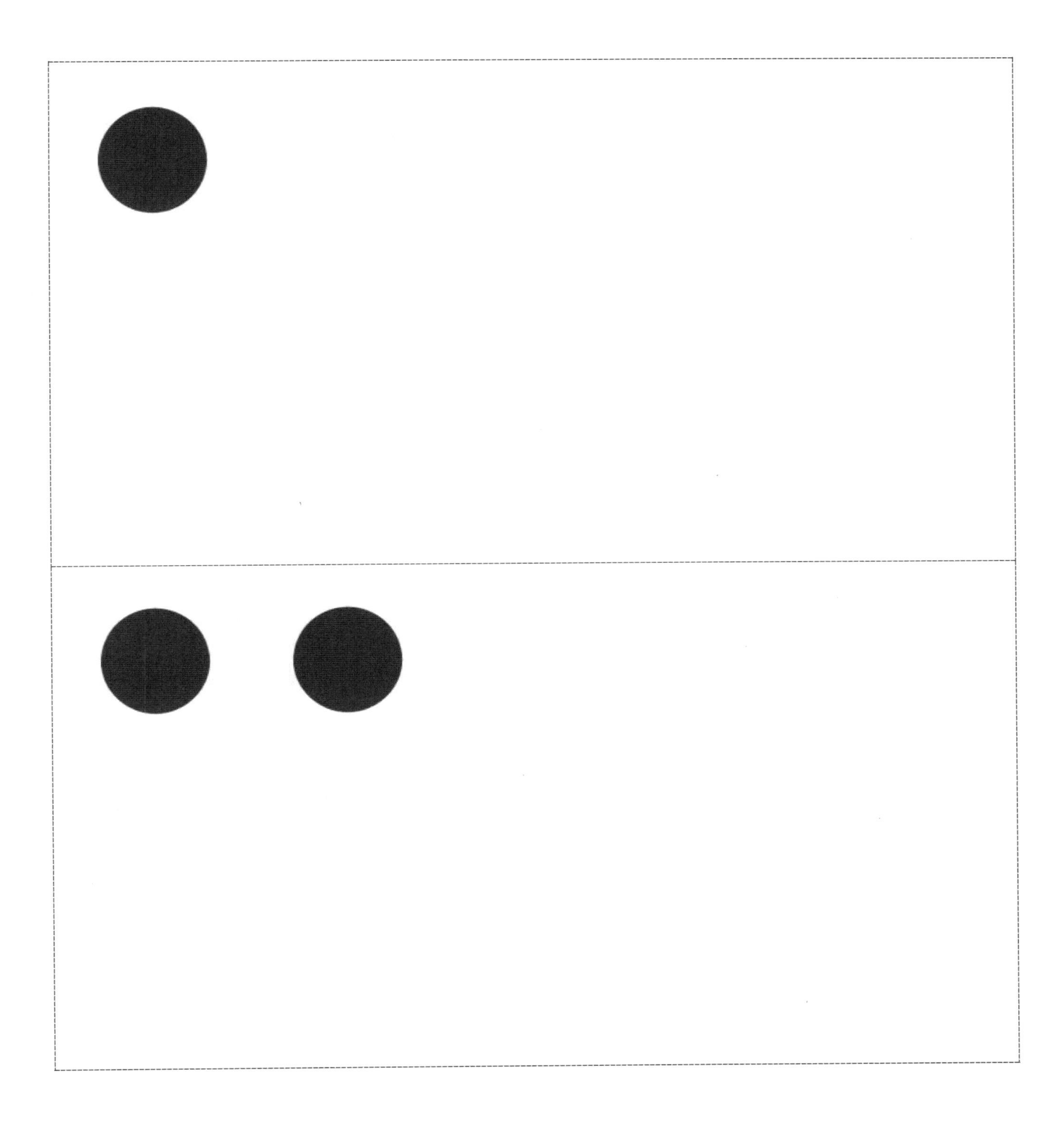

large 5-group cards (Copy on card stock and cut. Save the full set.)

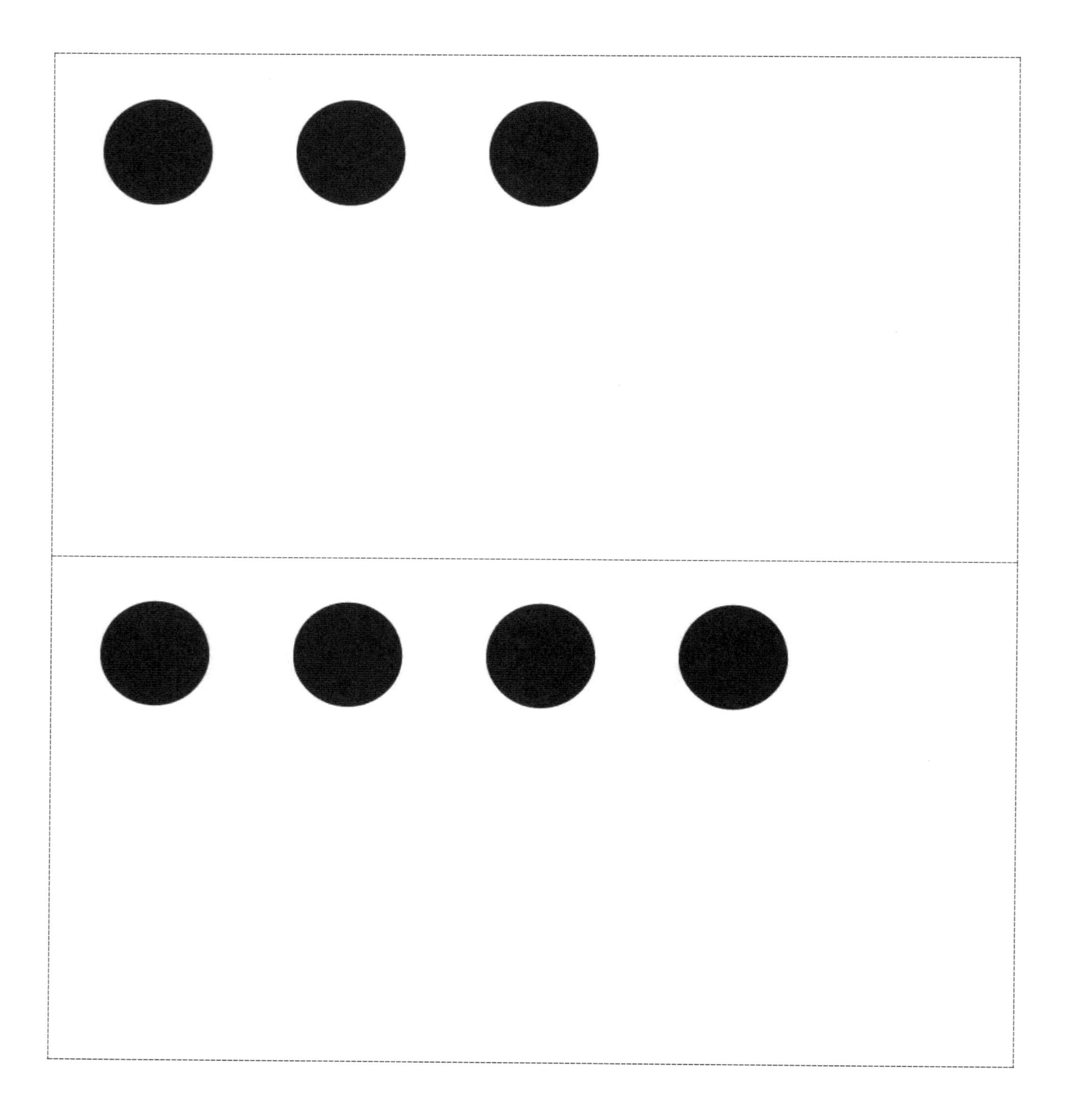

large 5-group cards (Copy on card stock and cut. Save the full set.)

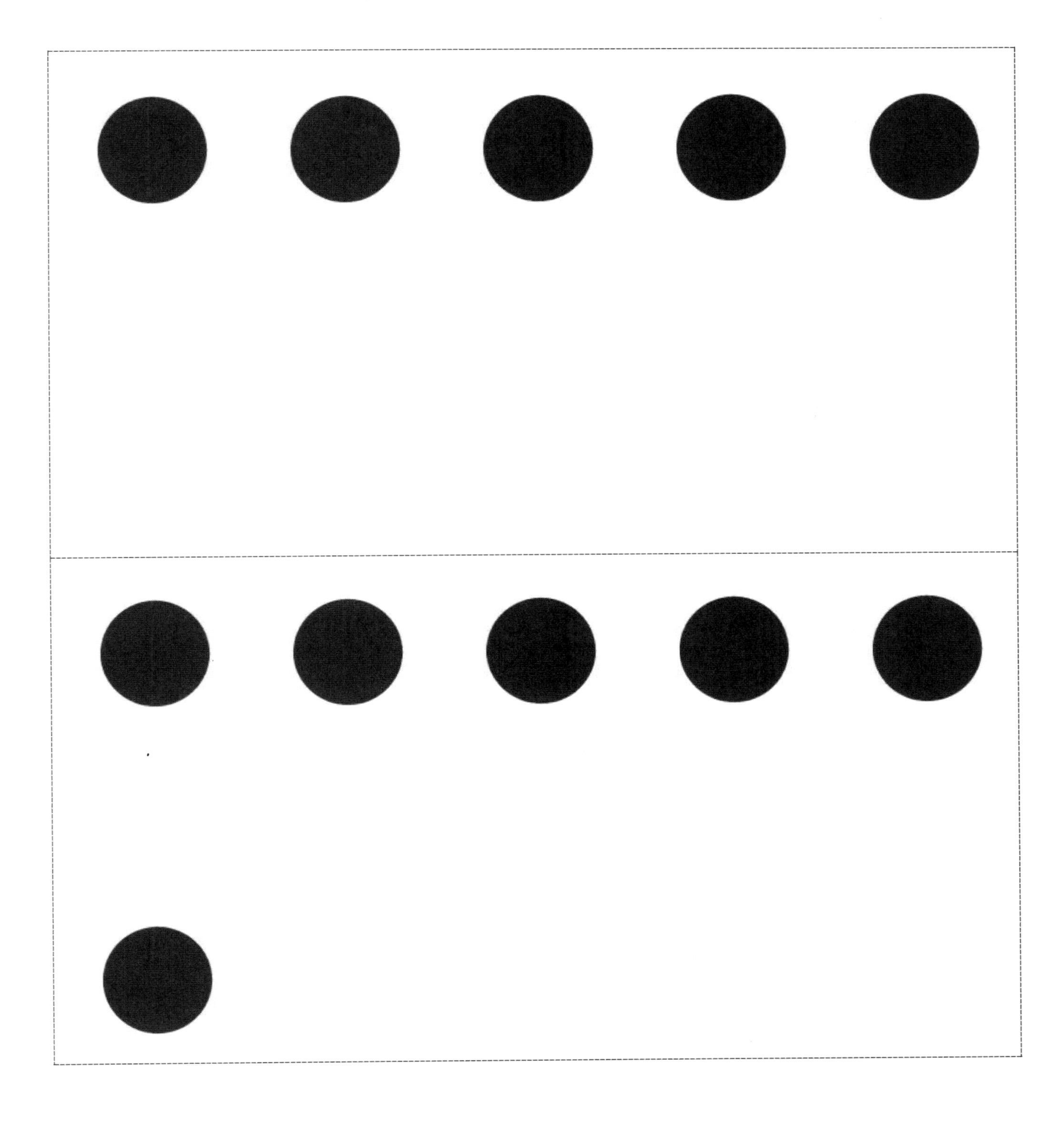

large 5-group cards (Copy on card stock and cut. Save the full set.)

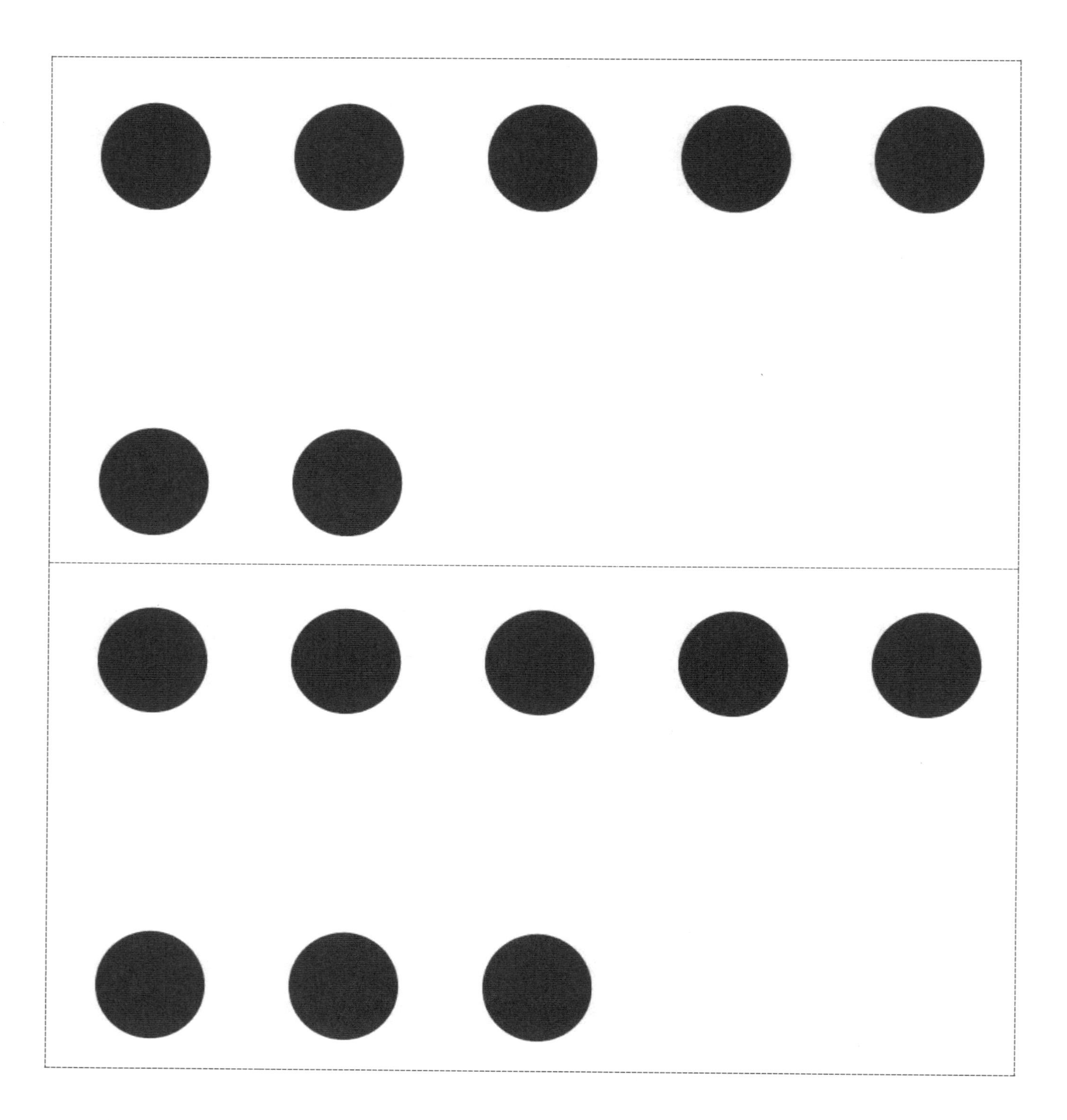

large 5-group cards (Copy on card stock and cut. Save the full set.)

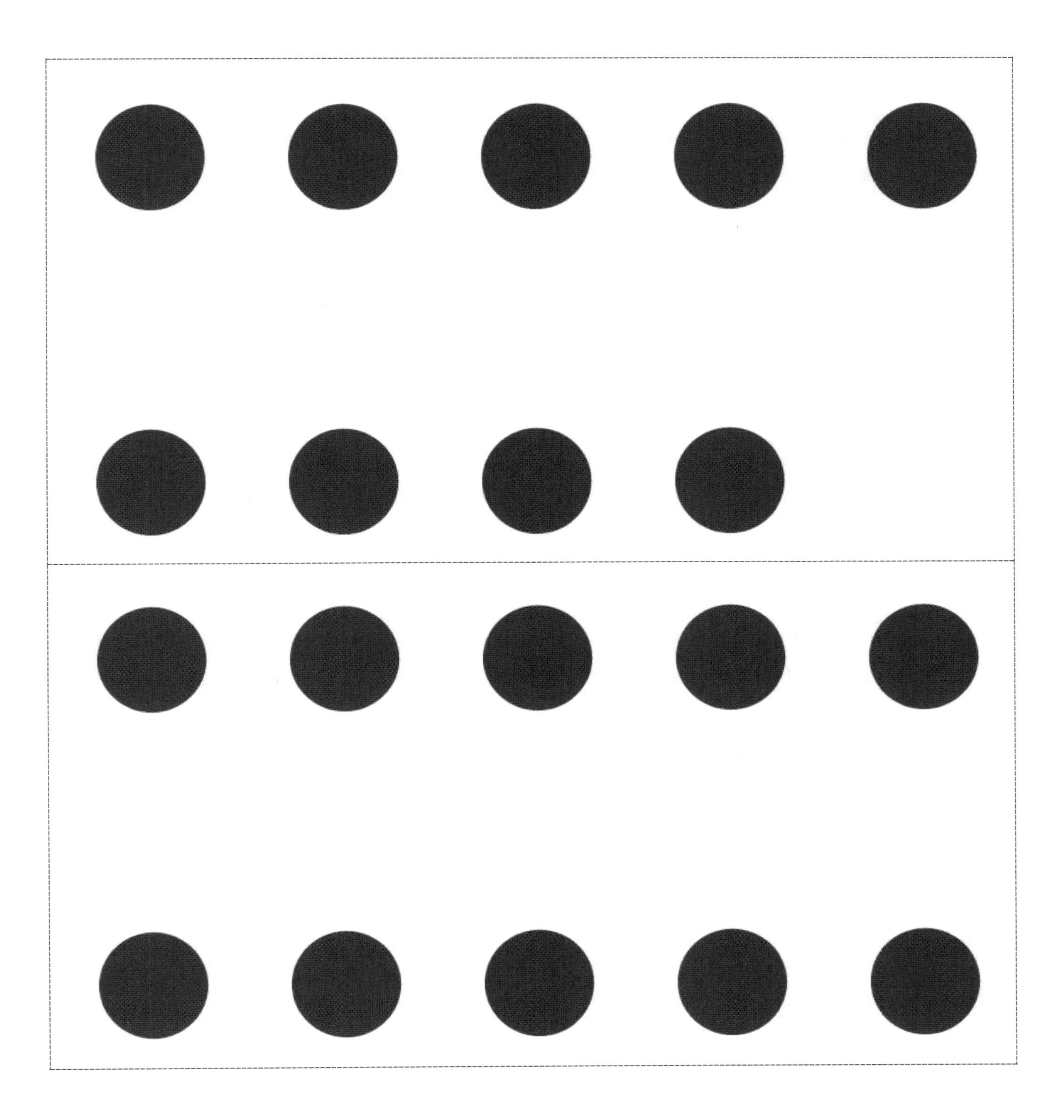

large 5-group cards (Copy on card stock and cut. Save the full set.)

0	1	2	3
4	5	5	6
7	8	9	10

Note: Consider copying on different colors of card stock for ease of organization.

5-group cards (numeral side) (Copy double-sided with 5-groups on card stock, and cut.)

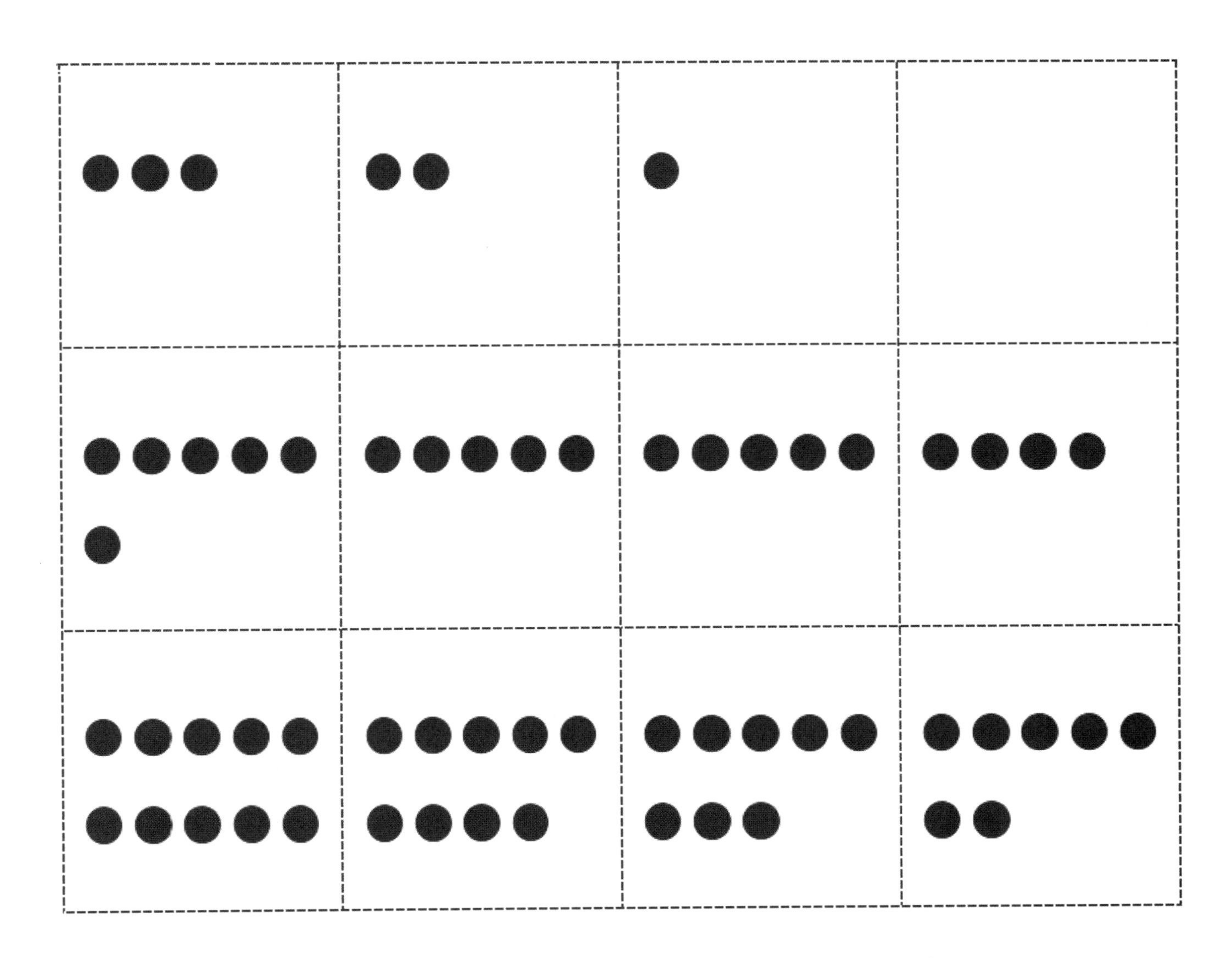

5-group cards (5-group side) (Copy double-sided with numerals on card stock, and cut.)

Lesson 2

Objective: Count 10 objects within counts of 10 to 20 objects, and describe as 10 ones and ___ ones.

Suggested Lesson Structure

Fluency Practice	(9 minutes)
Application Problem	(5 minutes)
Concept Development	(30 minutes)
Student Debrief	(6 minutes)
Total Time	**(50 minutes)**

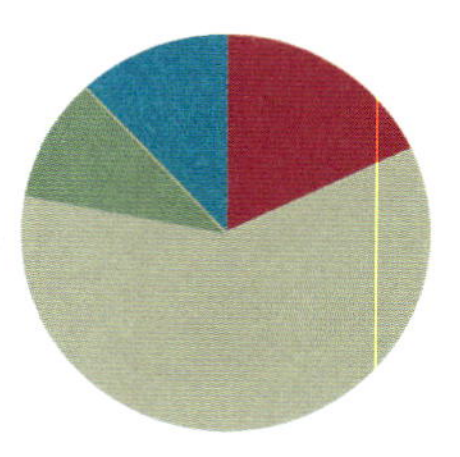

Fluency Practice (9 minutes)

- How Many Is One More? **K.CC.2** (3 minutes)
- Show One More on Fingers **K.CC.2** (3 minutes)
- Count Piles of Ten **K.CC.2, K.CC.4** (3 minutes)

How Many Is One More? (3 minutes)

Materials: (T) Large 5-group cards (Lesson 1 Fluency Template 1) (S) 5-group cards (Lesson 1 Fluency Template 2)

Note: This fluency activity advances the familiar work with the pattern of *1 more* as it requires students to visualize an additional dot on the 5-groups.

T: (Show 3.) How many dots?

S: 3.

T: What's one more than 3?

S: 4 is one more than 3.

Continue with the following possible sequence: 1, 4, 2, 4, 5, 6, 7, 9, 5, 8, 7. Eliminate asking them to identify the base number as quickly as possible. Students then continue this activity with each other in pairs.

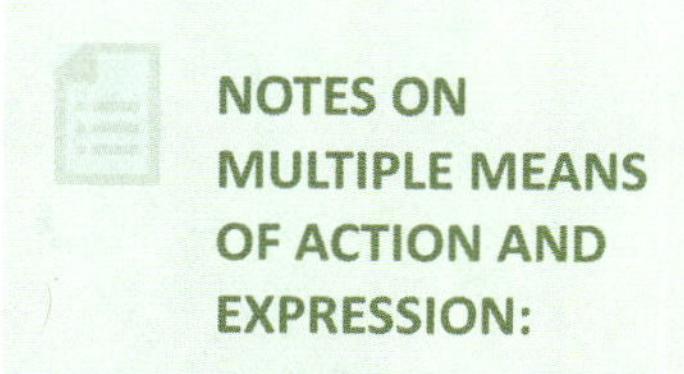

NOTES ON MULTIPLE MEANS OF ACTION AND EXPRESSION:

Deepen the understanding of students working above grade level by asking students to explain strategies for identifying *one more*. Then, have them apply their strategies through practice with a partner.

Ask students:

Could you use the same strategy for solving *two more* and *three more*?

Show One More on Fingers (3 minutes)

Materials: (T) 20-bead Rekenrek

Note: This fluency activity maintains students' proficiency with the pattern of *1 more* and connects two 5-group models, the Rekenrek, and counting the Math Way.

T: (Show 5 beads.) Count the number of beads.
S: 1, 2, 3, 4, 5.
T: Count one more on your fingers left to right.
S: (Hover hands as if playing the piano. Drop a finger or *play a note*, starting with the left pinky.) 1, 2, 3, 4, 5, 6.

Continue with the following possible sequence:
6, 4, 7, 9, 8, 7, 6.

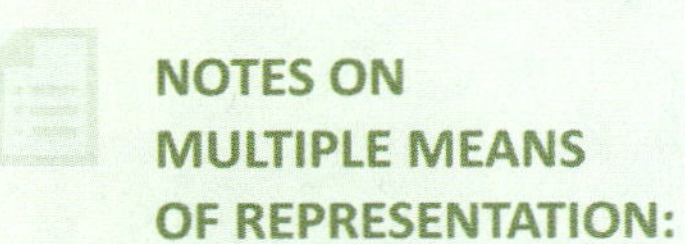

Access prior knowledge. Remind students of what a ten looks like by providing them with empty ten-frames. Students might then draw sets of ten sticks in the ten-frames.

Count Piles of Ten (3 minutes)

Materials: (S) About 40 straws for each pair of students

Note: Making groups of ten objects calls students' attention to the number 10 as a significant number in today's lesson.

Have students see how many piles of 10 straws they can count.

Application Problem (5 minutes)

Lisa counted some sticks into one pile of 10. She counted 5 other sticks into another pile. Draw a picture to show Lisa's piles of sticks.

Note: For now, just focus on the pile of 10 sticks and the pile of 5 rather than composing the teen number.

(Extension: Have early finishers draw Lisa's piles on another day when she made one pile of 10 sticks and one pile of 8 sticks!)

Concept Development (30 minutes)

Materials: (T) 10 bags with different items in each (suggestions to the right) (S) 1 egg carton cut to have 10 compartments for each pair of students

T: Count to find out how many slots there are in your egg carton. Wait for the signal to tell me.

Bag Contents:
- 18 clothespins
- 20 pasta shells
- 13 beads
- 16 pennies
- 11 crayons
- 10 erasers
- 14 linking cubes
- 12 walnuts in the shell
- 10 play dollars
- 15 counting chips

T: (Pause. When all are ready, give the signal.)

S: 10.

T: Each team will count the objects in ten bags. To count the objects in your bag, start by placing the objects in the egg carton, and then put any extra objects next to the carton.

T: Tell your partner, "I have 10 ones and ____ ones."

T: We'll do one together first. (Demonstrate.)

Have pairs of students count out the given **teen number**, decomposing it as 10 ones and some more ones. After counting the objects, have pairs trade bags and count the new objects.

T: (Allow students time to count all 10 bags.) Let's see what you discovered! Count the clothespins with me.

S: (Show each one using the egg carton.) 1, 2, 3, 4, 5, 6, 7, 8, 9, 10, 11, 12, 13, 14, 15, 16, 17, 18.

T: How many clothespins are there?

S: 18.

T: (Write 10 ones and ___ ones.) Let's complete this sentence.

S: 10 ones and 8 ones.

T: Yes!

Have students, in pairs, count and then decompose the other quantities in the other bags using their egg cartons, allowing them to recognize and internalize the structure of teen numbers as 10 ones and some more ones. Continue to encourage statements following the pattern "12 is 10 ones and 2 ones."

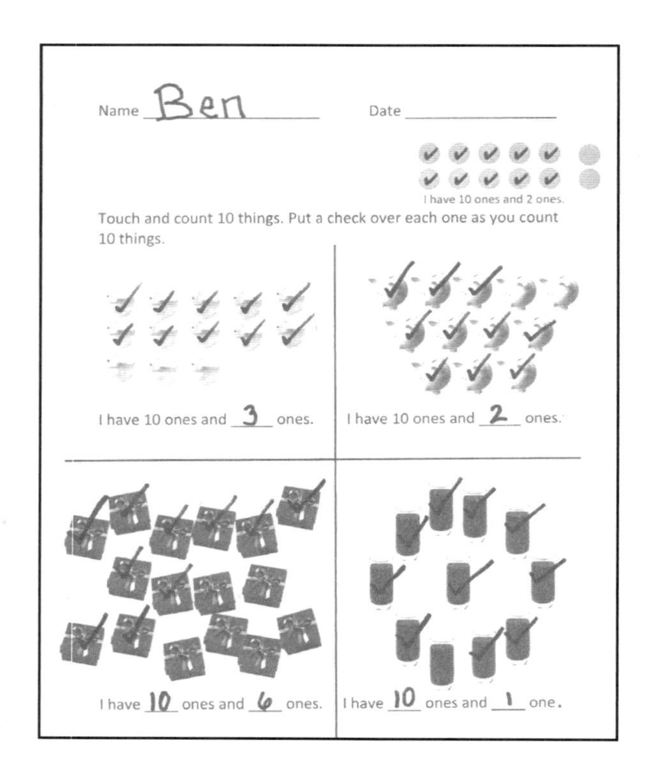

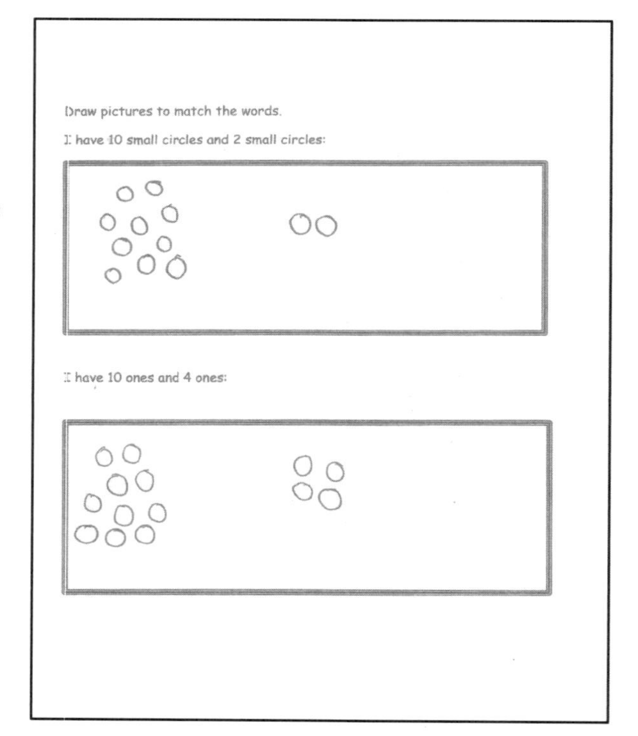

Problem Set (8 minutes)

Students should do their personal best to complete the Problem Set within the allotted time.

Note: Students use the method of checking off one object each time they count. This is an easier strategy than circling 10 items, which is part of the next lesson.

Student Debrief (6 minutes)

Lesson Objective: Count 10 objects within counts of 10 to 20 objects, and describe as 10 ones and ___ ones.

The Student Debrief is intended to invite reflection and active processing of the total lesson experience. Have students bring their Problem Set to the carpet and work with a partner to check their count of 10 ones and some more ones. Have them say the teen number as 10 ones and some more ones.

S: There are 1, 2, 3, 4, 5, 6, 7, 8, 9, 10, 11, 12, 13 ducks.

S: 13 is 10 ones and 3 ones.

Ask students to look at the picture of the ducks. Guide students in a conversation to debrief the Problem Set and process the lesson. Any combination of the questions below may be used to lead the discussion.

- Is it easy to see 10 ones in this picture? Why?
- How is this picture the same and different from counting using the egg carton?
- Which was easier to count, the ducks or the glasses of juice? Why? Show your friend how you counted the glasses of juice.
- Does your drawing of 10 ones and 2 ones look exactly the same as your friend's? How is it the same? How is it different?
- Write the number 17 on the board. Can someone come up and draw 17 squares on the board? Can someone come up and circle 10? Fill in this sentence for me: 17 is 10 ones and ____ ones.
- 14 is 10 ones and ____ ones. Fourteen is a **teen number**. What is another teen number?
- Eleven and twelve don't have *teen*, but most grown-ups call them teen numbers. What have you noticed today about teen numbers?

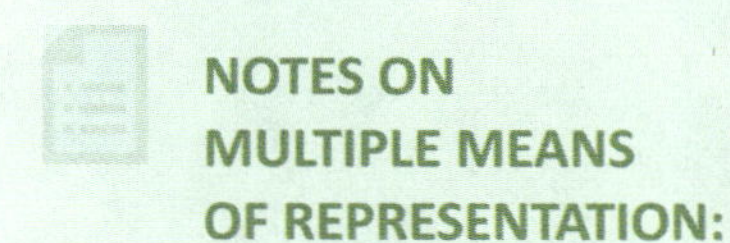

For students with developing language skills, review academic vocabulary. Before beginning student sharing during the Debrief, count to 20 with the Rekenrek to practice pronouncing numbers.

Exit Ticket (3 minutes)

After the Student Debrief, instruct students to complete the Exit Ticket. A review of their work will help with assessing students' understanding of the concepts that were presented in today's lesson and planning more effectively for future lessons. The questions may be read aloud to the students.

Name ______________________________ Date ______________

I have 10 ones and 2 ones.

Touch and count 10 things. Put a check over each one as you count 10 things.

I have 10 ones and ___ ones.

I have 10 ones and ___ ones.

I have ___ ones and ___ ones.

I have ___ ones and ___ ones.

Draw pictures to match the words.

I have 10 small circles and 2 small circles:

I have 10 ones and 4 ones:

Name ______________________________ Date ______________

10 ones and 3 ones

10 ones and 1 one

Circle the correct numbers that describe the pictures.

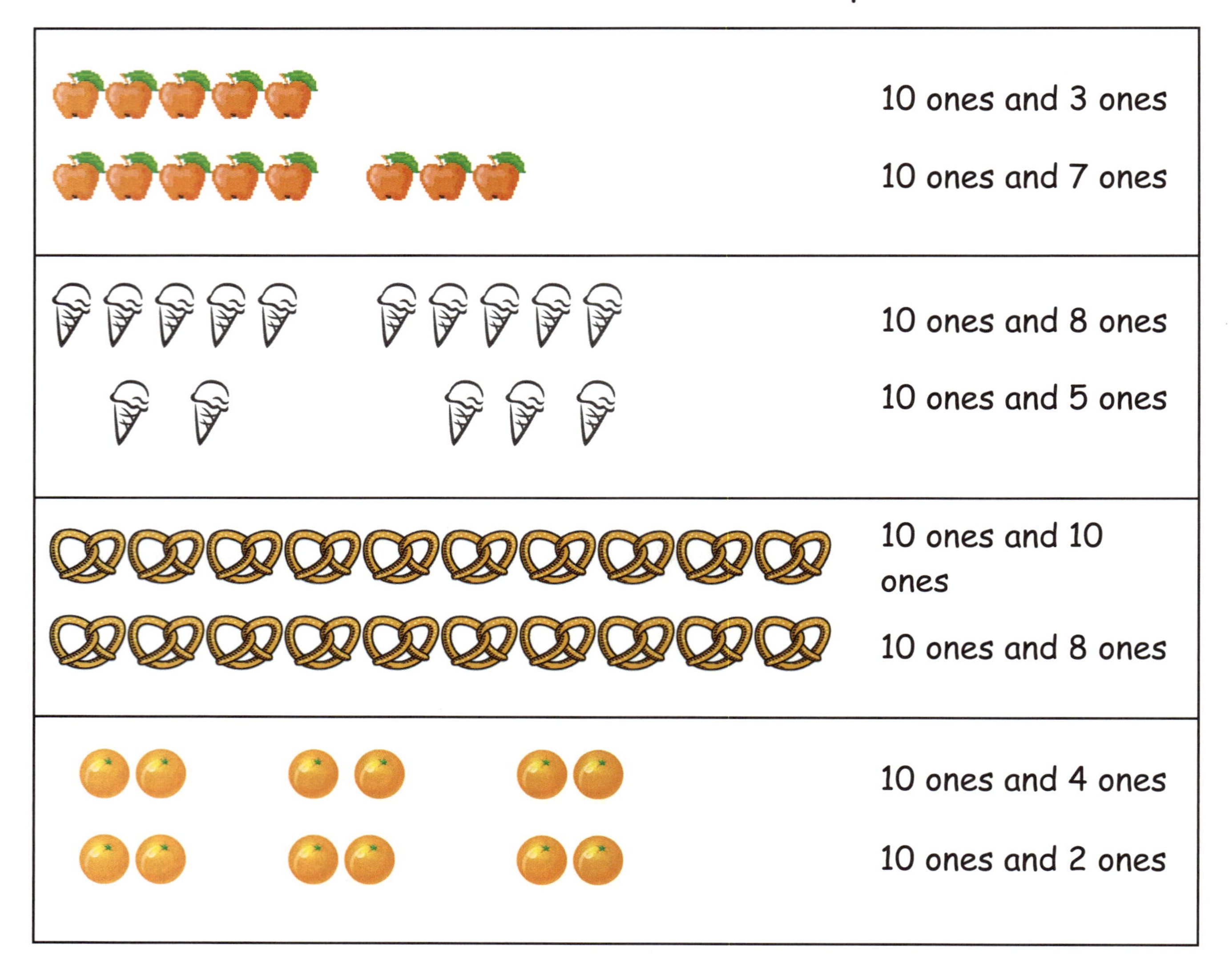

Name ______________________ Date ____________

10 ones and 3 ones

Draw more to show the number.

10 ones and 2 ones

10 ones and 5 ones

10 ones and 7 ones

10 ones and 4 ones

Lesson 3

Objective: Count and circle 10 objects within images of 10 to 20 objects, and describe as 10 ones and ___ ones.

Suggested Lesson Structure

Fluency Practice	(10 minutes)
Application Problem	(7 minutes)
Concept Development	(26 minutes)
Student Debrief	(7 minutes)
Total Time	**(50 minutes)**

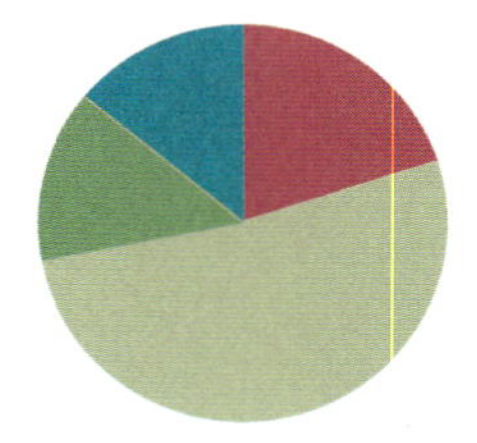

Fluency Practice (10 minutes)

- Hide 1 **K.OA.1** (4 minutes)
- How Many Do You See? **K.CC.2** (3 minutes)
- Grouping 10 Objects **K.NBT.1** (3 minutes)

Hide 1 (4 minutes)

Materials: (T) Large 5-group cards (Lesson 1 Fluency Template 1) (S) 5-group cards (Lesson 1 Fluency Template 2)

Note: This fluency activity advances the familiar work with the pattern of *1 less* as it requires students to visualize removing a dot from the 5-group card.

T: (Show 5.) Use your imagination to hide 1. How many are left?

S: 4.

T: (Show 10.) Use your imagination to hide 1. How many are left?

S: 9.

Continue with the following possible sequence: 1, 6, 2, 7, 3, 8, 4, 9. Have students repeat the activity in pairs if there is time.

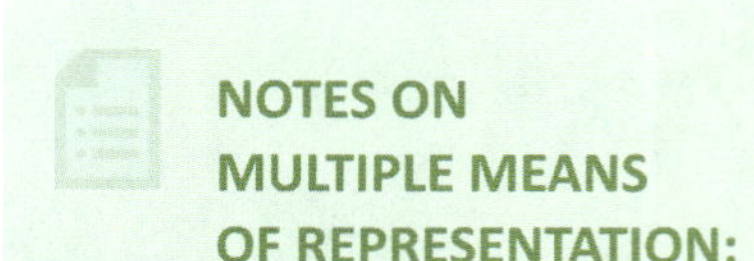

NOTES ON MULTIPLE MEANS OF REPRESENTATION:

Make instructions visual as well as oral for English language learners. When instructing students, "Use your imagination to hide 1," illustrate this process by covering one dot on the 5-group card. Repeat for the first few numbers.

How Many Do You See? (3 minutes)

Materials: (T) Large 5-group cards (Lesson 1 Fluency Template 1)

Note: This fluency activity advances students' ability to rapidly recognize quantities on 5-group cards by requiring them to visualize.

T: (Show dots for several seconds, and then hide the card.) Wait for the signal. How many dots did you see?

S: 7.

T: Who can explain how they see 7?

S: I see a 5 group on top and 2 more on the bottom. (Draw as the student speaks.)

Continue with the following possible sequence: 3, 9, 1, 8, 7, 4.

Grouping 10 Objects (3 minutes)

Materials: (S) Bag with about 20 small objects for each student

Note: Making groups of 10 ones in varied configurations brings attention to the number as significant in today's lesson and allows students to experience conservation of the number.

T: Place the items from your bag on your work mat. Count out 10 ones, and move them together into a bunch.

T: (Wait while they work.) By counting, prove to your partner there are 10 things in your bunch.

S: (Count.)

T: Push all your things back together. Mix them up. Count out 10 ones again, and move them together into a bunch.

Repeat process two or three more times. Ask students if the same 10 things are in the bunch each time.

Application Problem (7 minutes)

Each gingerbread man got 10 sprinkles as buttons with 2 sprinkles to show the eyes. Draw to show the 12 sprinkles as 10 buttons and 2 eyes.

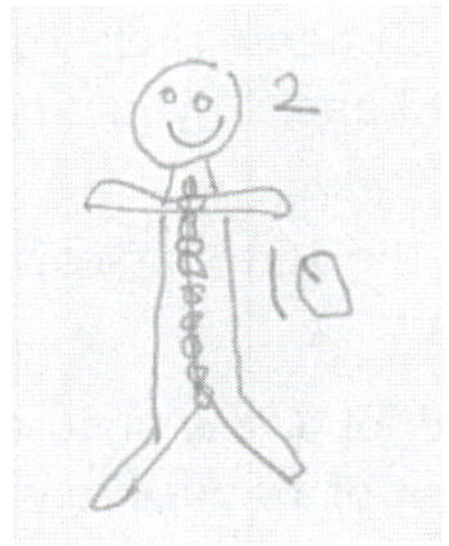

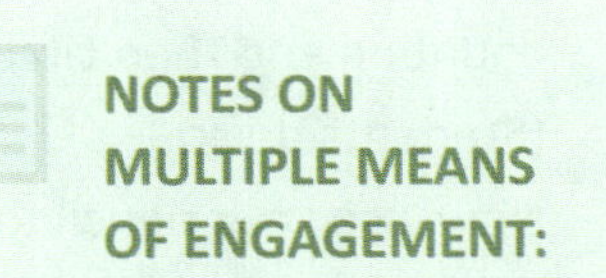

NOTES ON MULTIPLE MEANS OF ENGAGEMENT:

Challenge students working above grade level during the Application Problem by asking them to draw a 5-group that represents this problem. Ask: "What if each gingerbread man got 1 more sprinkle for the nose?"

Concept Development (26 minutes)

Materials: (S) Find 10 (Template) cut into strips

T: (Draw two rows of five circles with three more off to the side.)

T: Let's count all the circles.

S: 1, 2, ..., 13.

T: Talk to your elbow partner. Can you count 10 ones in my picture?

S: (Students talk with their partners. Watch for pointing and counting. Expect students to count one at a time. Do not insist they recognize the 2 fives as 10 automatically.)

T: Who can come to the board and show us how they counted 10 ones?

S: (Student comes to the board and designates his 10.)

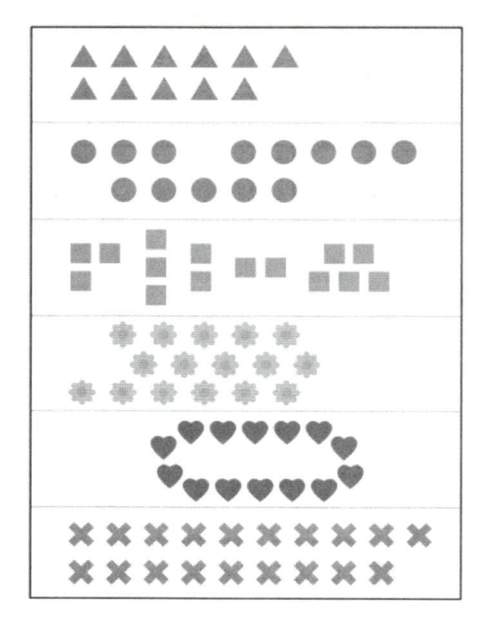

T: Let's count with him while he points.

S: 1, 2, 3, 4, 5, 6, 7, 8, 9, 10.

T: Are there more?

S: Yes!

T: How many more?

S: 3 more.

T: Use your finger to circle the 10 ones from your seat.

S: (Make circles around the 10 ones with fingers.)

T: Can you see the 3 ones without counting?

S: Yes!

T: Now, find 10 triangles inside this group of triangles. (Distribute the template strip of triangles pictured to the right.) Find 10 ones, and circle them carefully with your finger.

S: (Count and circle 10 ones with finger.)

T: Show your partner how you found and circled 10 ones with your finger. Prove to him that it is 10 by counting and then circling.

S: (Students do so.)

T: Now, use your pencil to find and circle your 10 ones. (Students circle 10 ones.) Trade papers with your partner, and count to be sure he circled exactly 10 ones. If you disagree, tell your partner why you think the answer should be different.

T: How many extra ones did you have after you counted the 10 triangles?

S: 1.

T: When you and your partner are ready, raise your hand for a new picture. Find and circle 10 ones with your finger and then with your pencil. Prove your count of 10 ones to your partner. Trade papers with your partner, and check his count. (Continue distributing additional strips of teen items from the template.)

Problem Set (8 minutes)

Students should do their personal best to complete the Problem Set within the allotted time.

Note: Ask students to find and circle 10 objects with their fingers before circling them with their pencils. They are *finding* an embedded number; just as when they were *seeing* seven, they may have seen a 5-group and 2 more. The difference here is that they must count to find 10 ones. Later, in Grade 1, they will recognize certain configurations of 10 ones (such as the ten-frame) as 1 ten.

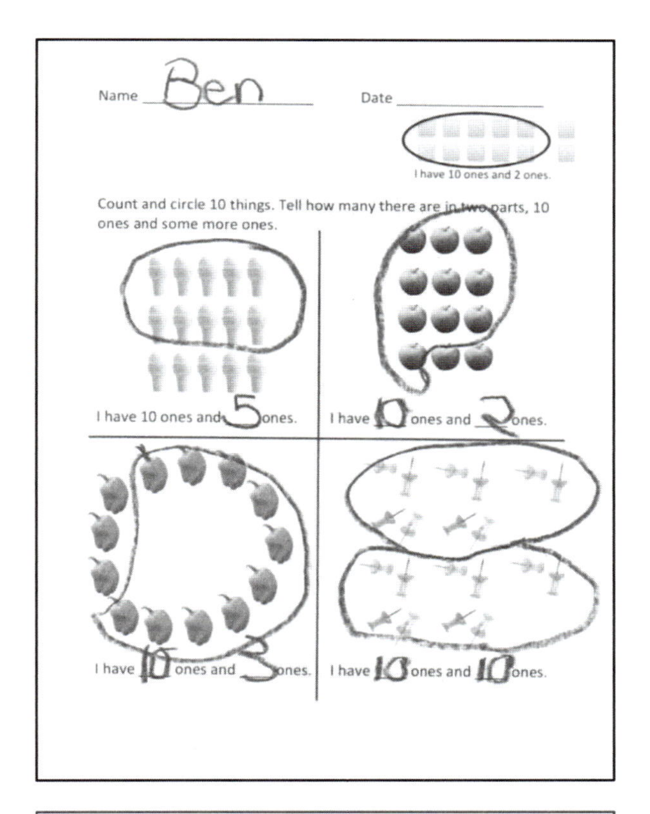
Name Ben Date

I have 10 ones and 2 ones.

Count and circle 10 things. Tell how many there are in two parts, 10 ones and some more ones.

I have 10 ones and 5 ones. I have 10 ones and 2 ones.

I have 10 ones and 3 ones. I have 10 ones and 10 ones.

Student Debrief (7 minutes)

Lesson Objective: Count and circle 10 objects within images of 10 to 20 objects, and describe as 10 ones and __ ones.

The Student Debrief is intended to invite reflection and active processing of the total lesson experience.

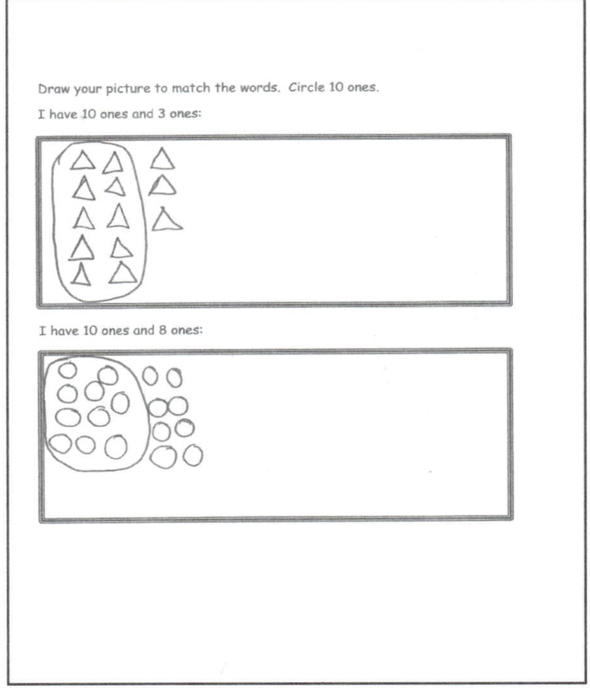
Draw your picture to match the words. Circle 10 ones.

I have 10 ones and 3 ones:

I have 10 ones and 8 ones:

Invite students to review their solutions for the Problem Set. They should check work by comparing answers with a partner before going over answers as a class. Look for misconceptions or misunderstandings that can be addressed in the Debrief. Guide students in a conversation to debrief the Problem Set and process the lesson.

Any combination of the questions below may be used to lead the discussion.

- Did your friend circle the exact same ice-cream cones? Apples? Peppers? Tacks?
- Were both your answers correct? Why?
- How did your friend represent 10 ones in his picture?
- How do we say 10 ones and 5 ones (and the other numbers represented) as one number? (The students have been counting to higher numbers during Fluency Practice since early in the year. Pre-K standards call for counting to 20.)
- Which pictures were the easiest for you to count? Why?

- What do all these examples have in common? Do 10 ones always look the same? What other things in our classroom could we make into a bunch or pile of 10 ones?

Exit Ticket (3 minutes)

After the Student Debrief, instruct students to complete the Exit Ticket. A review of their work will help with assessing students' understanding of the concepts that were presented in today's lesson and planning more effectively for future lessons. The questions may be read aloud to the students.

Name ______________________________ Date ______________

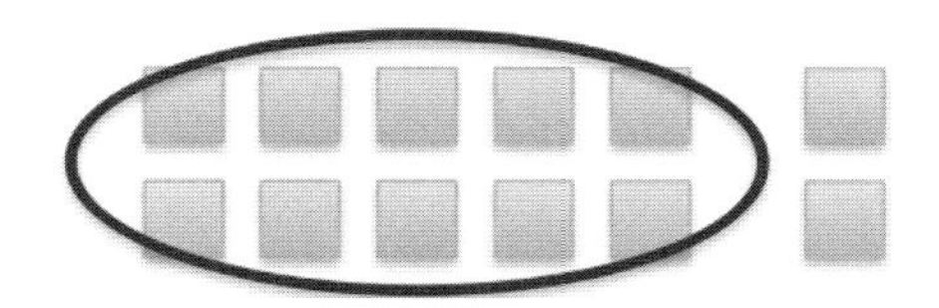

I have 10 ones and 2 ones.

Count and circle 10 things. Tell how many there are in two parts, 10 ones and some more ones.

I have 10 ones and ___ ones.

I have ___ ones and ___ ones.

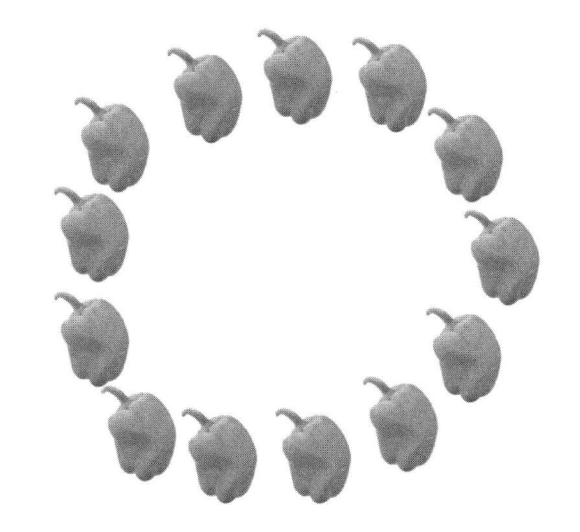

I have ___ ones and ___ ones.

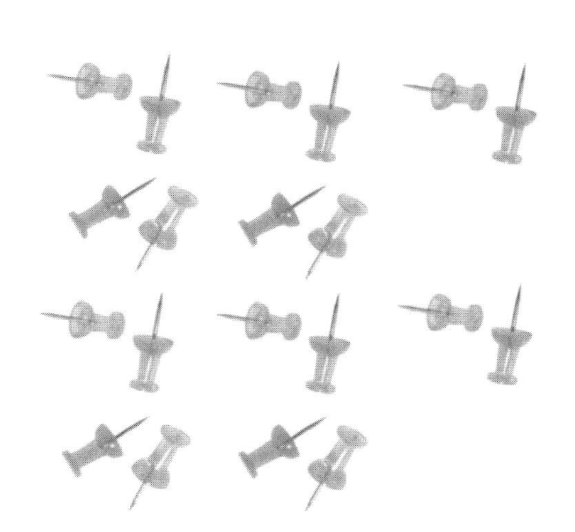

I have ___ ones and ___ ones.

Draw your picture to match the words. Circle 10 ones.

I have 10 ones and 3 ones:

I have 10 ones and 8 ones:

GK-M5-TE-B5-1.3.1-01.2016

Name ______________________________ Date ______________

Circle 10 ones.

I have 10 ones and ___ ones.

Draw 10 ones and 6 ones.

I have 10 ones and 6 ones.

Name ______________________________ Date ______________

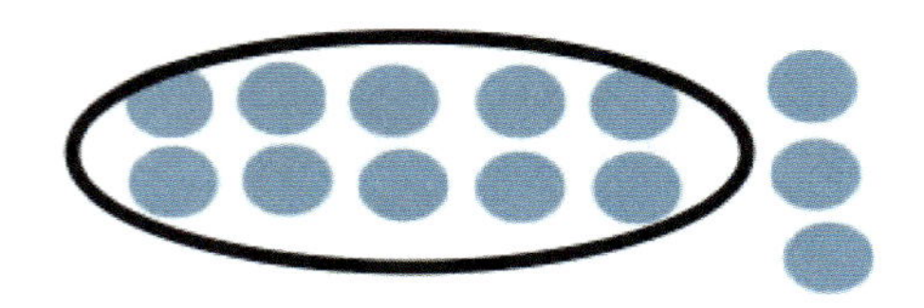

I have 10 ones and 3 ones.

Circle 10 things. Tell how many there are in two parts, 10 ones and some more ones.

I have 10 ones and ___ ones.

I have 10 ones and ___ ones.

I have ___ ones and ___ ones.

I have ___ ones and ___ ones.

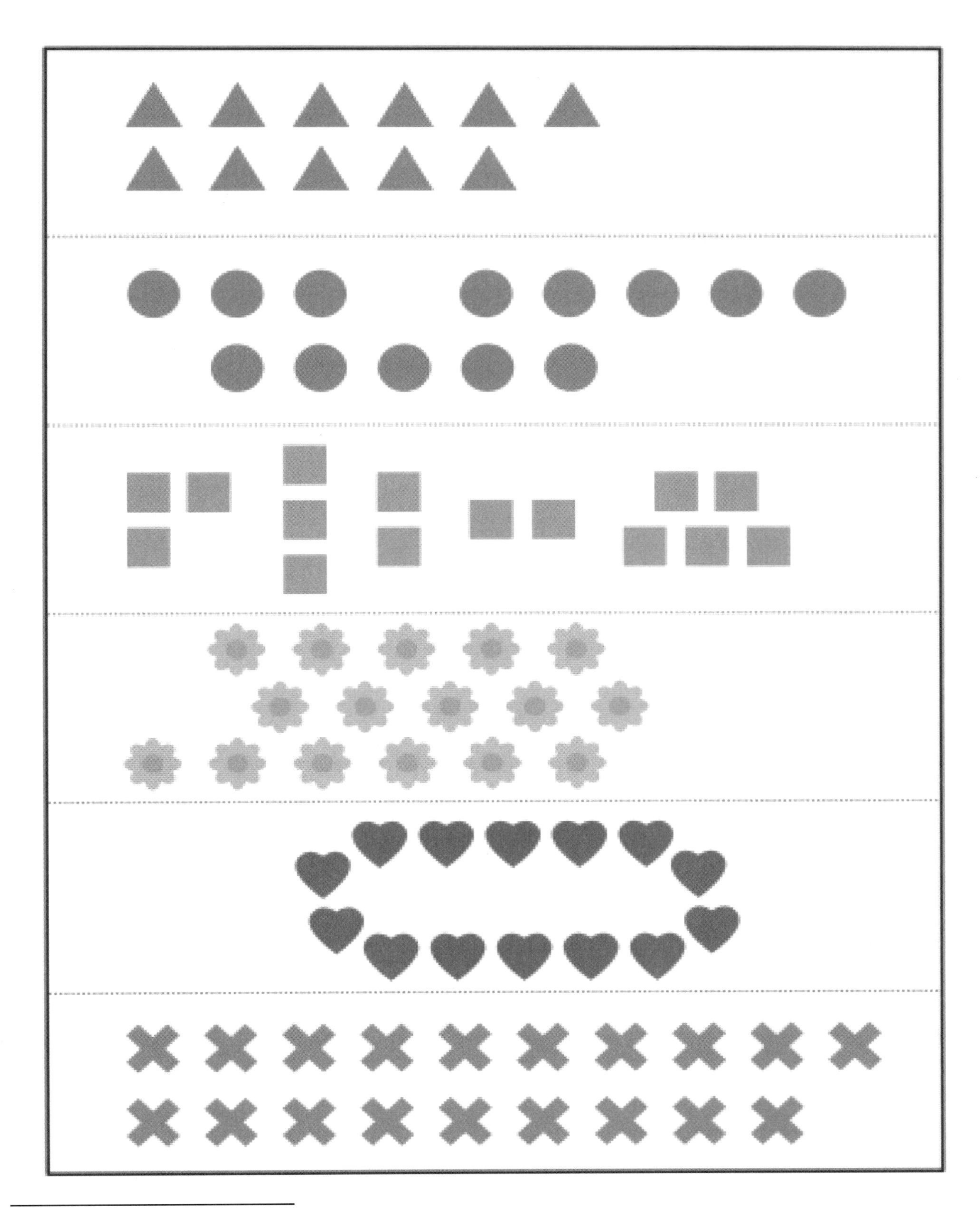

find 10

Lesson 4

Objective: Count straws the Say Ten way to 19; make a pile for each ten.

Suggested Lesson Structure

Fluency Practice	(12 minutes)
Application Problem	(6 minutes)
Concept Development	(26 minutes)
Student Debrief	(6 minutes)
Total Time	**(50 minutes)**

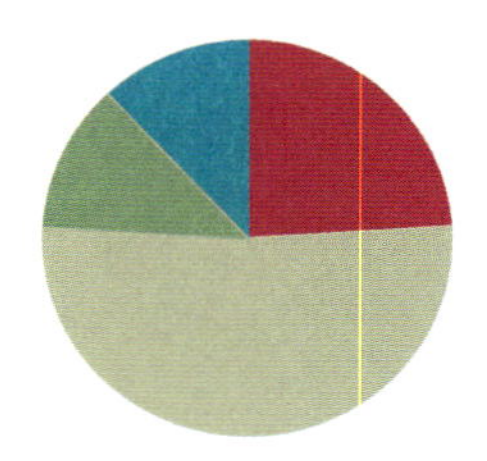

Fluency Practice (12 minutes)

- Dot Cards of Six **K.CC.2, K.CC.5** (4 minutes)
- Number Pairs of Six **K.CC.2** (4 minutes)
- Circle 10 Objects **K.NBT.1** (4 minutes)

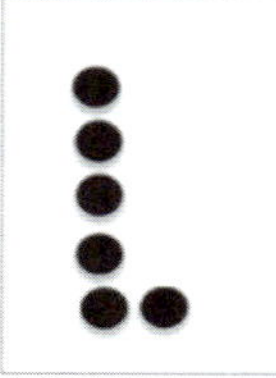

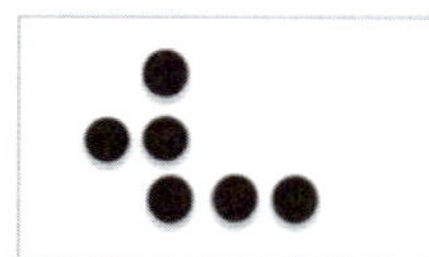

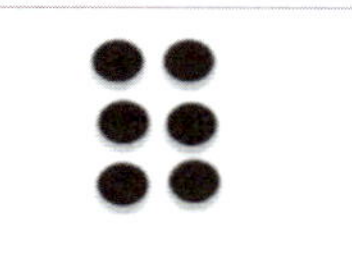

Dot Cards of Six (4 minutes)

Materials: (T/S) Dot cards of 6 (Fluency Template 1)

Note: This fluency activity gives students an opportunity to develop increased familiarity with decompositions of six and practice seeing part–whole relationships.

T: (Show 6 dots.) How many do you see? (Give students time to count).

S: 6.

T: How can you see 6 in two parts?

S: (Come up to the card.) 5 here and 1 here. I see 3 here and 3 here.

Continue with other cards of six. Distribute the cards to the students for partner sharing time. Have them *pass on* the card to a different set of partners at a signal.

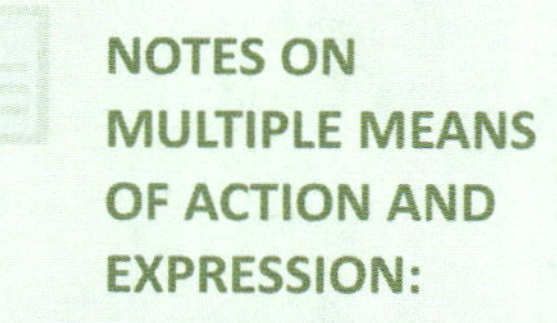

NOTES ON MULTIPLE MEANS OF ACTION AND EXPRESSION:

Provide students with disabilities with extra minutes to process questions before giving the signal to respond. When students are responding chorally, ask them to "show thumbs up when ready" to ensure ample think time.

Number Pairs of Six (4 minutes)

Materials: (T) Linking cube sticks or dot cards of 6 (Fluency Template 1) (S) Personal white board

Note: This fluency activity gives students an opportunity to develop increased familiarity with compositions of six and practice seeing part–whole relationships. Do not expect automaticity from most students, but make note of advanced thinking. Allow time to count all if necessary.

Show a stick of linking cubes or the dot cards with 5 and 1 indicated as parts.

T: Say the larger part. (Give students time to count.)
S: 5.
T: Say the smaller part.
S: 1.
T: What is the total number of dots? (Give them time to recount.)
S: 6.
T: Show the number bond on your personal white board.

Continue with 4 and 2, 3 and 3, and 6 and 0.

Circle 10 Objects (4 minutes)

Materials: (S) Circle 10 (Fluency Template 2)

Note: This activity requires students to locate 10 as an embedded number within a pictorial group of 10 ones and some ones.

Distribute circle 10 template. Please note that this template will be used in the Student Debrief.

Application Problem (6 minutes)

At recess, 17 students were playing. 10 students played handball while 7 students played tetherball. Draw to show the 17 students as 10 students playing handball and 7 students playing tetherball.

Note: In this Application Problem, students are not adding to solve, but rather they are being guided to decompose the 17 as 10 ones and 7 ones. This is not asking *how many* but rather separating 17 into 10 ones and some ones (**K.NBT.1**). The problem is not asking them to count the total but is instead telling them the total.

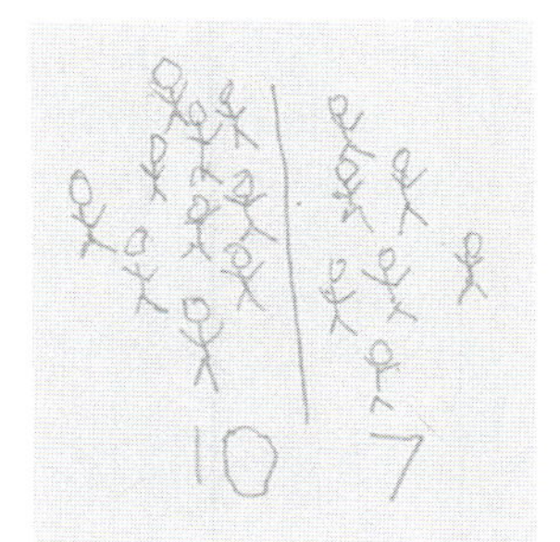

NOTES ON MULTIPLE MEANS OF ENGAGEMENT:

Support your English language learners' developing academic vocabulary; as students count the Say Ten way, ask them to also tell the standard number name.

T: How many?
S: Ten 1.
T: Right. And the regular way?
S: Eleven.

Concept Development (26 minutes)

Materials: (T) 19 linking cubes (S) bag of 19 small counting objects such as pennies or beans; 19 straws (per pair)

T: Come sit with me on the carpet. (Choose a student helper to sit next to you on the left.)
T: (Place a linking cube on each of your fingers.) How many cubes do you see?
S: 10.
T: (Ask your helper to place a cube on her right pinky finger.) Now, how many cubes do you see?
S: Eleven! → I see 10 and 1.
T: You're all correct! Eleven is 10 and 1. I'm going to teach you to count the Say Ten way!
T: (With a linking cube on each finger, raise your hands again.) How many linking cubes is this?
S: Ten.
T: Every time Lucy adds another cube to her fingers, we'll say, "Ten" (show your hands) and the number of ones you see on her fingers. Ready?
S: (Have helper add cubes on her fingers from right to left in sequential order up to 19.) Ten 1, ten 2, ten 3, ten 4, ten 5, ten 6, ten 7, ten 8, ten 9.
T: Excellent! Now, go back to your seats, and we'll practice counting the Say Ten way using straws.
T: (Pass out 19 straws to each pair of students.) One student, Partner A, will count out 10 straws into a pile. The other student, Partner B, will place one straw next to the pile, and we'll say, "Ten 1." Ready?
S: (Show a pile of 10 straws and 1 more straw.) Ten 1.
T: Partner B, place another straw next to the pile of 10. How many straws now?
S: Ten 2, ten 3, ten 4, ...(continue to ten 9).
T: Put all the straws back into one pile, and switch roles. Partner B, count out 10 straws into a pile. Partner A, place 1 straw next to the pile, and let's practice counting again the Say Ten way.
S: (Count up to ten 9.)

Problem Set (7 minutes)

Students should do their personal best to complete the Problem Set within the allotted time.

Begin by having students use concrete materials on the ten-frames of the Problem Set. Have them count the Say Ten way as they work. Direct students to fill the ten-frame on the left, first with one row of 5 from left to right, and then the row below from left to right. Remind them that these are like their egg cartons. After doing some examples with materials, have students draw and count the specified amounts while they count the Say Ten way.

Student Debrief (6 minutes)

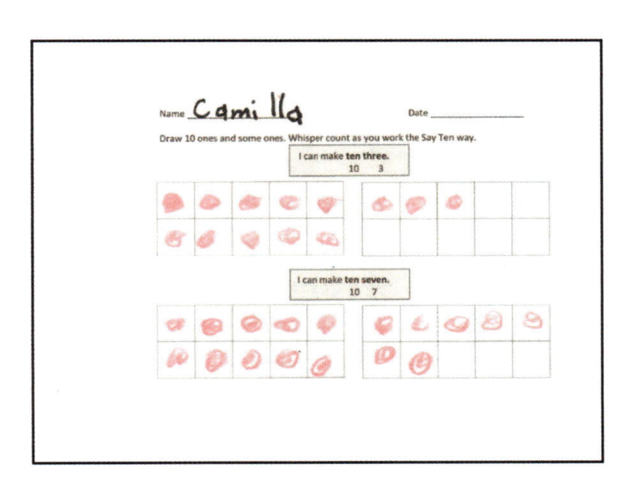

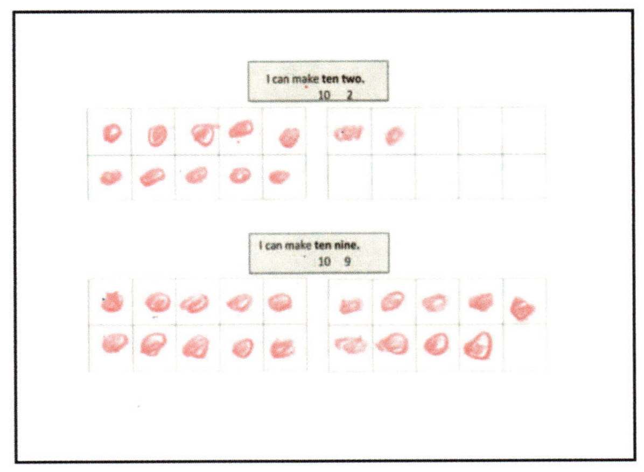

Lesson Objective: Count straws the Say Ten way to 19; make a pile for each ten.

The Student Debrief is intended to invite reflection and active processing of the total lesson experience.

The following is a suggested list of questions to invite reflection and active processing of the total lesson experience. Use what best supports students' ability to articulate the focus of the lesson.
Have students bring their circle 10 template to the carpet. This is the template from the Fluency Practice.

Suggestions for the Debrief:

- Look at your circle 10 template. Can you say the numbers the Say Ten way?
- Did your friend circle 10 objects the same way you did?
- Were both of your answers correct? Why?
- How do we say ten 9 as one number?
- How do we say 16 the Say Ten way?
- Which pictures were the easiest for you to count? Why?
- What do all the pictures have in common?

Exit Ticket (3 minutes)

After the Student Debrief, instruct students to complete the Exit Ticket. A review of their work will help with assessing students' understanding of the concepts that were presented in today's lesson and planning more effectively for future lessons. The questions may be read aloud to the students.

Name ____________________ Date ________________

Draw 10 ones and some ones. Whisper count as you work the Say Ten Way.

I can make **ten three.**
10 3

I can make **ten seven.**
10 7

I can make **ten two.**
10 2

I can make **ten nine.**
10 9

Name ______________________________ Date ________________

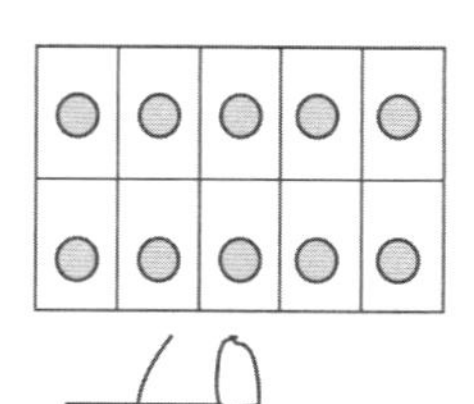

10 3

Count and write how many the Say Ten way.

10 ________

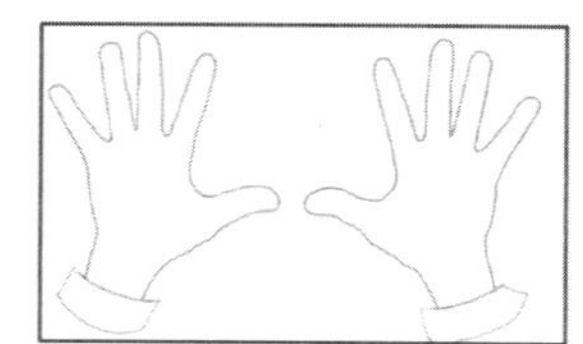
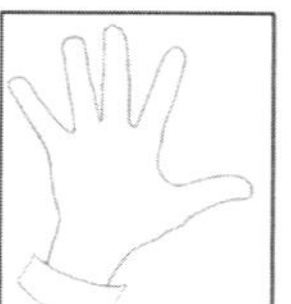

10 ________

________ ________

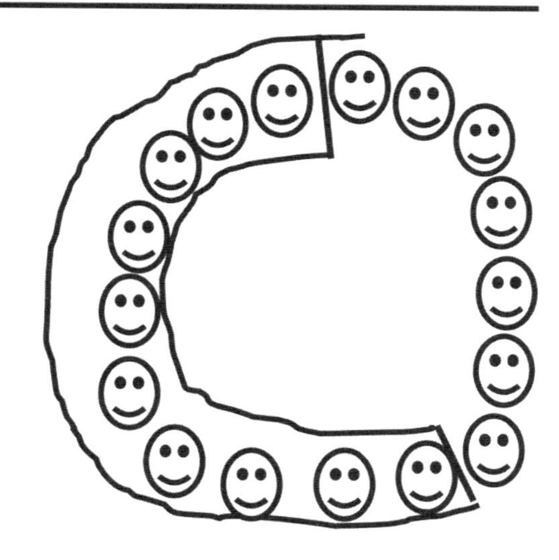

________ ________

EUREKA MATH

Name ______________________________ Date ______________

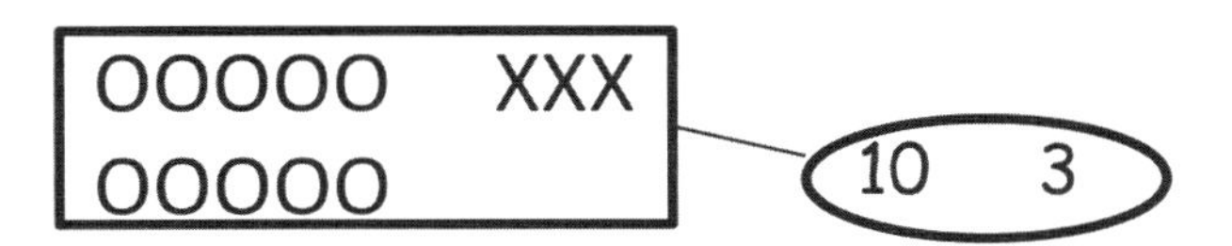

Draw a line to match each picture with the numbers the Say Ten way.

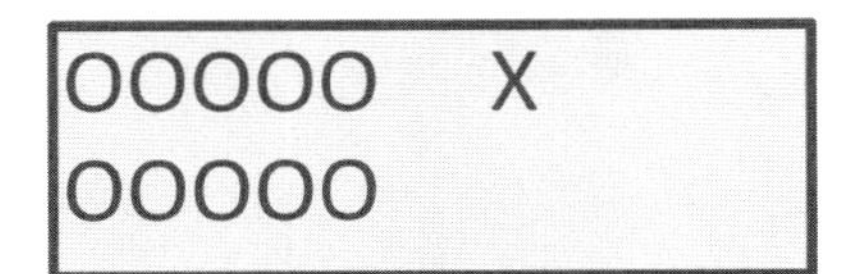

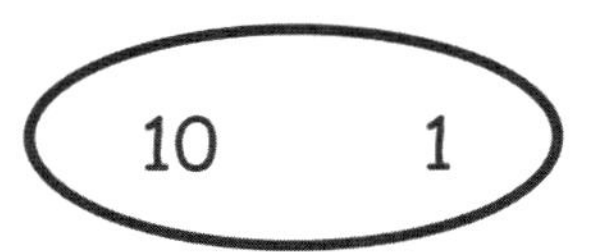

OOOOO XX
OOOOO

10 6

OOOOO XXX
OOOOO

10 10

OOOOO XXXXX
OOOOO X

10 2

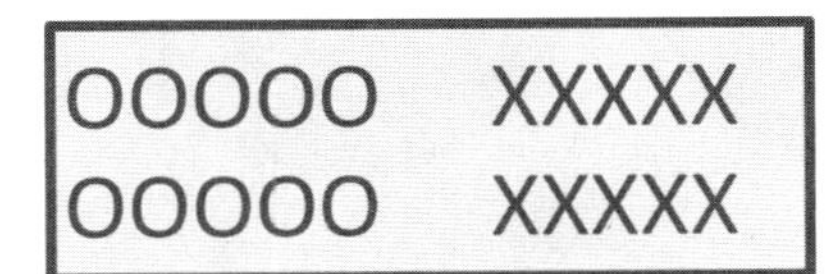

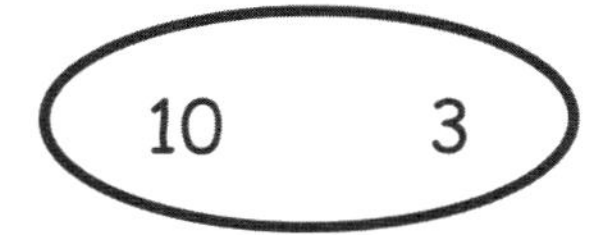

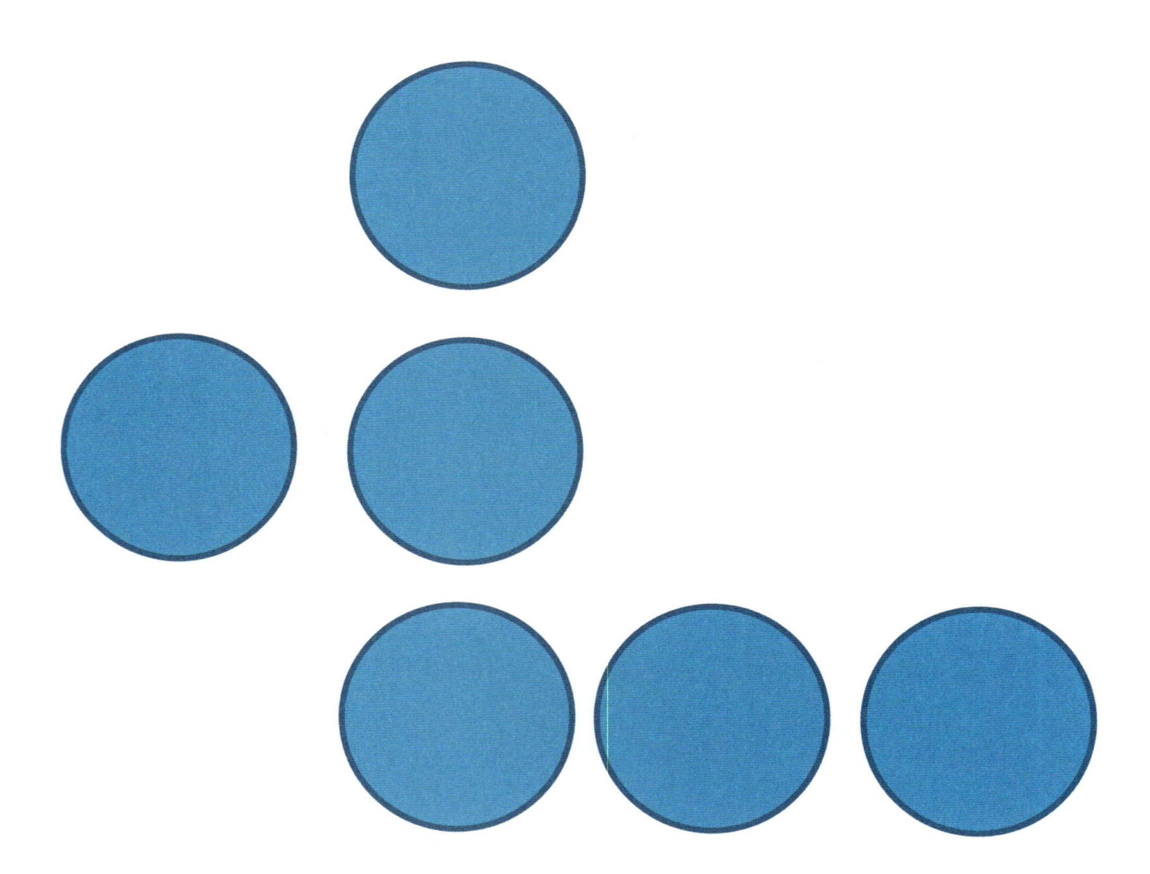

dot cards of 6

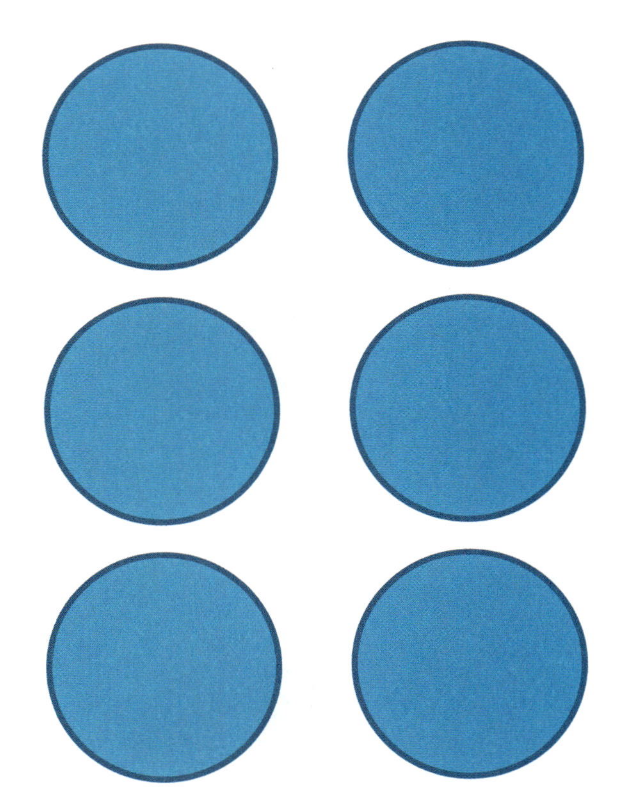

dot cards of 6

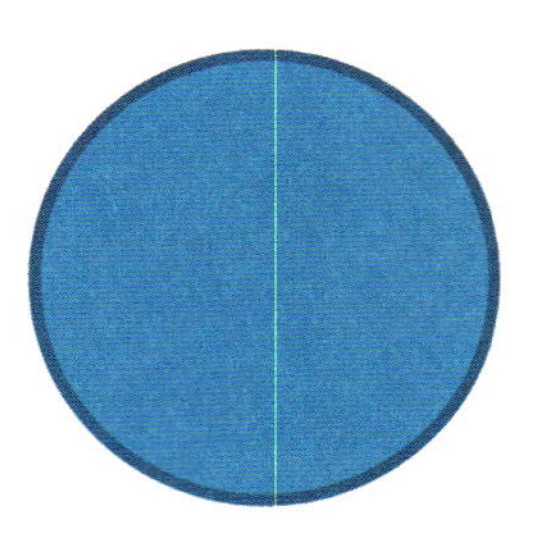

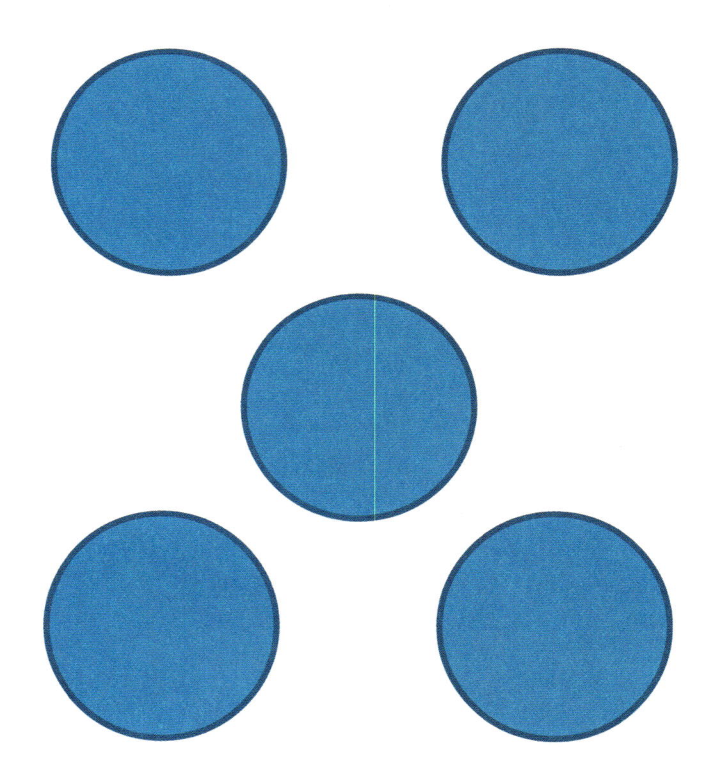

dot cards of 6

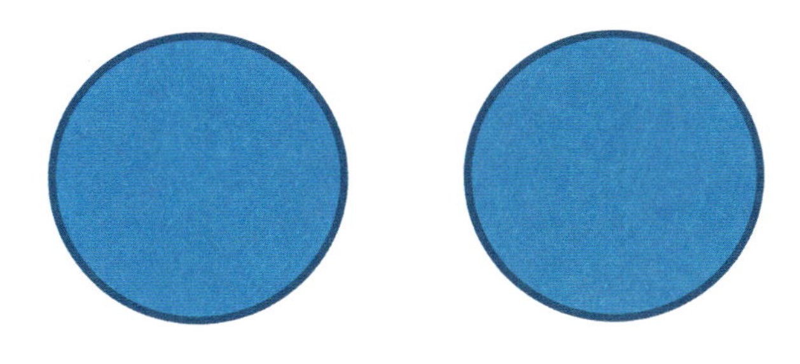

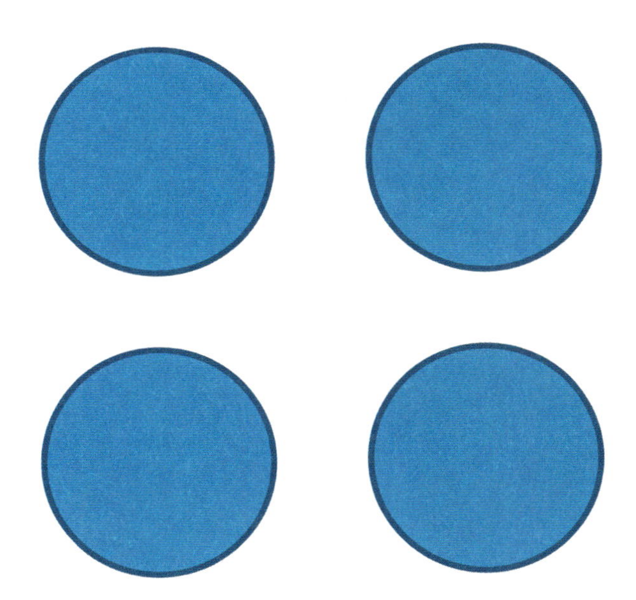

dot cards of 6

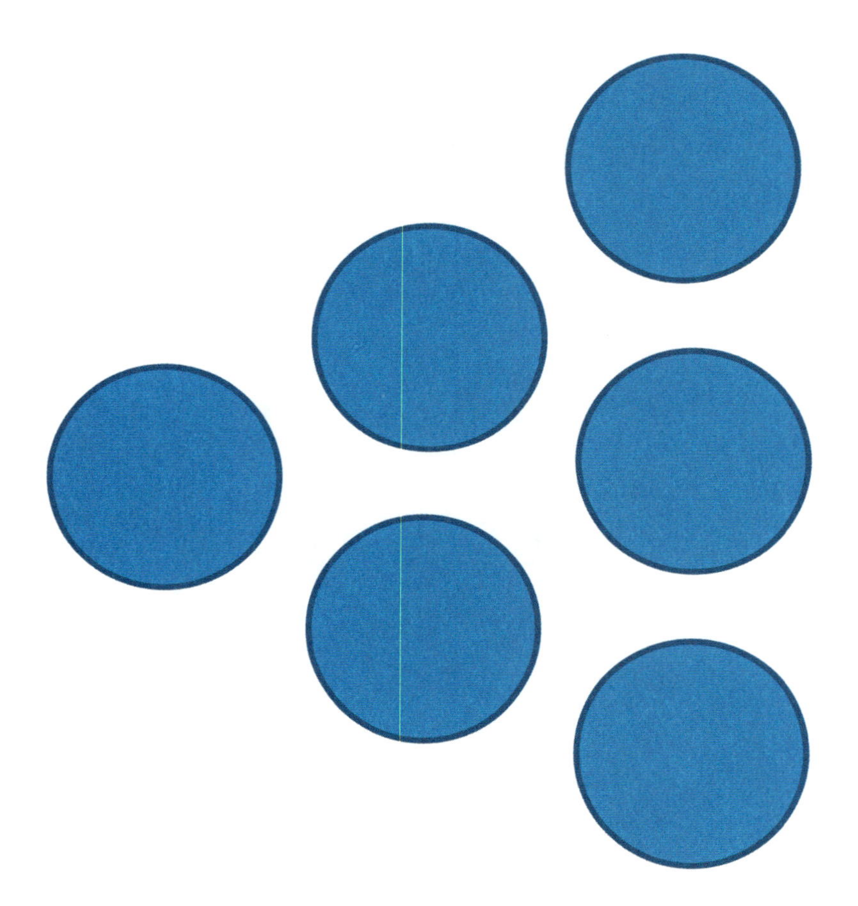

dot cards of 6

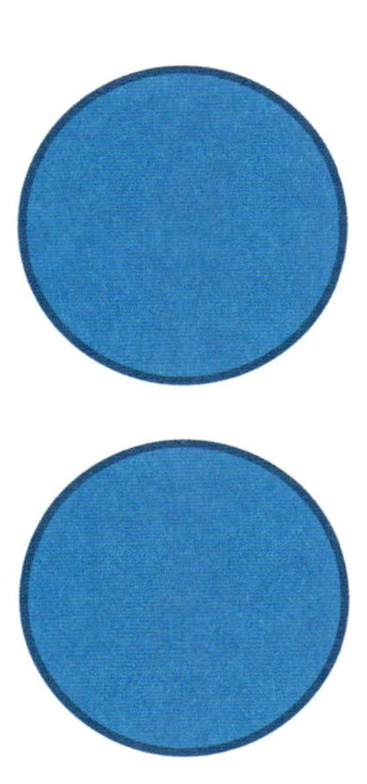

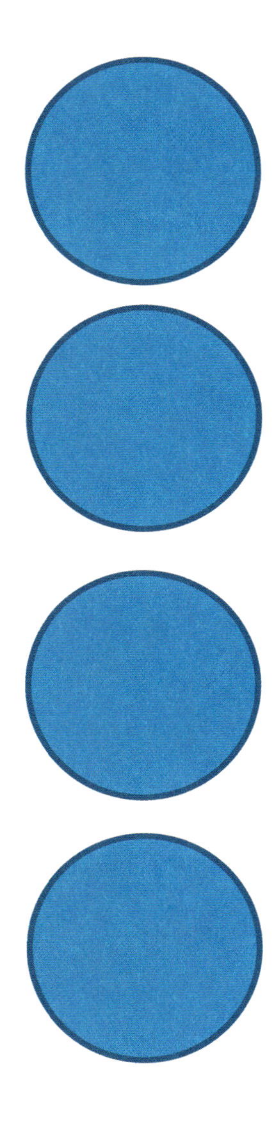

dot cards of 6

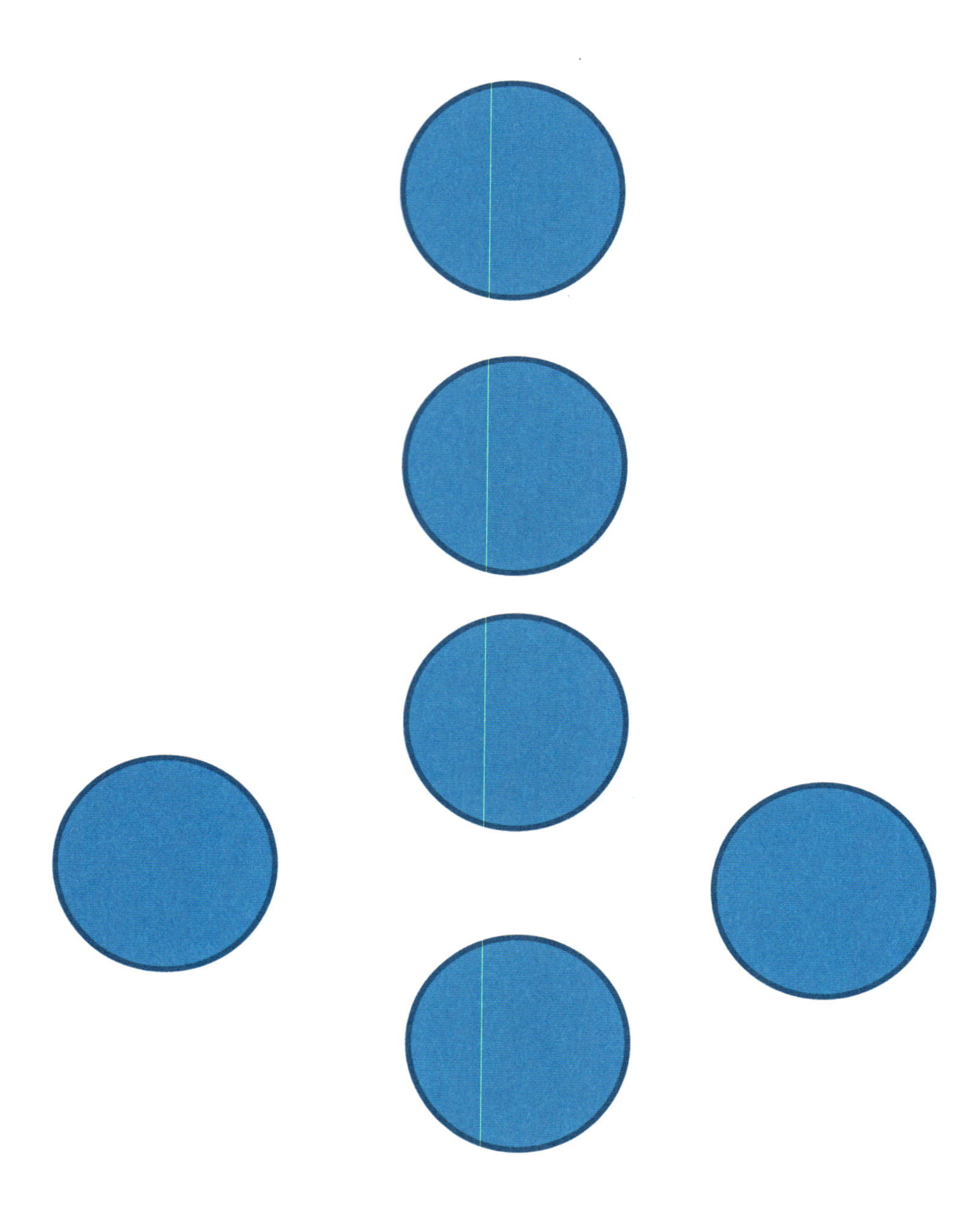

dot cards of 6

dot cards of 6

Name ______________________ Date ____________

Circle 10.

circle 10

Lesson 5

Objective: Count straws the Say Ten way to 20; make a pile for each ten.

Suggested Lesson Structure

Fluency Practice	(12 minutes)
Application Problem	(5 minutes)
Concept Development	(25 minutes)
Student Debrief	(8 minutes)
Total Time	**(50 minutes)**

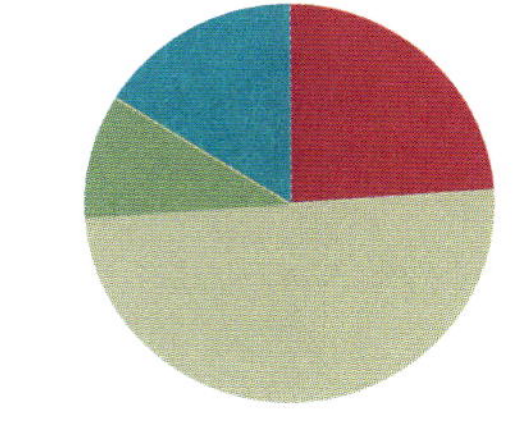

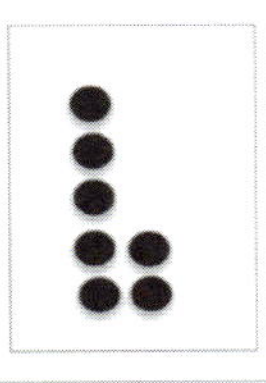

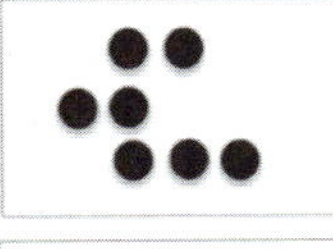

Fluency Practice (12 minutes)

- Dot Cards of Seven **K.CC.2, K.CC.5** (4 minutes)
- Number Pairs of Seven **K.CC.2** (4 minutes)
- Circling 10 Ones **K.NBT.1** (4 minutes)

Dot Cards of Seven (4 minutes)

Materials: (T/S) Dot cards of 7 (Fluency Template 1)

Note: This fluency activity gives students an opportunity to develop increased familiarity with decompositions of seven and practice seeing part–whole relationships.

T: (Show 7 dots.) How many do you see? (Give students time to count.)

S: 7.

T: How can you see 7 in two parts?

S: (Come up to the card.) 5 here and 2 here. I see 3 here and 4 here.

Continue with other cards of seven. Distribute the cards to students for partner sharing time. Have them *pass on* the card to a different set of partners at a signal.

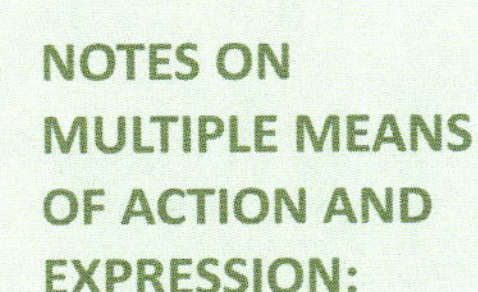

NOTES ON MULTIPLE MEANS OF ACTION AND EXPRESSION:

Students who are working below grade level will need to do more counting. They need more time and may benefit from working with the cards one at a time while you move more rapidly through the cards with the majority of the class.

NOTES ON MULTIPLE MEANS OF ENGAGEMENT:

Let students who are working above grade level work in a small group with more of a flashing approach. Assign one student or classroom helper to be the teacher. Engage them in sharing the different ways they saw the subsets.

Number Pairs of Seven (4 minutes)

Materials: (S) Dot cards of 7 (Fluency Template 1), personal white board

Note: This fluency activity gives students an opportunity to develop increased familiarity with decompositions of seven and practice seeing part–whole relationships.

T: (Indicate 6 and 1 as parts.) Say the larger part.
S: 6.
T: Say the smaller part.
S: 1.
T: What is the total number of dots? (Give students time to recount.)
T: Write the number bond on your personal white board. Continue with 5 and 2, 4 and 3, and 7 and 0.

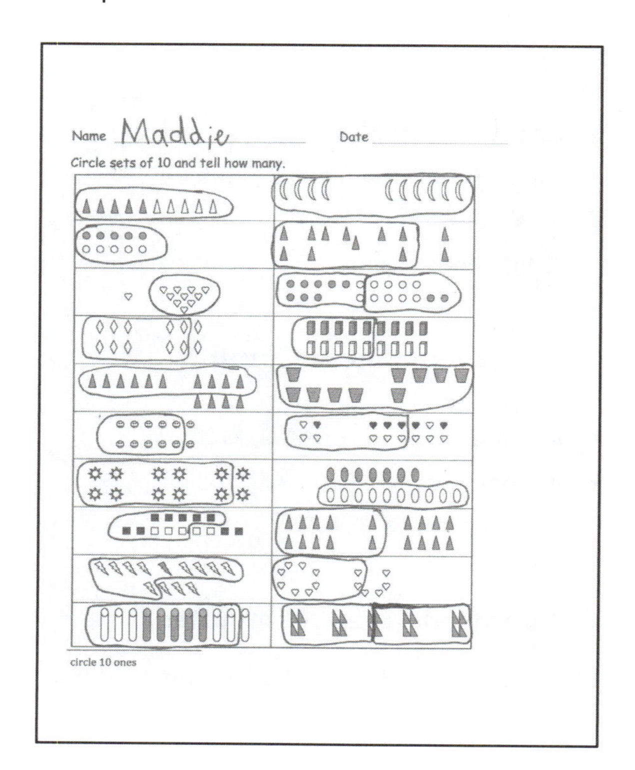
Name Maddie Date

Circle sets of 10 and tell how many.

circle 10 ones

Circling 10 Ones (4 minutes)

Materials: (S) Circle 10 ones (Fluency Template 2) (pictured to the right)

Note: This activity gives students repeated experience in locating 10 ones embedded within a pictorial group of 10 ones and some ones. Challenge students working above grade level to circle a different group of 10 than last time.

Application Problem (5 minutes)

Pat covered 16 holes when playing the flute. She covered 10 holes with her fingers on the first note she played. She covered 6 holes on the next note she played. Draw the 10 holes. Draw the 6 holes. Use your drawing to count all the holes the Say Ten way.

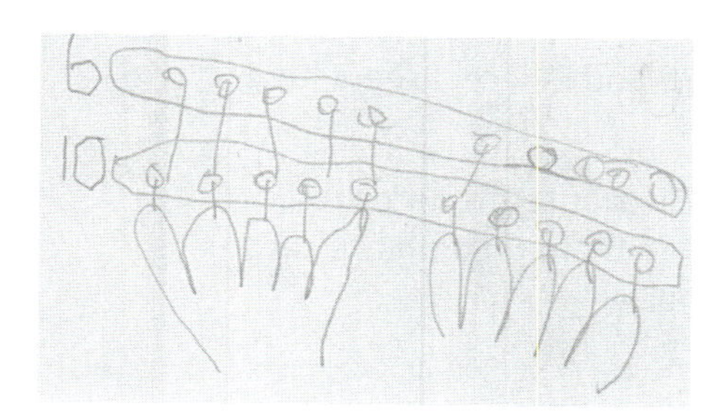

Note: The focus here is on counting to find the total rather than on addition. They are also seeing the embedded 10 and 6 as they count to 16 the Say Ten way.

Concept Development (25 minutes)

NOTES ON MULTIPLE MEANS OF ACTION AND EXPRESSION:

Support English language learners by using gestures during the lesson. Flash 10, and gesture with your hands for the word. Flash 1. Gesture again for the word. This engages students to figure out the intent and bypasses all the potential confusion in oral directions.

Materials: (S) 20 straws (per pair)

T: Come sit with me on the carpet.

T: I'm going to flash numbers with my hands. Tell me the number, and then tell me the number the Say Ten way. Let's do one as an example.

MP.7

T: (Hold out both hands, palms out, to show 10. Then, show your right hand with the pinky extended.)

S: Eleven.

T: The Say Ten way?

S: Ten 1.

T: Perfect. (Show 10 again, and then show 2 on your right hand with the pinky and ring finger.)

S: Twelve! Ten 2.

T: Yes!

T: (Continue this way up to ten nine.) What comes after 19? (Flash 2 tens.)

S: Twenty! 2 tens!

T: Very good! Please return to your seats, and we'll practice counting the Say Ten way using straws. Partner A will count 10 straws into a pile. The other student, Partner B, will place one straw next to the pile and say, "Ten 1." Ready?

S: (Show a pile of 10 straws and 1 one.) Ten 1.

T: Partner B, place another straw. How many straws are there now?

S: Ten 2.

T: (Continue this way up to 2 tens.) How many straws are there?

S: 2 tens!

T: You are all correct! There are 2 piles of 10 straws. We say, "2 tens."

T: Put all the straws back into one pile, and switch roles. Partner B, count out 10 straws into a pile. Partner A, place one straw next to the pile, and let's practice counting again the Say Ten way.

S: (Count up to 2 tens.)

Problem Set (7 minutes)

Students should do their personal best to complete the Problem Set within the allotted time.

Direct students to circle 10 objects and check the extra ones. Have them count the total using the Say Ten way. Watch to see that they count the 10 ones within the circle first from left to right, row by row. They then match the drawing to its numerical representation.

Student Debrief (8 minutes)

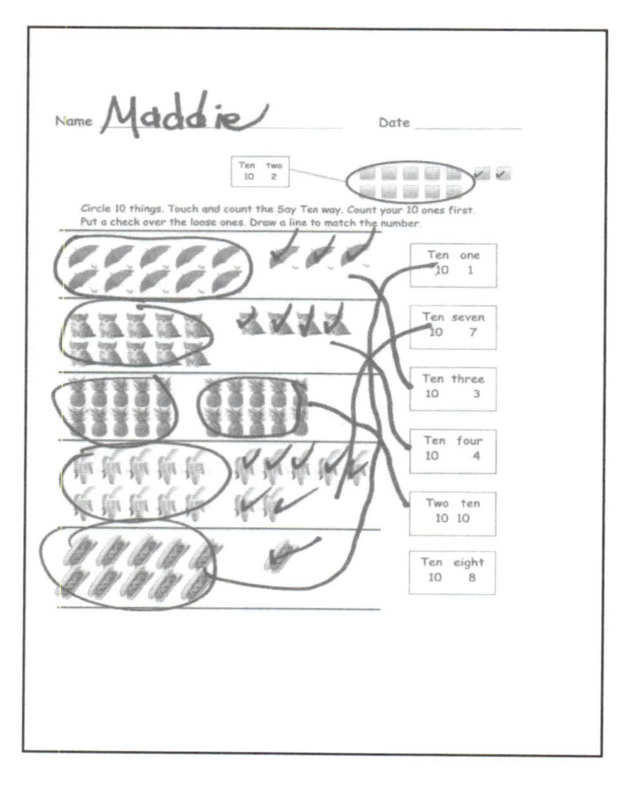

Lesson Objective: Count straws the Say Ten way to 20; make a pile for each ten.

The Student Debrief is intended to invite reflection and active processing of the total lesson experience.

Invite students to review their solutions for the Problem Set. They should check work by comparing answers with a partner before going over answers as a class. Look for misconceptions or misunderstandings that can be addressed in the Debrief. Guide students in a conversation to debrief the Problem Set and process the lesson.

Any combination of the questions below may be used to lead the discussion.

- Look at your circle 10 ones template. Can you say the numbers the Say Ten way?
- Did your friend circle 10 ones the same way you did?
- Were both your answers correct? Why?
- How do we say 2 tens as one number?
- How do we say 17 the Say Ten way?
- Which pictures were the easiest for you to count? Why?
- Look at your Problem Set. Tell your partner what makes it easy for you to count.
- What is the same about all of the pictures? What is different?

Exit Ticket (3 minutes)

After the Student Debrief, instruct students to complete the Exit Ticket. A review of their work will help with assessing students' understanding of the concepts that were presented in today's lesson and planning more effectively for future lessons. The questions may be read aloud to the students.

Name ______________________________ Date ______________

Ten two
10 2

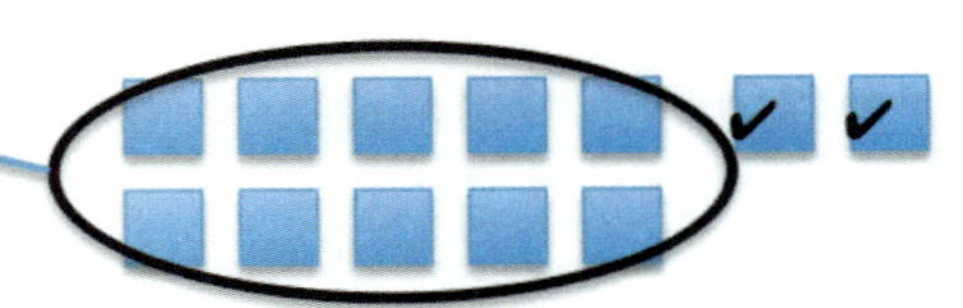

Circle 10 things. Touch and count the Say Ten way. Count your 10 ones first. Put a check over the loose ones. Draw a line to match the number.

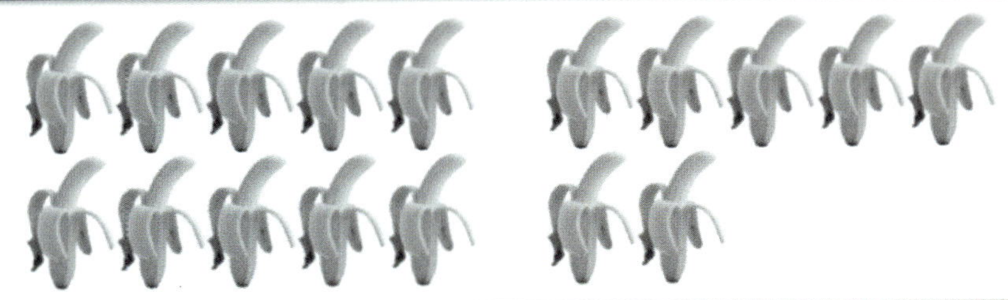

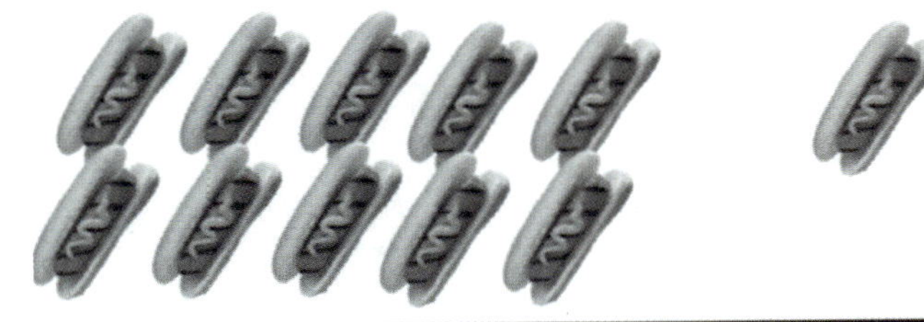

Ten one
10 1

Ten seven
10 7

Ten three
10 3

Ten four
10 4

Two ten
10 10

Ten eight
10 8

Name ________________________________ Date _______________

Write and whisper the missing numbers.
Count the Say Ten way from 11 to 20.

10 and 1	10 and 2	10 and ___	10 and 4	10 and ___
10 and 6	___ and ___	___ and ___	___ and ___	10 and 10

Name ______________________________ Date ______________

Write the numbers that go before and after, counting the Say Ten way.

BEFORE	NUMBER	AFTER
10 and 3	10 and 4	10 and 5
and	10 and 2	and
and	10 and 5	and
and	10 and 6	and
and	10 and 1	and
and	10 and 9	and

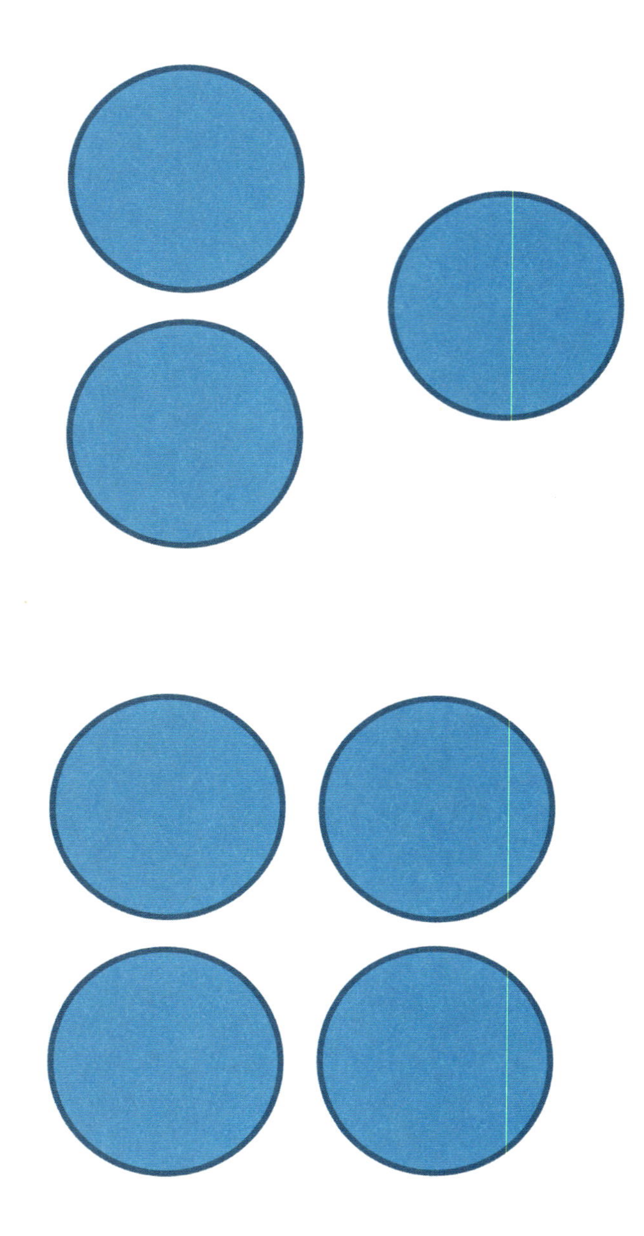

dot cards of 7

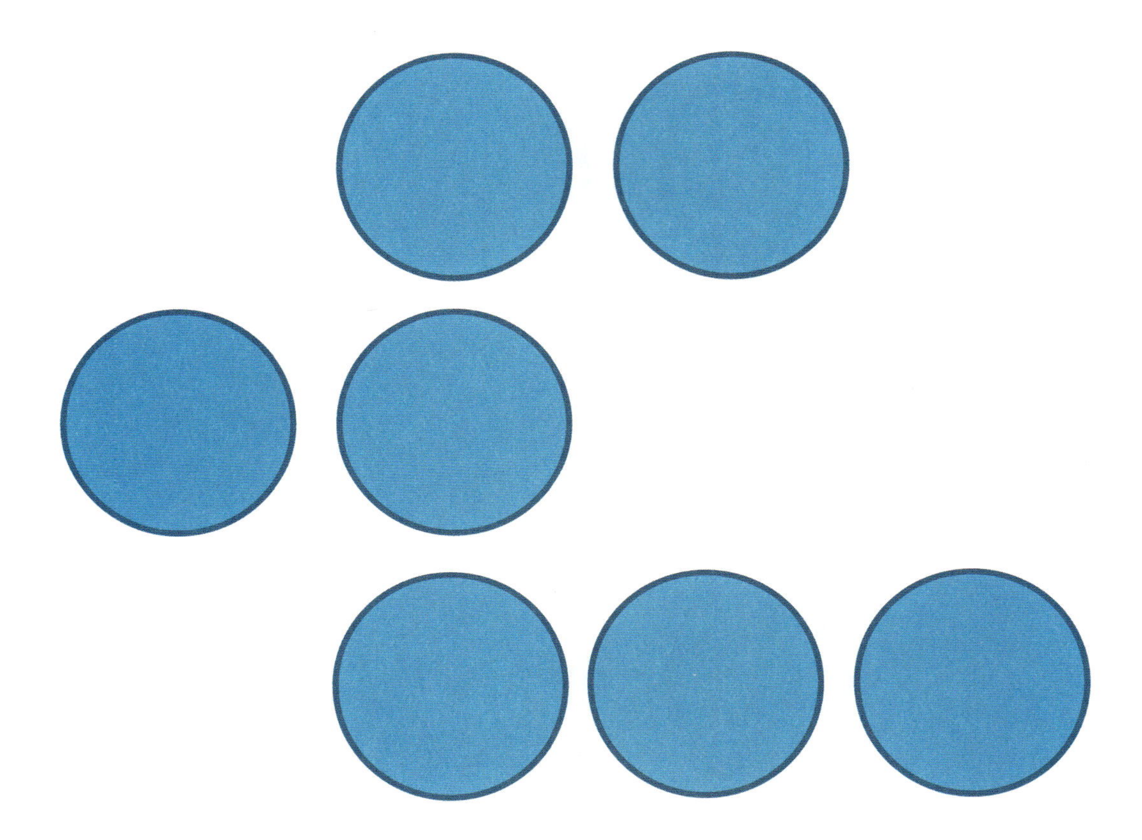

dot cards of 7

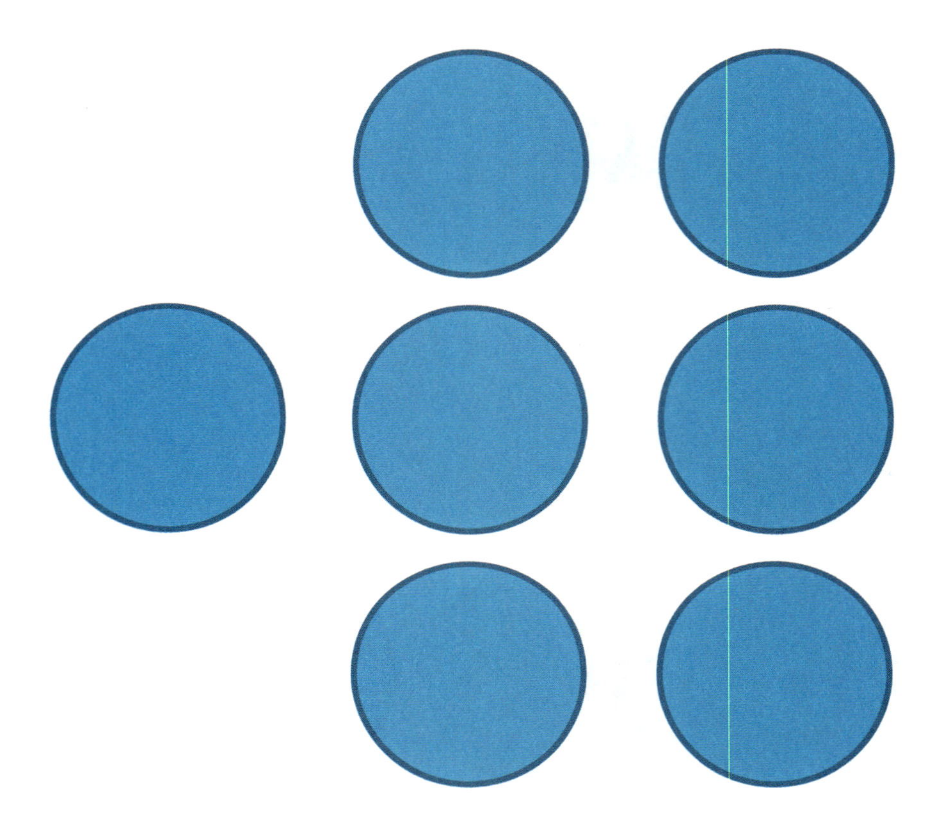

dot cards of 7

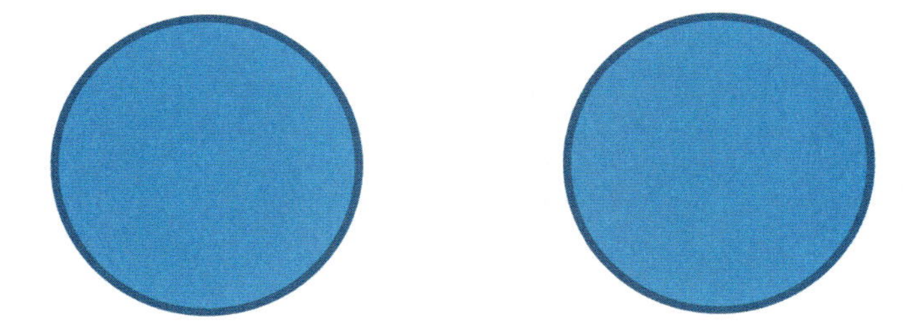

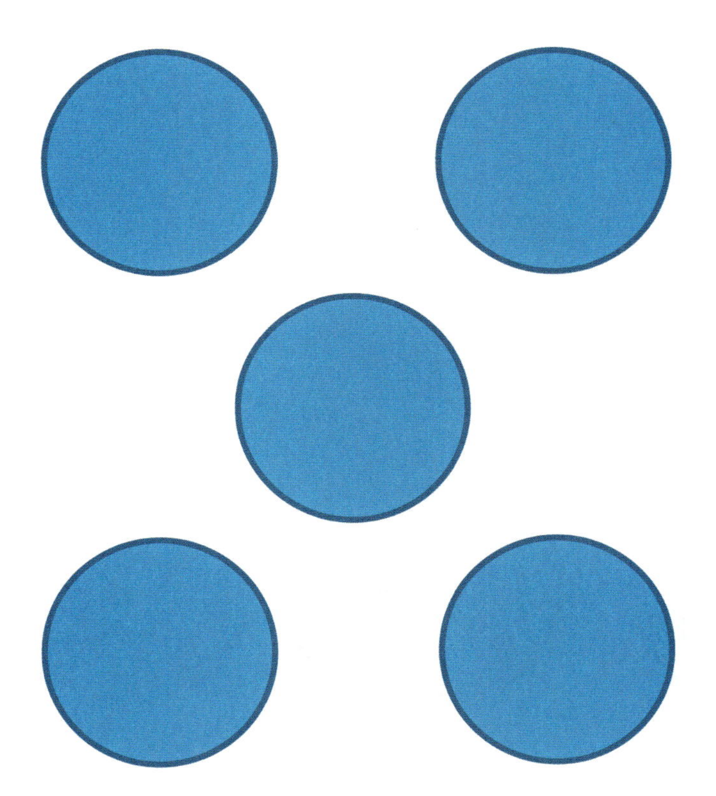

dot cards of 7

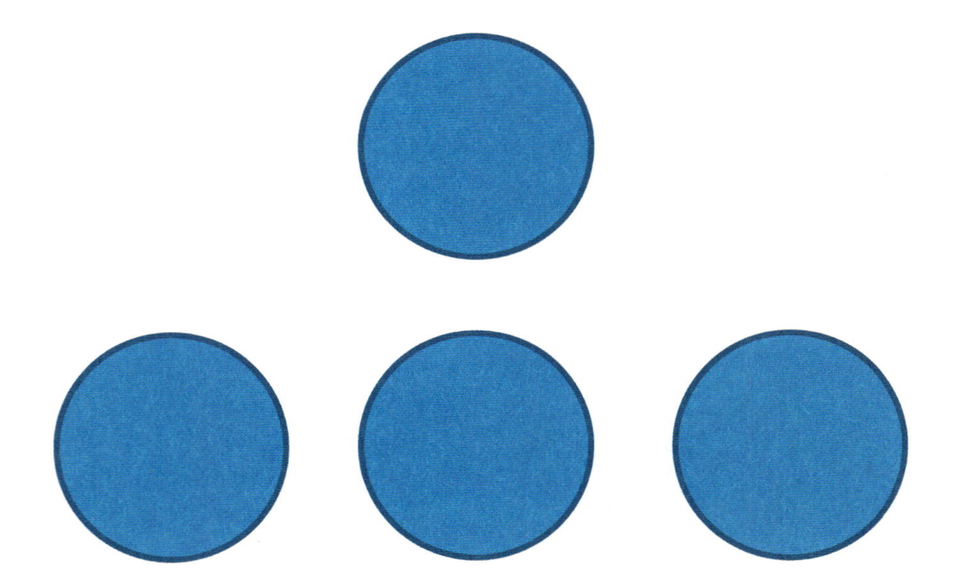

dot cards of 7

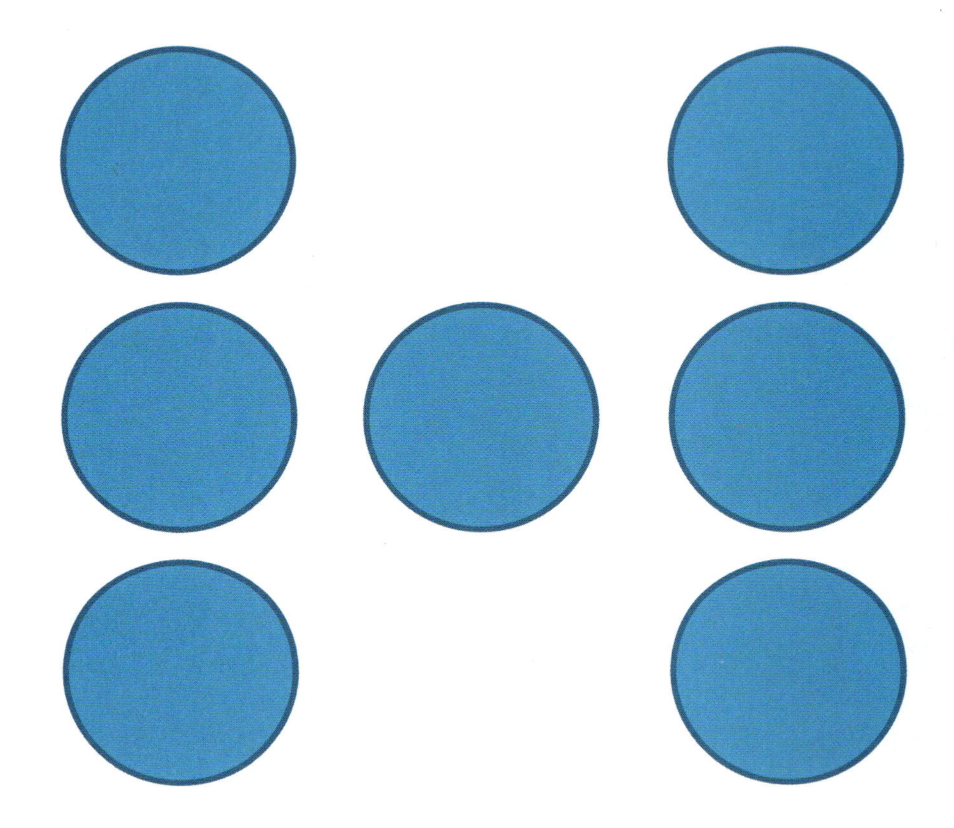

dot cards of 7

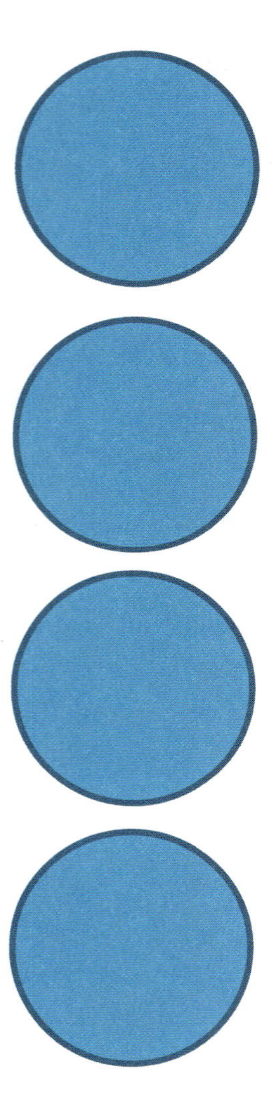

dot cards of 7

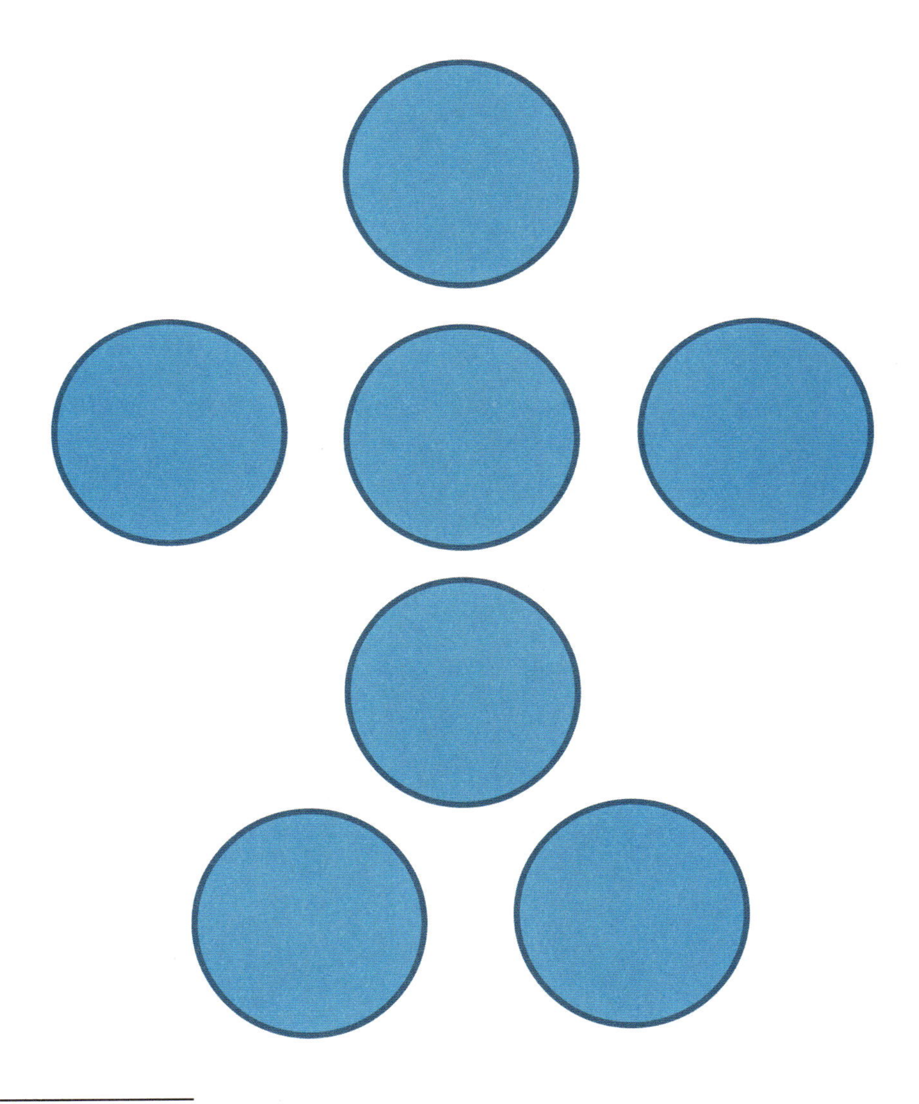

dot cards of 7

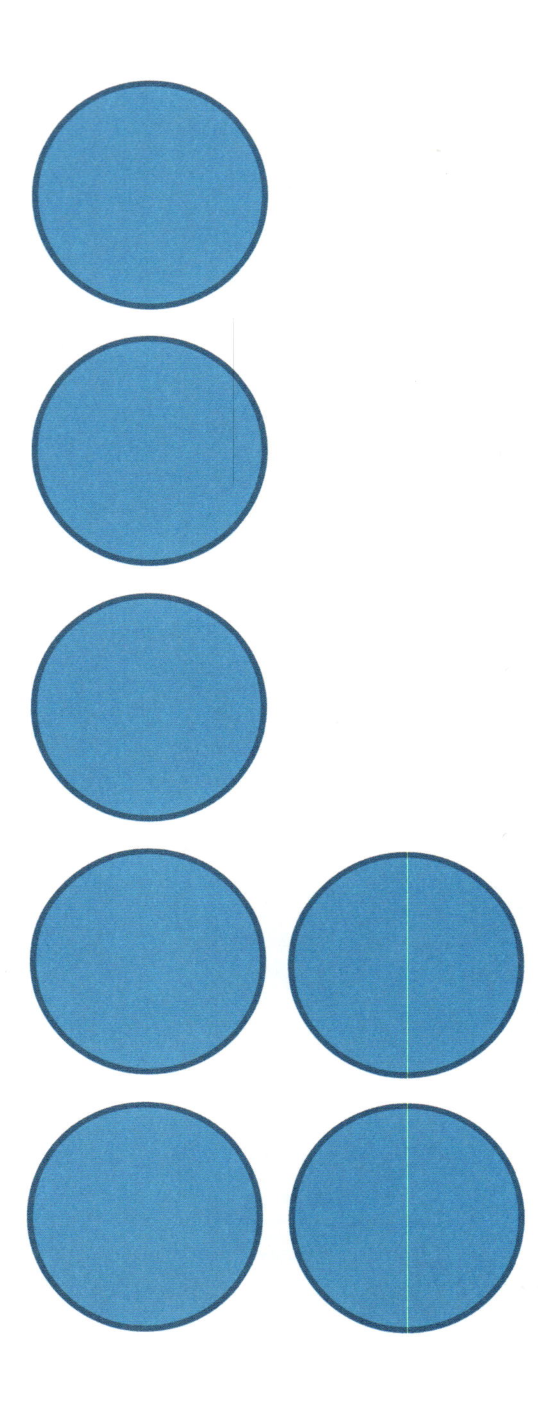

dot cards of 7

Name ______________________________ Date ______________

Circle sets of 10, and tell how many.

circle 10 ones

Topic B

Compose Numbers 11–20 from 10 Ones and Some Ones; Represent and Write Teen Numbers

K.CC.3, K.NBT.1, K.CC.1, K.CC.2, K.CC.4a, K.CC.4b, K.CC.4c, K.CC.5

Focus Standards:		K.CC.3	Write numbers from 0 to 20. Represent a number of objects with a written numeral 0–20 (with 0 representing a count of no objects).
		K.NBT.1	Compose and decompose numbers from 11 to 19 into ten ones and some further ones, e.g., by using objects or drawings, and record each composition or decomposition by a drawing or equation (e.g., 18 = 10 + 8); understand that these numbers are composed of ten ones and one, two, three, four, five, six, seven, eight, or nine ones.
Instructional Days:		4	
Coherence	**-Links from:**	GPK–M5	Addition and Subtraction Stories and Counting to 20
	-Links to:	G1–M2	Introduction to Place Value Through Addition and Subtraction Within 20

In Topic B, students advance to a more abstract level, representing the decomposition of teen numbers first with Hide Zero cards (place value cards) and in Lesson 7 with number bonds. They then work from the abstract to the concrete and pictorial in Lessons 8 and 9 as they are directed to "show (and in Lesson 9 draw) me this many cubes (as teacher displays 13)."

Application Problems in Topic B are experiences with decomposition and composition of teen numbers (**K.NBT.1**) rather than word problems (**1.OA.1**). For example, in Lesson 7, the problem reads, "Gregory drew 10 smiley faces and 5 smiley faces. He put them together and had 15 smiley faces. Draw his 15 smiley faces as 10 smiley faces and 5 smiley faces." In this instance, there is no unknown. We do not ask, "How many in all?" or "How many?" as within a word-problem setting. The students represent 15 with their Hide Zero cards, both when the zero is hiding and when it is not hiding, as they apply all their experiences from Topic A to deeply understand the meaning of the digit 1 in the tens place in teen numbers.

GK-M5-TE-B5-1.3.1-01.2016

A Teaching Sequence Toward Mastery of Composing Numbers 11–20 from 10 Ones and Some Ones; Representing and Writing Teen Numbers

Objective 1: **Model with objects and represent numbers 10 to 20 with place value or Hide Zero cards.** **(Lesson 6)**

Objective 2: **Model and write numbers 10 to 20 as number bonds.** **(Lesson 7)**

Objective 3: **Model teen numbers with materials from abstract to concrete.** **(Lesson 8)**

Objective 4: **Draw teen numbers from abstract to pictorial.** **(Lesson 9)**

Lesson 6

Objective: Model with objects and represent numbers 10 to 20 with place value or Hide Zero cards.

Suggested Lesson Structure

Fluency Practice	(12 minutes)
Application Problem	(6 minutes)
Concept Development	(24 minutes)
Student Debrief	(8 minutes)
Total Time	**(50 minutes)**

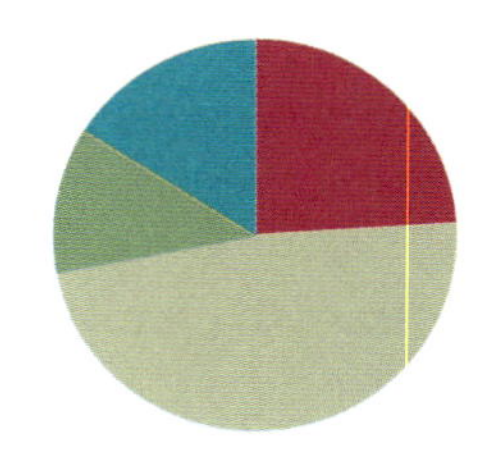

Fluency Practice (12 minutes)

- How Many More to Make 10? **K.CC.2** (4 minutes)
- Dot Cards of Eight **K.CC.2, K.CC.5** (4 minutes)
- Counting Straws the Say Ten Way **K.CC.2** (4 minutes)

How Many More to Make 10? (4 minutes)

Materials: (T/S) Large 5-group cards (Lesson 1 Fluency Template 1) (S) 5-group cards (Lesson 1 Fluency Template 2)

Note: This activity helps students develop automaticity with partners to 10 through visualizing with the 5-group model.

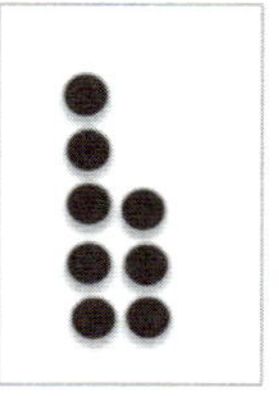

T: (Show 5.) How many dots do you see?
S: 5.
T: How many more does 5 need to make 10?
S: (Full sentence.) 5 needs 5 more to make 10.

Continue with the following possible sequence: 9, 8, 7, 6, 1, 4, 3, 9, 2, 5.
Allow students to play with a partner briefly.

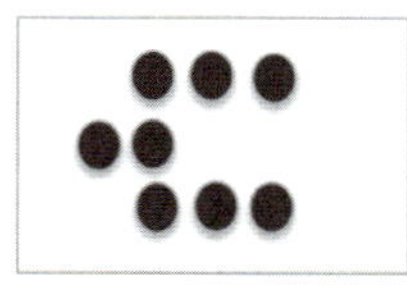

Dot Cards of Eight (4 minutes)

Materials: (T/S) Dot cards of 8 (Fluency Template)

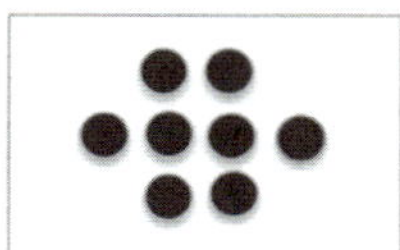

Note: This fluency activity gives students an opportunity to develop increased familiarity with decompositions of eight and practice seeing part–whole relationships.

T: (Show a card with 8 dots.) How many dots do you count? Wait for the signal to tell me.

S: 8.

T: How can you see them in 2 parts?

S: (Students come up to the card.) I saw 4 here and 4 here. → I saw 5 here and and 3 here. → I saw 6 here and 2 here.

Repeat with other cards. Pass out the cards for students to work with a partner.

Counting Straws the Say Ten Way (4 minutes)

Materials: (T) Large 5-group cards (Lesson 1 Fluency Template 1) (S) 5-group cards (Lesson 1 Fluency Template 2); 20 straws (per pair)

Note: Counting the Say Ten way prepares students to think of ten as part of a teen number in today's Concept Development.

T: (Show 10 and 3.) Say the number the Say Ten way.

S: Ten 3.

T: Count out that many straws with your partner.

Repeat the process with other teen numbers. Give students time to practice this exercise with a partner briefly.

Application Problem (6 minutes)

There are 18 students: 10 girls and 8 boys. Show the 18 students as 10 girls and 8 boys.

Note: Remember that the focus is on counting all to find the total rather than counting on or addition.

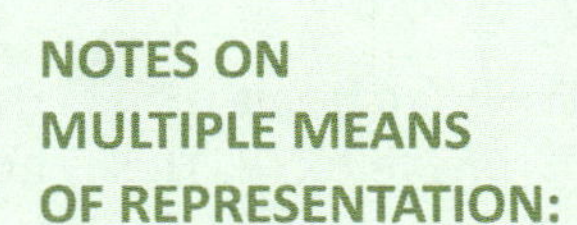

NOTES ON MULTIPLE MEANS OF REPRESENTATION:

Support English language learners by matching the linking cubes to the quantity and picture of the girls and boys from the Application Problem. This way, when asked, "What color is represented by the girls?" and "What color is represented by the boys?" students will already know the answer and can focus on answering mathematical questions.

Concept Development (24 minutes)

Materials: (T) Large Hide Zero cards (Template 1) (S) Hide Zero cards: 1 Hide Zero 10 card (Template 2) with 5-group cards 1–9 (Lesson 1 Fluency Template 2), two sets of 10 linking cubes (10 in one color and 10 in another color), personal white board (per pair)

T: Have one color of your cubes represent the boys and another one the girls from the story in the Application Problem. Show me the boys and girls that were in school. When you are done, check your partner's work to be sure you agree.

T: (Allow students time to finish.) Everyone hold up the stick that represents the girls. (Students do so.) Hold up the stick that represents the boys. (Students do so.)

T: How many girls are there?

S: 10 girls.

T: Show the girls. (Students show again.) Here is the number 10. (Show the 10 card.)

T: How many boys are there?

S: 8 boys.

T: Show the boys. (Students show again.) Here is the number 8. (Show the 8 card.)

T: Put the boys together with the girls. Count with your partner the Say Ten way to see how many students you have.

S: 1, 2, 3, 4, 5, 6, 7, 8, 9, 10, ten 1, ten 2, ten 3, ten 4, ten 5, ten 6, ten 7, ten 8. (Have early finishers count down to 1 from 18.)

T: How do we say the number of students the Say Ten way?

S: Ten 8.

T: Watch this magic. Here is my 10. Here is my 8. I push them together, and I have ten 8. This is how we write ten 8. (Pull the cards apart, and push them together a few times.)

1 0

8

T: Talk to your partner. What happened to the 0 of the 10 ones?

S: It went under the 8. → It disappeared. → It isn't there anymore. → It is hiding.

T: Yes! It is hiding. I'm going to write the number without the cards. (Write 18.) It is like there is a 0 hiding under this 8.

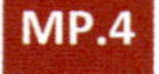

T: I want each of you to write this number on your personal white board. When I say to show me your board, show me.

S: (Write 18 on the personal white board.)

T: Show me!

S: (Hold up personal white board.)

T: Here is a bag with a set of these cards for you. Partner A, open the bag, and put all the numbers on your work mat. With your partner, put them in order from 1 to 10. (Wait.)

T: Partner B, show me ten 8 with your cards. Be sure to hide the zero!

T: Partner A, on this first turn, you will use the cubes. Partner B, you will use the cards and write the number on your personal white board.

T: Partners, show me ten 1.

T: Partner B, use the cubes, and Partner A, use the cards. Show me ten 5.

Continue the activity using other numbers. Different groups might work at varying speeds.

After about four different numbers, change the mode of representation from linking cubes to the dot side of the Hide Zero cards. Have students place the cards in decreasing order from 10 to 1 (for variety), and then match them with the corresponding numeral side. Repeat the process with about four more numbers.

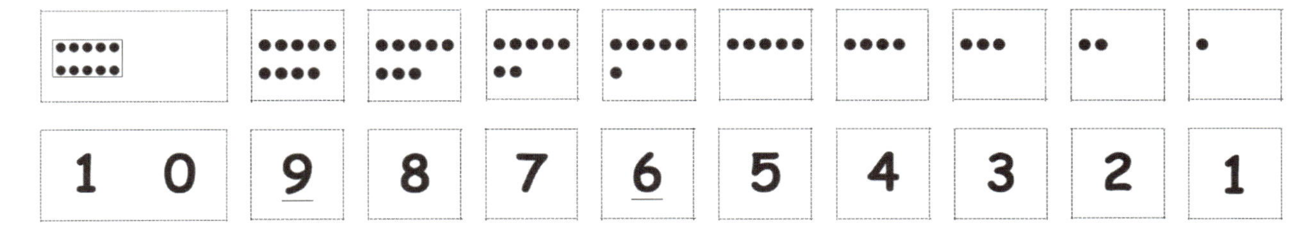

Problem Set (7 minutes)

Students should do their personal best to complete the Problem Set within the allotted time.

Have students use their Hide Zero cards while doing the Problem Set, drawing the number represented and then writing the teen number.

Early finishers can be given another number to represent both pictorially and with cards on the back.

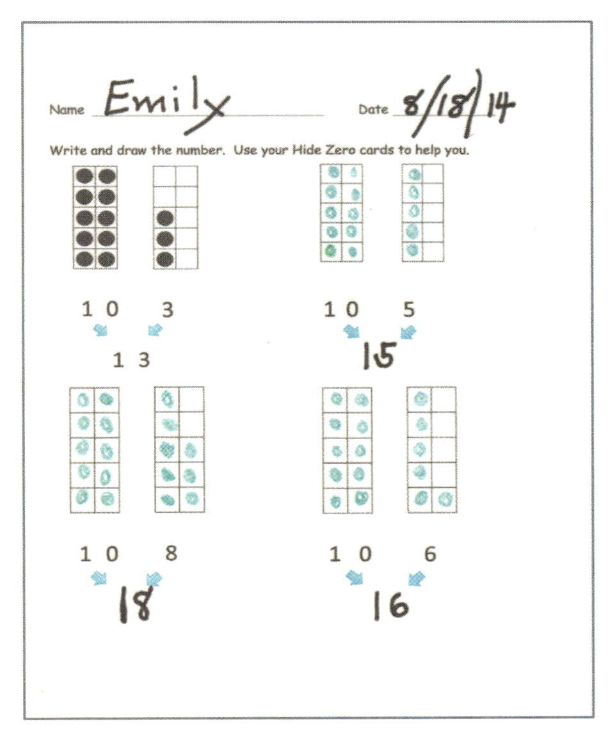

Student Debrief (8 minutes)

Lesson Objective: Model with objects and represent numbers 10 to 20 with place value or Hide Zero cards.

The Student Debrief is intended to invite reflection and active processing of the total lesson experience.

Invite students to review their solutions for the Problem Set. They should check work by comparing answers with a partner before going over answers as a class. Look for misconceptions or misunderstandings that can be addressed in the Debrief. Guide students in a conversation to debrief the Problem Set and process the lesson.

Introduce the cards as **Hide Zero cards**. Then, possibly discuss:

- Why do you think we call these cards Hide Zero cards?

- How is the number made by the Hide Zero cards different from, and the same as, the number written with pencil?
- How do the cards help you to understand the number 13? 18?
- If you didn't know the 0 was hiding, you might think the 1 in 13 was equal to 1 instead of 10. Then, the total value would be 4 because 1 + 3 is 4.

NOTES ON MULTIPLE MEANS OF ACTION AND EXPRESSION:

Students working below grade level will benefit from additional hands-on time with a Rekenrek. Look for opportunities to give them control of the movement of the beads. They may move the beads slowly or erratically. This allows students to hold a number in their minds and wait for the movement of the bead rather than simply rote count.

Exit Ticket (3 minutes)

After the Student Debrief, instruct students to complete the Exit Ticket. A review of their work will help with assessing students' understanding of the concepts that were presented in today's lesson and planning more effectively for future lessons. The questions may be read aloud to the students.

Name ______________________________ Date ______________

Write and draw the number. Use your Hide Zero cards to help you.

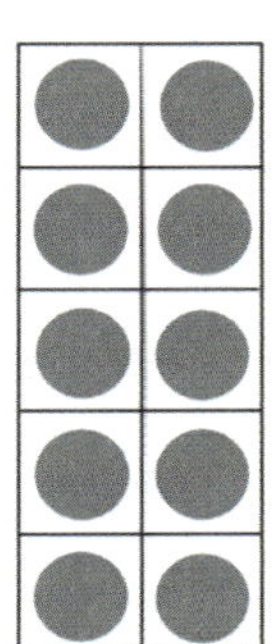
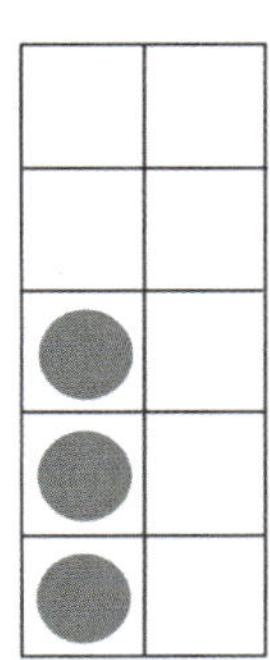

1 0   3

1 3

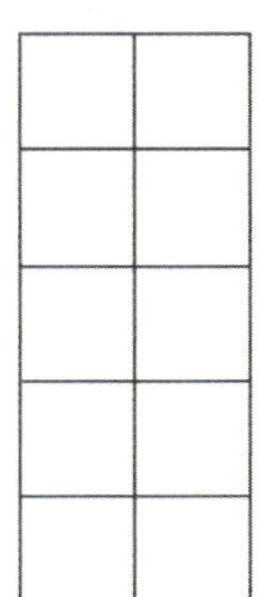
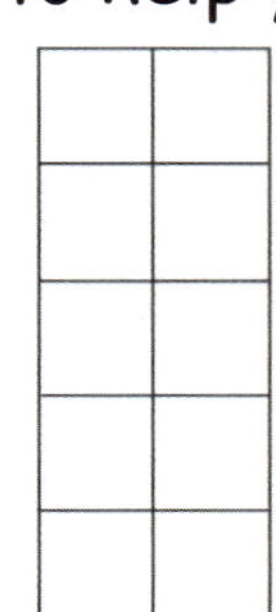

1 0   5

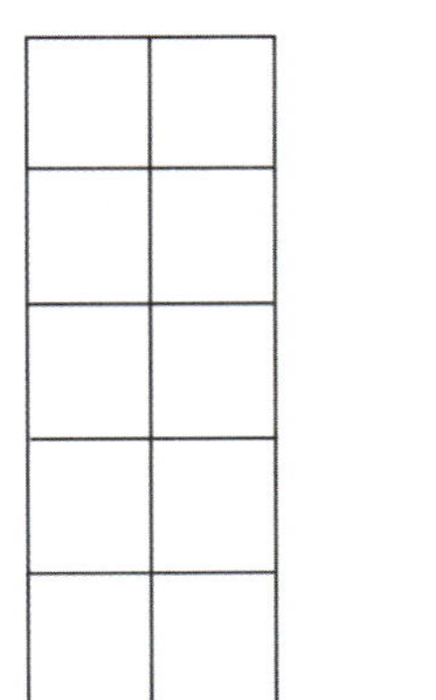
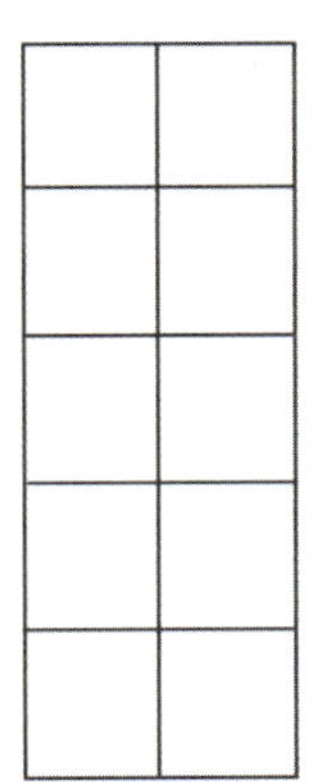

1 0   8

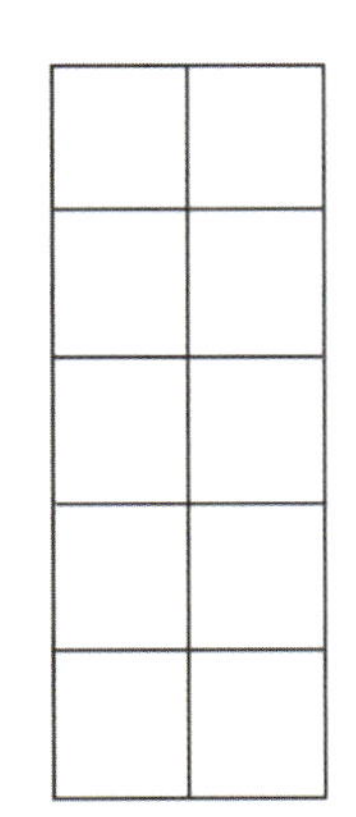
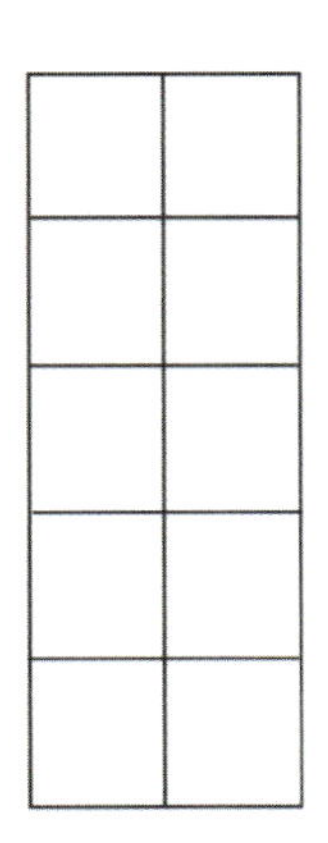

1 0   6

Name ______________________________ Date ______________

Draw the number shown on the Hide Zero cards with a drawing in the ten-frame. Write the number below after the 0 is hidden.

Show the number again on the right with a count of 10 ones and 4 ones. Circle the 10 ones.

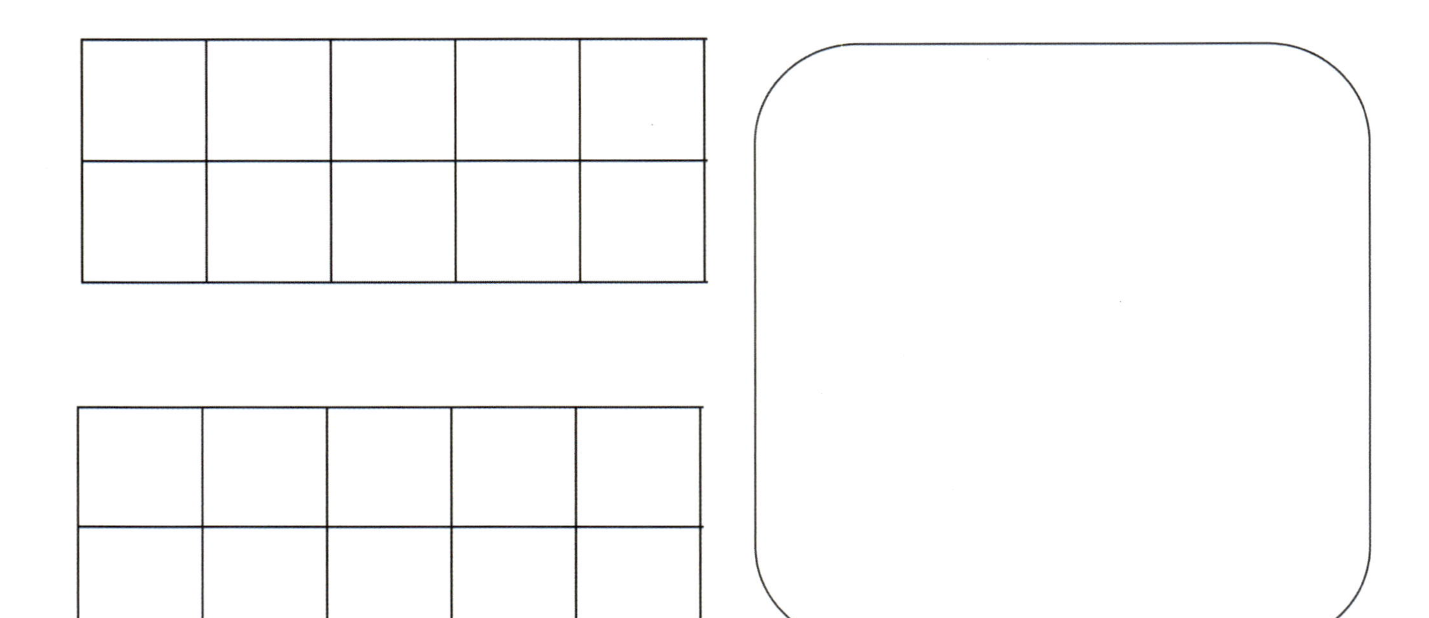

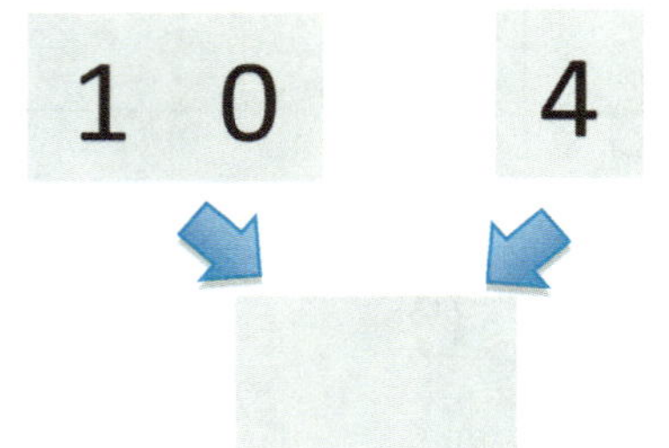

EUREKA MATH

Name ______________________ Date ____________

Write and draw the number. Use your Hide Zero cards to help you.

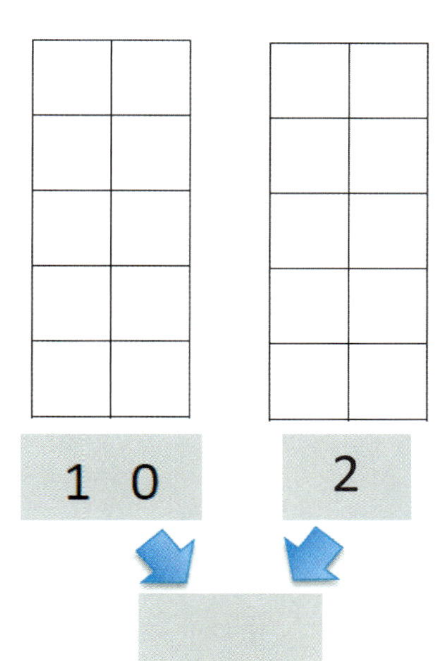

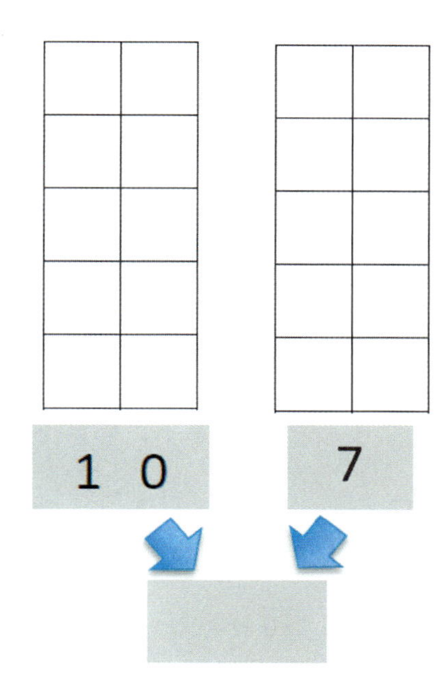

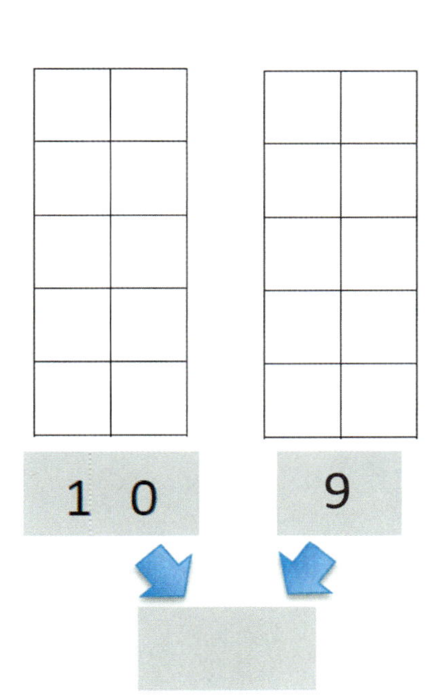

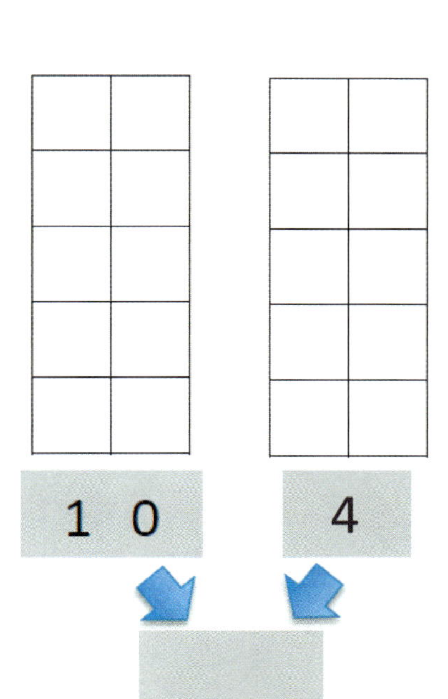

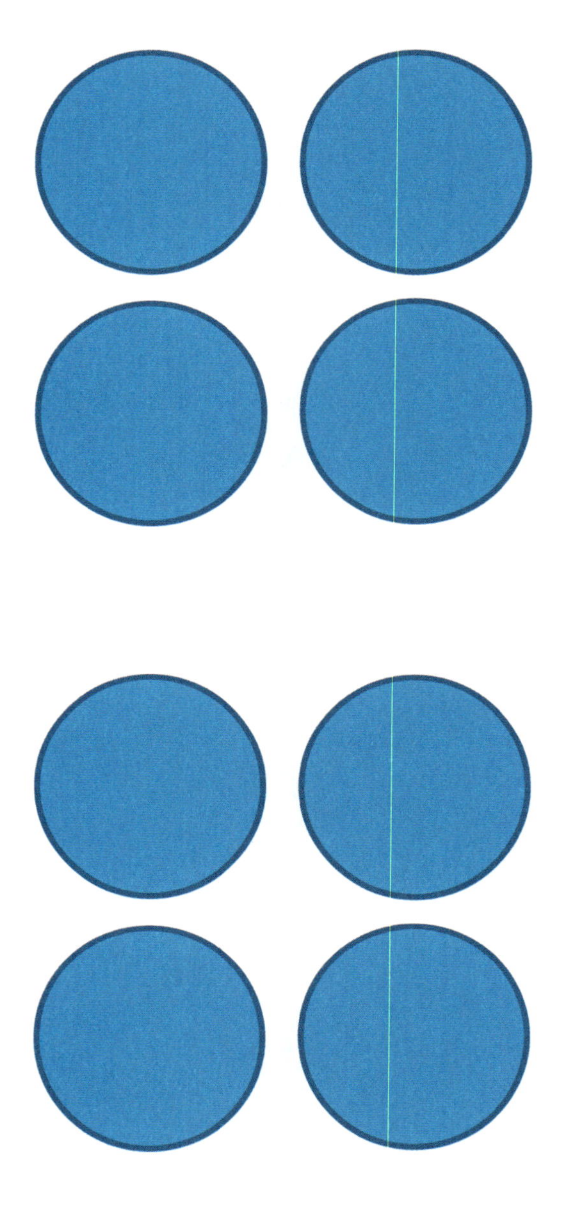

dot cards of 8

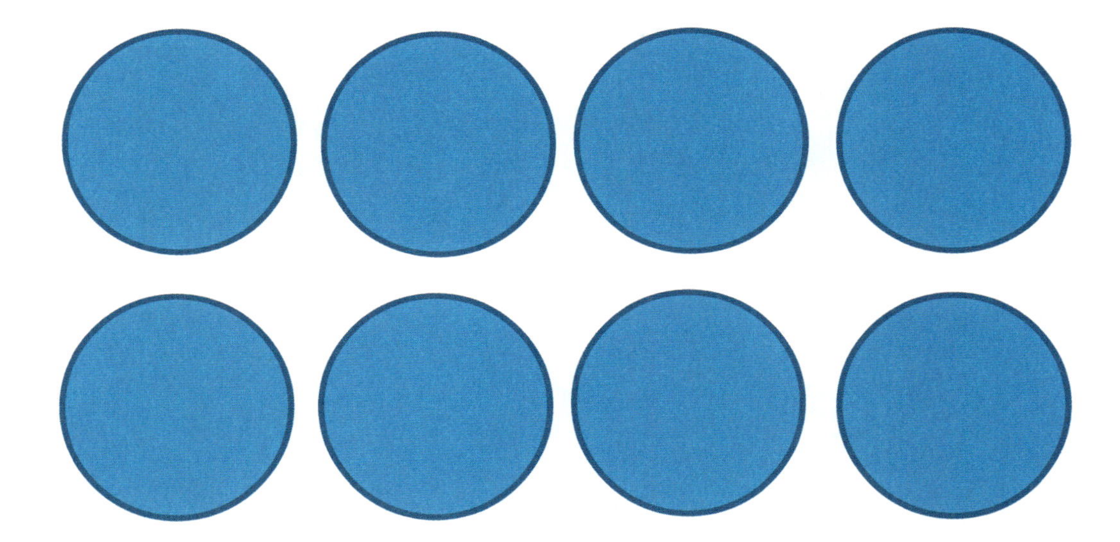

dot cards of 8

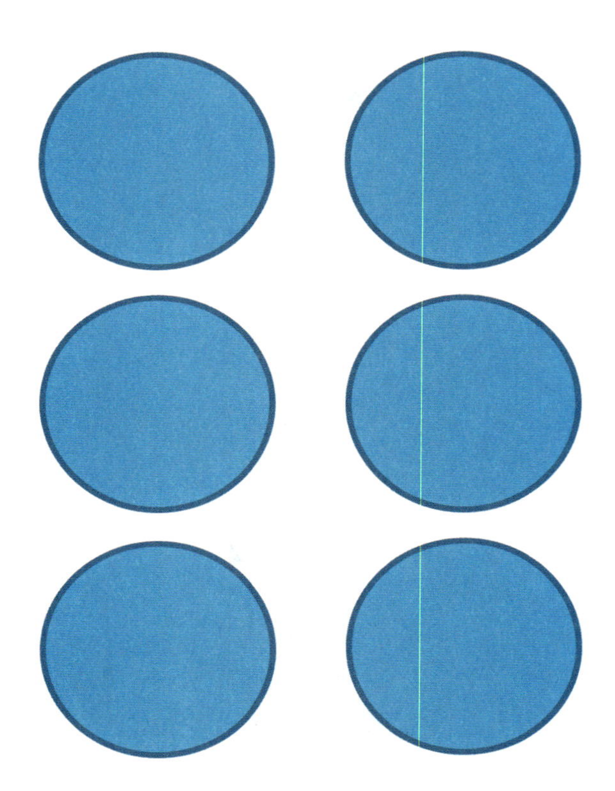

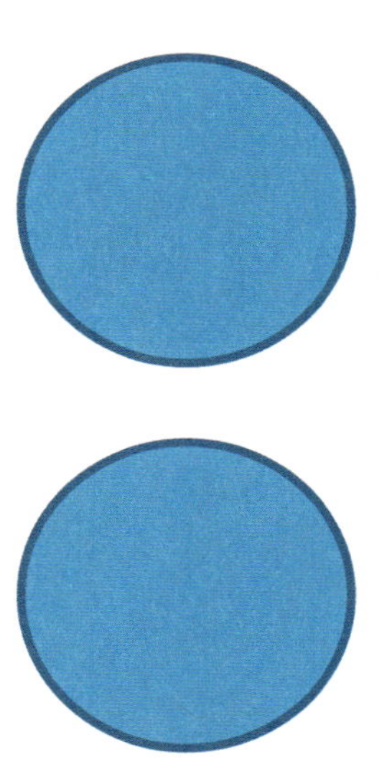

dot cards of 8

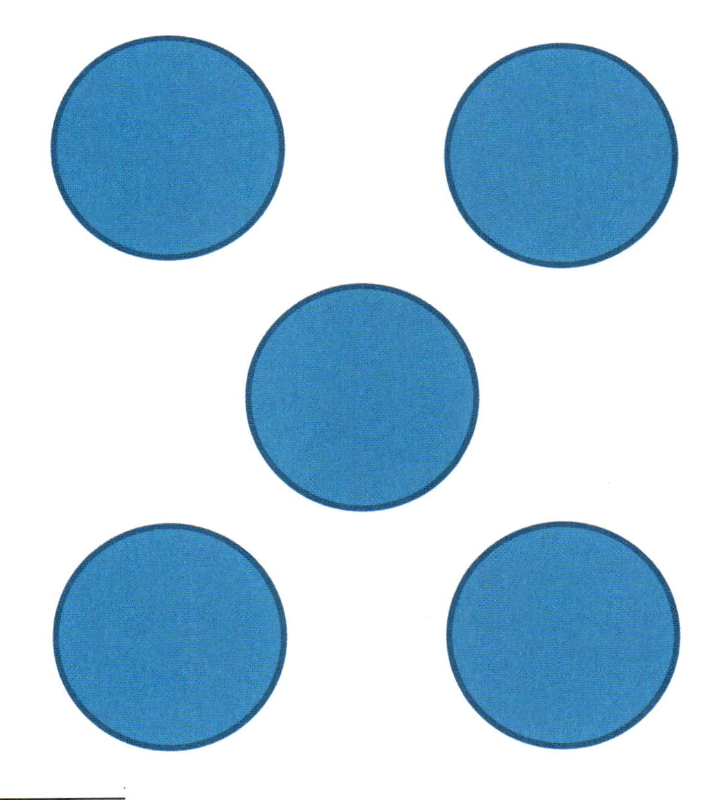

dot cards of 8

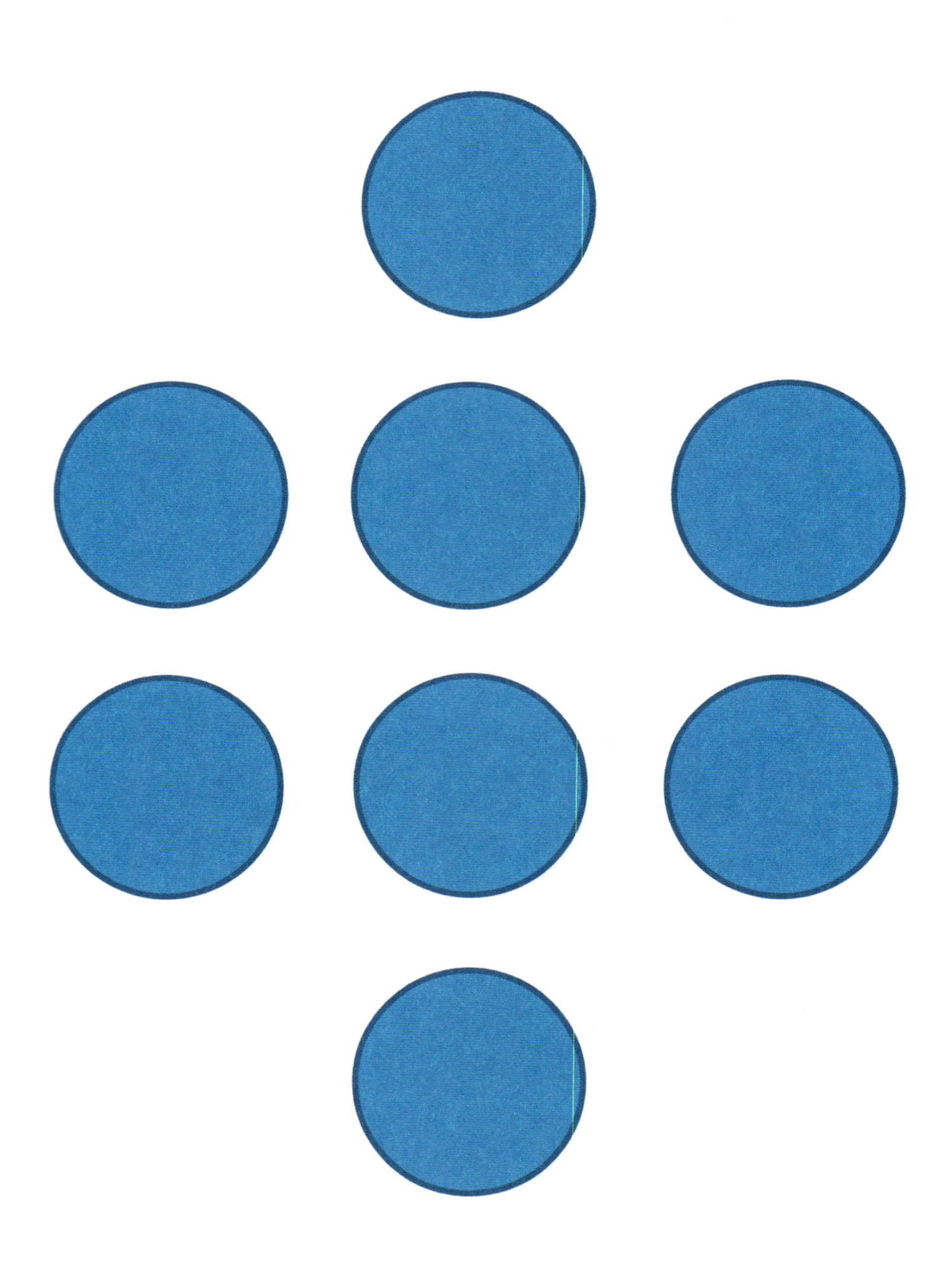

dot cards of 8

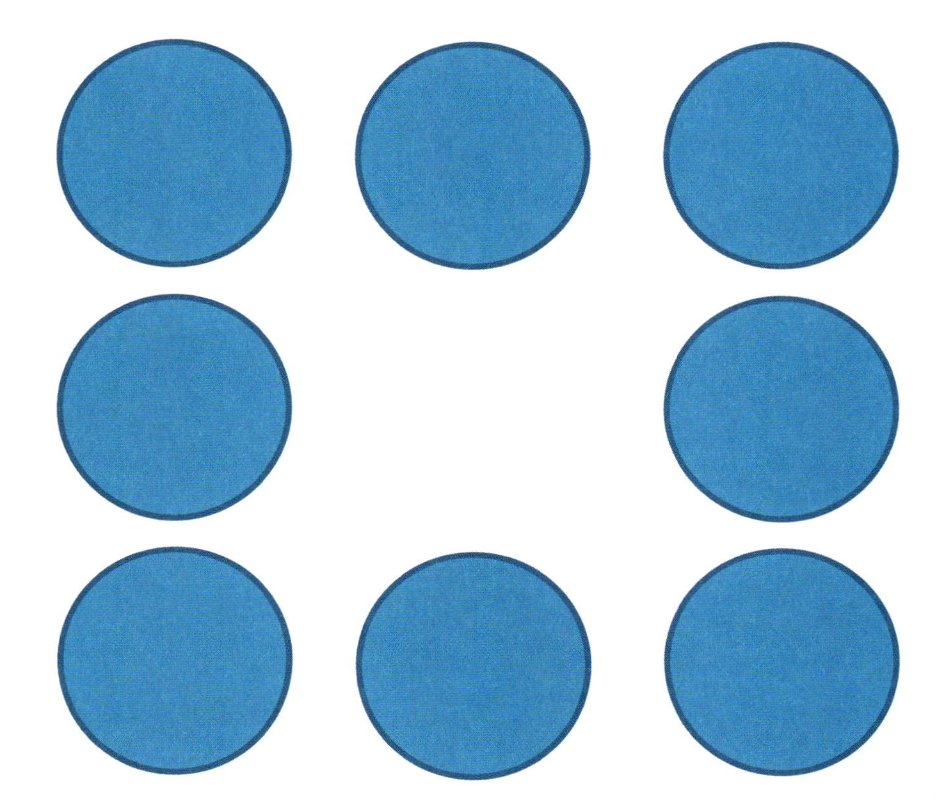

dot cards of 8

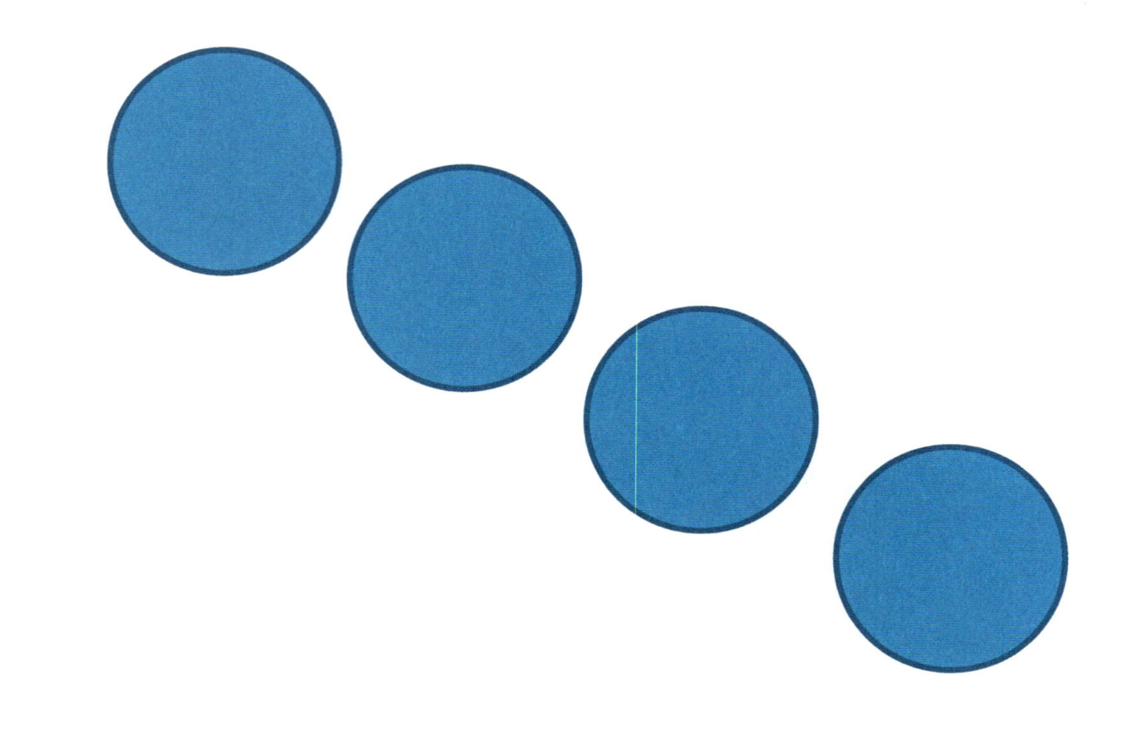

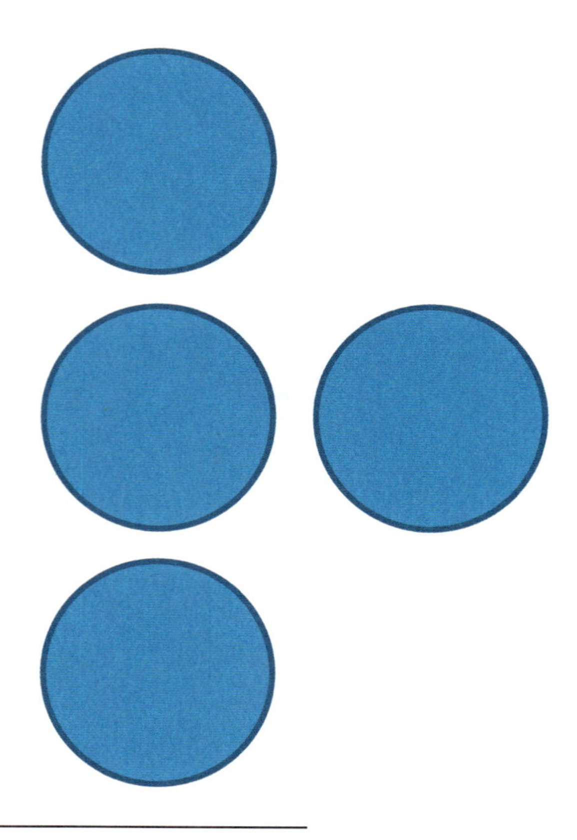

dot cards of 8

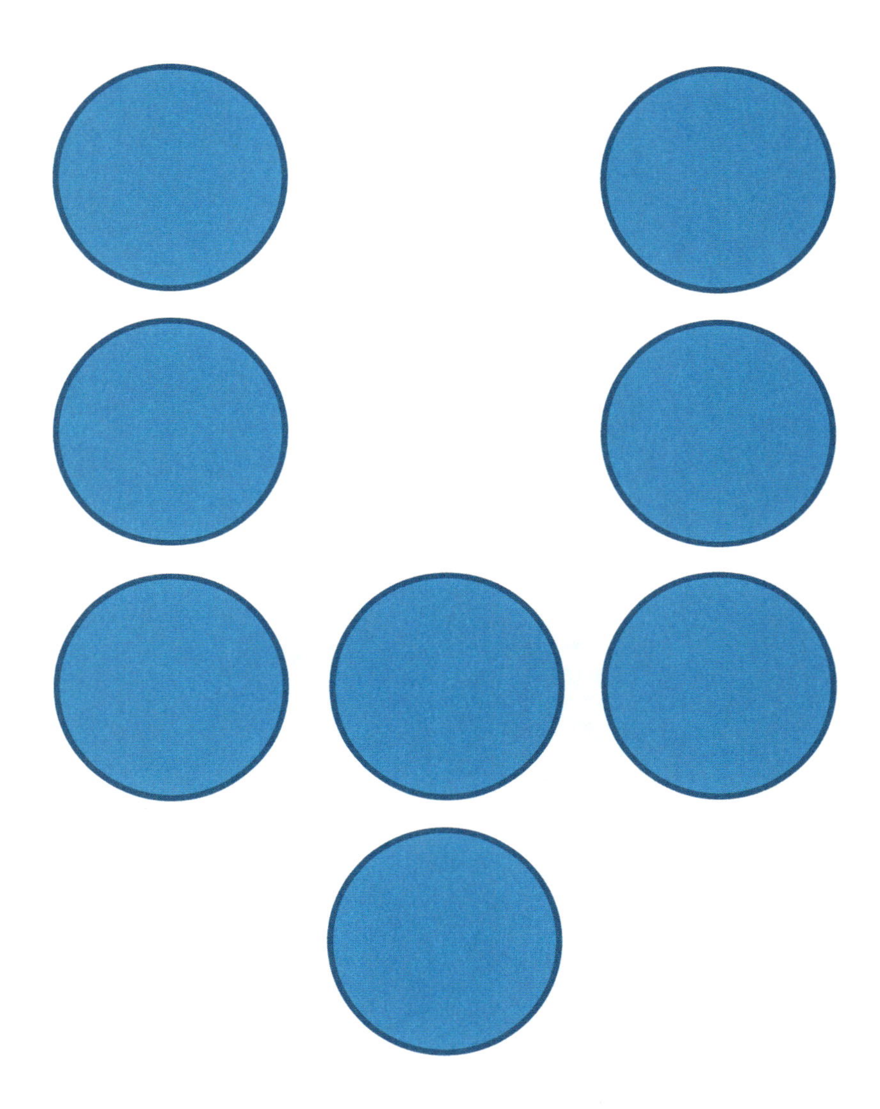

dot cards of 8

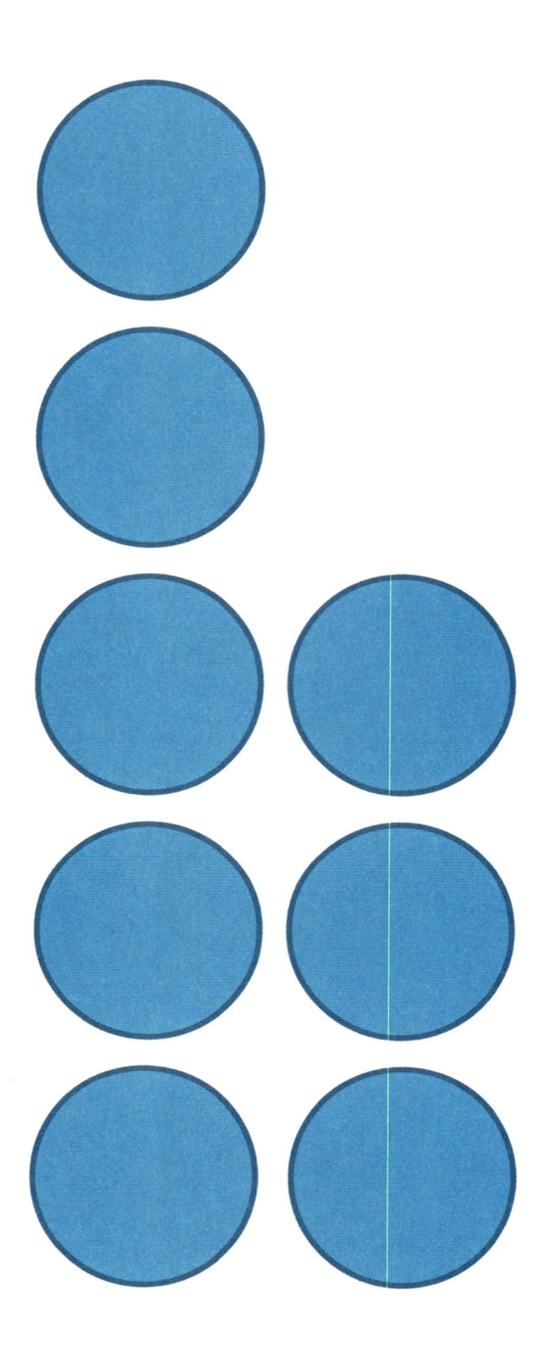

dot cards of 8

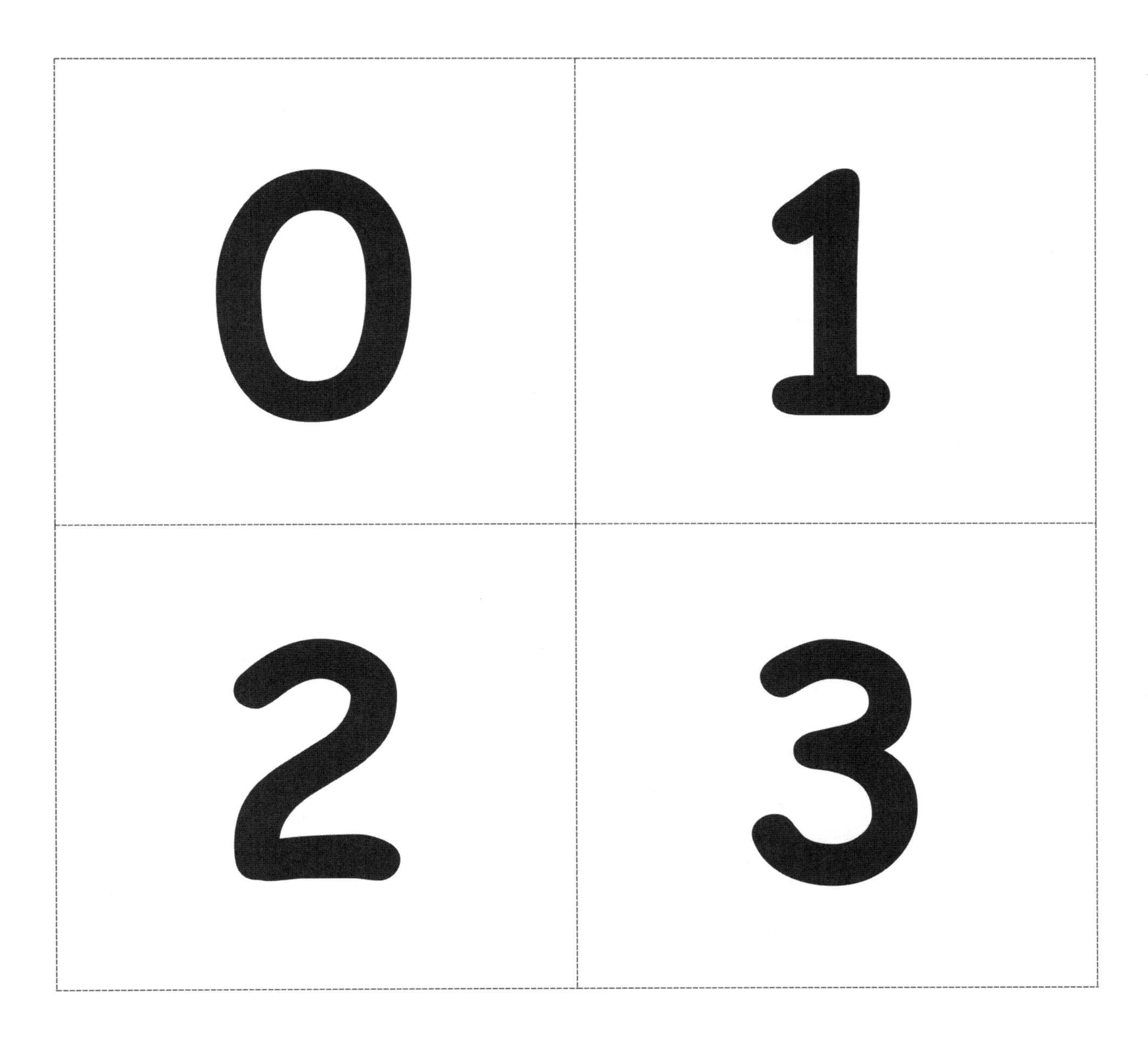

Note: Match to corresponding 5-group side and copy double-sided on card stock.

---

large Hide Zero cards (numeral side)

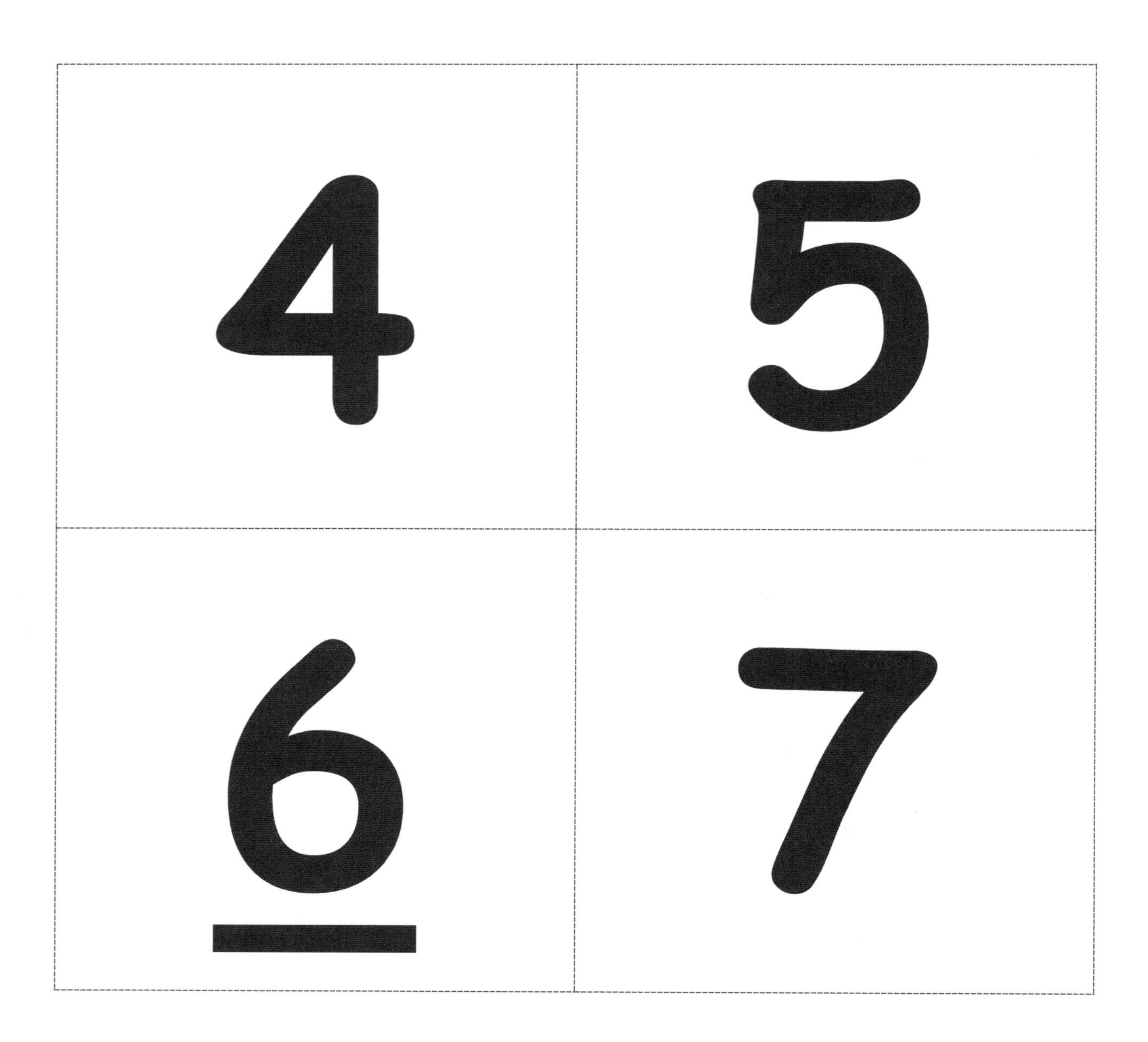

Note: Match to corresponding 5-group side and copy double-sided on card stock.

large Hide Zero cards (numeral side)

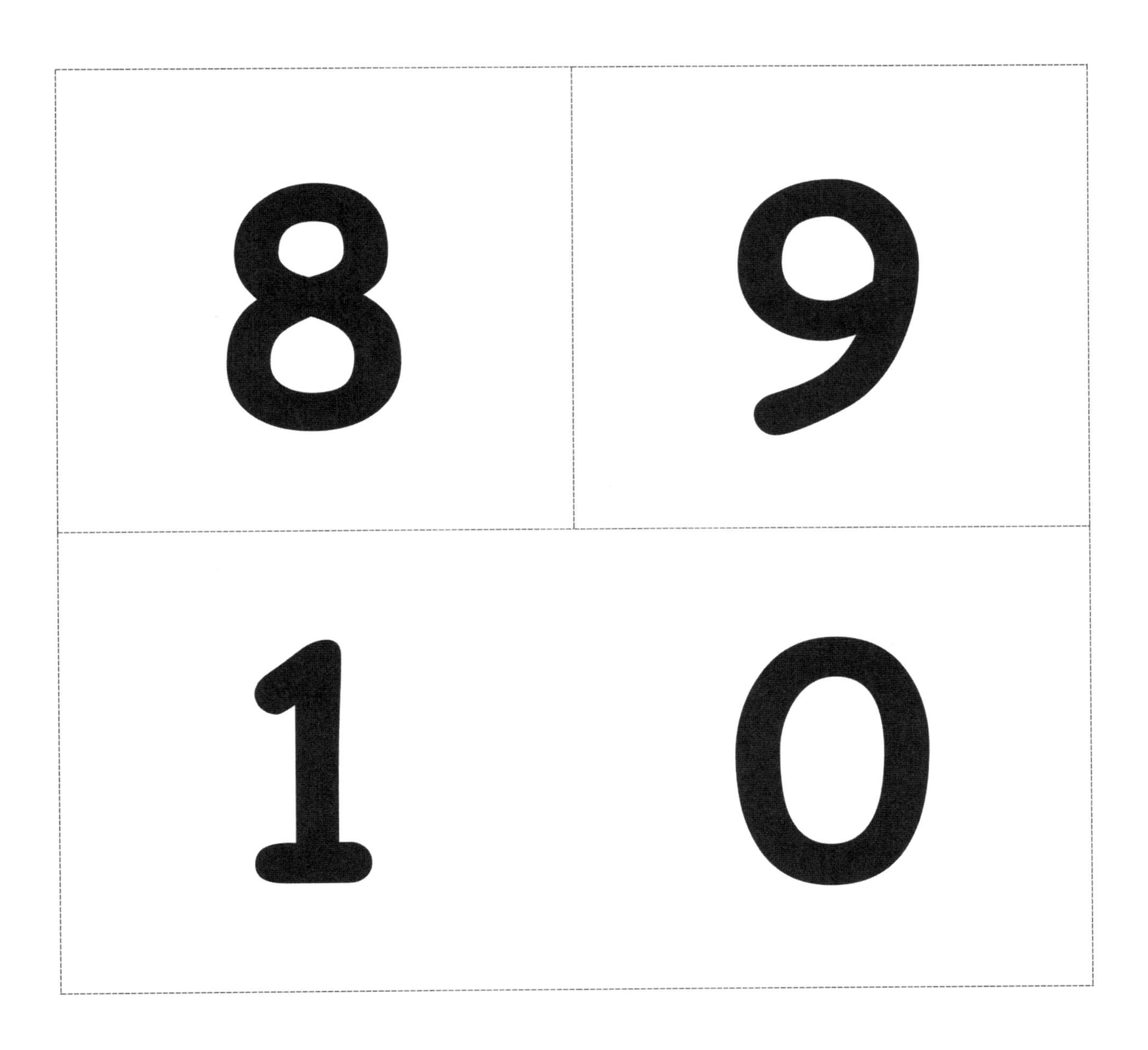

Note: Match to corresponding 5-group side and copy double-sided on card stock.

large Hide Zero cards (numeral side)

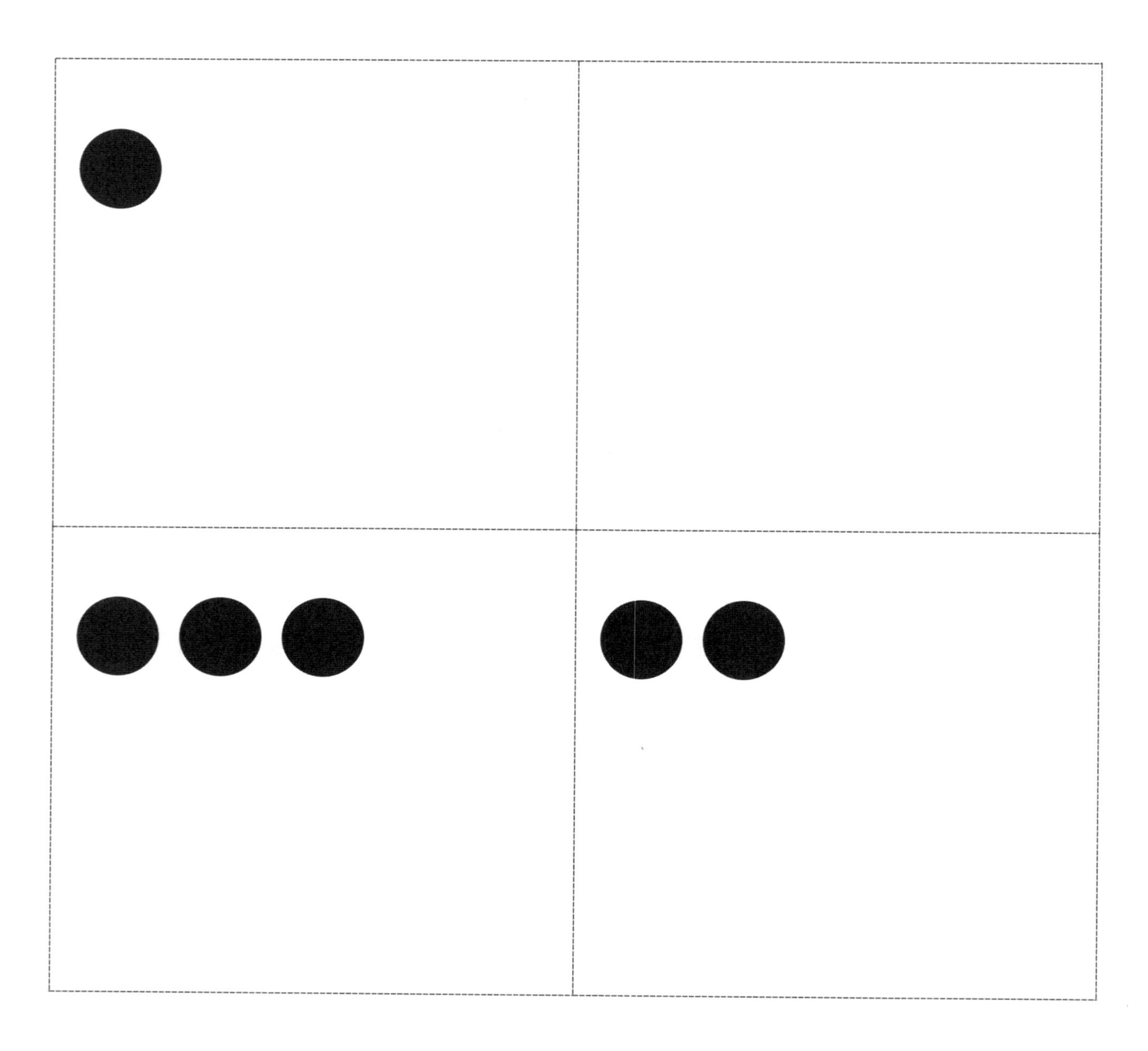

Note: Match to corresponding numeral side and copy double-sided on card stock.

large Hide Zero cards (5-group side)

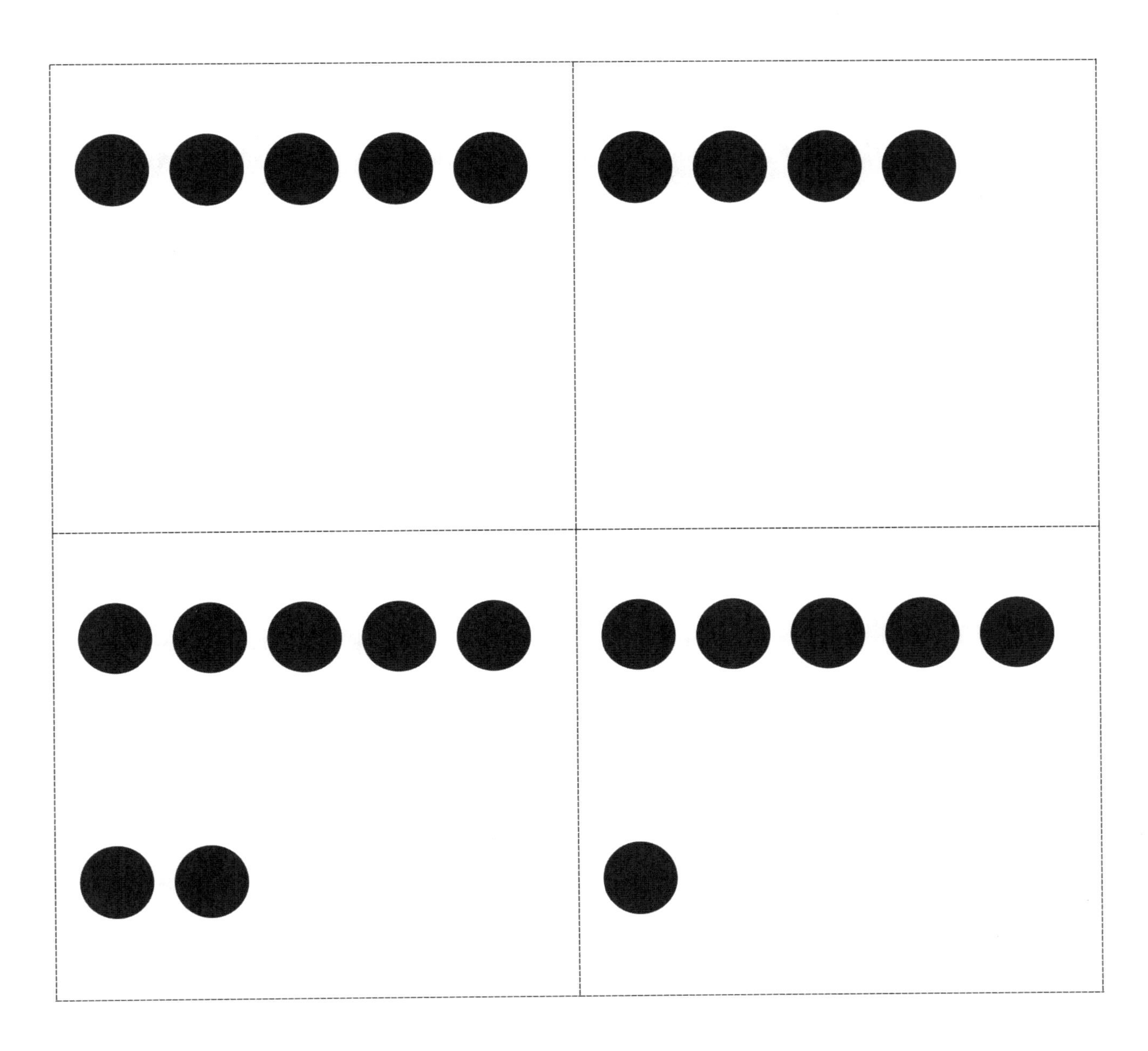

Note: Match to corresponding numeral side and copy double-sided on card stock.

large Hide Zero cards (5-group side)

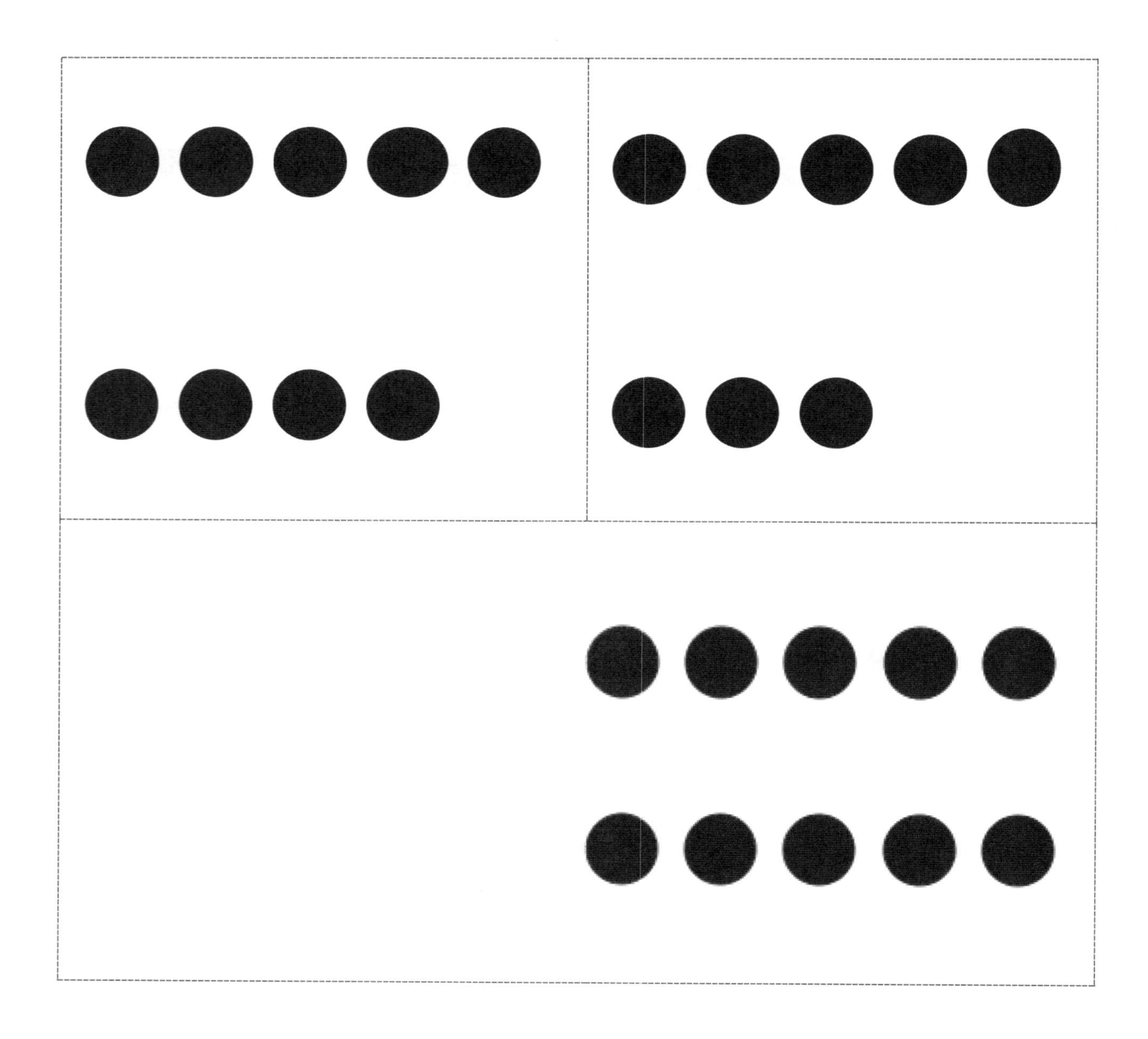

Note: Match to corresponding numeral side and copy double-sided on card stock.

large Hide Zero cards (5-group side)

1 0	1 0
1 0	1 0
1 0	1 0
1 0	1 0

Note: Copy double-sided with the Hide Zero 10 card (5-group side) on card stock. Each student needs one, double-sided Hide Zero 10 card. This card is used with 5-group cards 1–9 (Lesson 1 Fluency Template 2), which combined, make the full set of Hide Zero cards.

Hide Zero 10 card (numeral side)

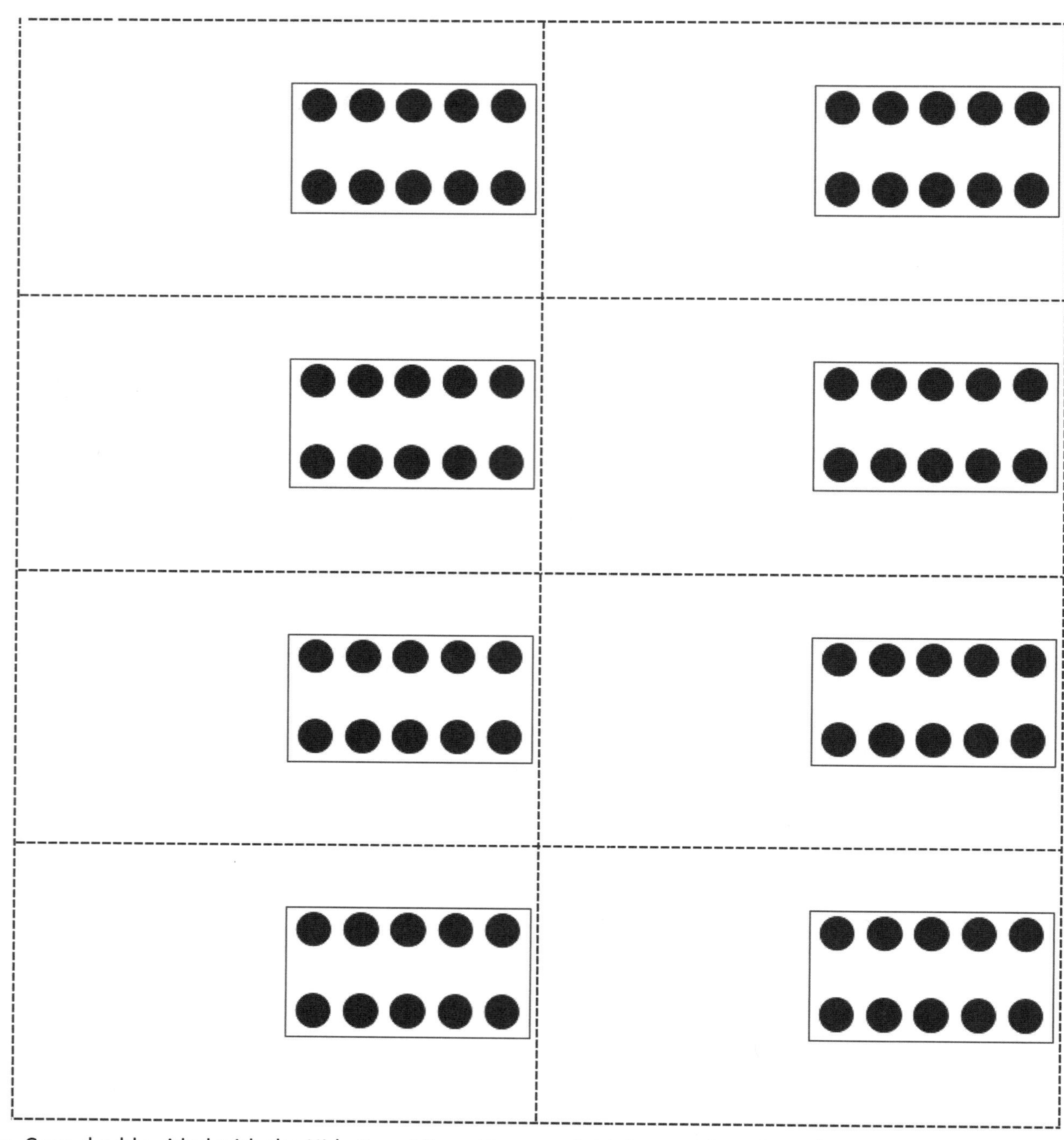

Note: Copy double-sided with the Hide Zero 10 card (numeral side) on card stock. Each student needs one, double-sided Hide Zero 10 card. This card is used with 5-group cards 1–9 (Lesson 1 Fluency Template 2), which combined, make the full set of Hide Zero cards.

Hide Zero 10 card (5-group side)

Lesson 7

Objective: Model and write numbers 10 to 20 as number bonds.

Suggested Lesson Structure

Fluency Practice	(10 minutes)
Application Problem	(5 minutes)
Concept Development	(28 minutes)
Student Debrief	(7 minutes)
Total Time	**(50 minutes)**

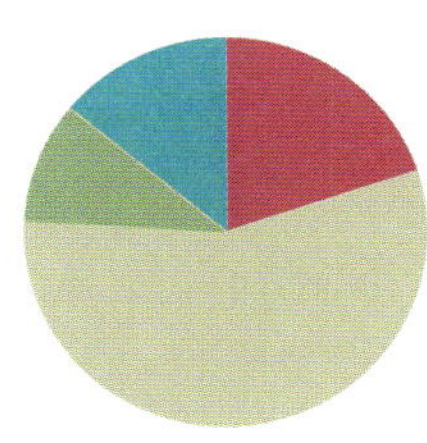
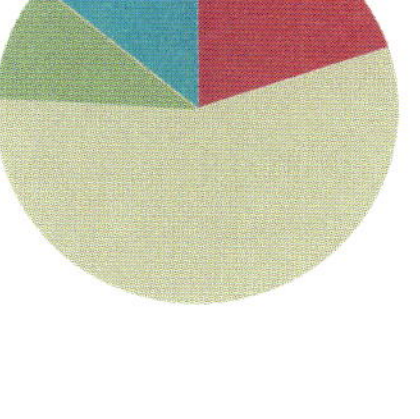

Fluency Practice (10 minutes)

- Dot Cards of Eight **K.CC.5, K.CC.2** (4 minutes)
- Counting **K.CC.2** (3 minutes)
- Decompose Teen Numbers **K.NBT.1** (3 minutes)

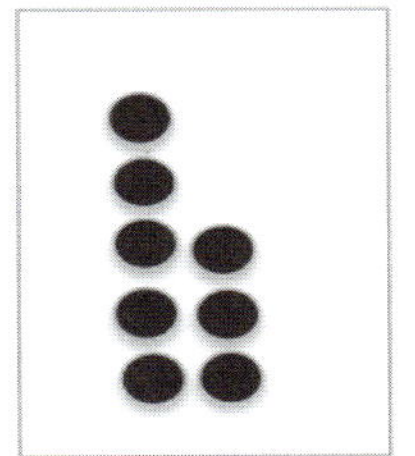

Dot Cards of Eight (4 minutes)

Materials: (T/S) Dot cards of 8 (Lesson 6 Fluency Template)

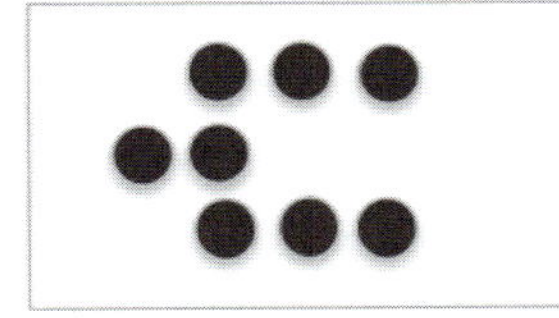

Note: This fluency activity gives students an opportunity to develop increased familiarity with decompositions of eight and practice seeing part–whole relationships.

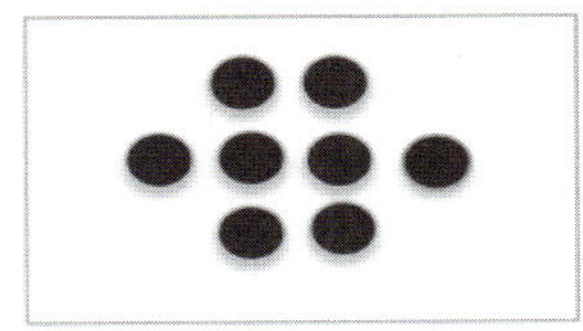

T: (Show a card with 8 dots.) How many dots do you count? Wait for the signal to tell me.

S: 8.

T: How can you see them in 2 parts?

S: (Students come up to the card.) I saw 4 here and 4 here. → I saw 5 here and 3 here. → I saw 6 here and 2 here.

Repeat with other cards. Pass out the cards for students to work with a partner.

Counting (3 minutes)

Note: Extending the counting sequence on partners' fingers prepares students to model teen numbers as 10 ones and some ones.

Partners hover their hands as if playing the piano. Student on the teacher's right begins by "playing" the pinky of the left hand and continuing from left to right. Once a finger is counted, it remains down on the keyboard.

Students count their own and their partner's fingers first the Say Ten way, ten 1, ten 2, etc., and then in standard form. Have them count down from 20 to 0 if they finish early.

NOTES ON MULTIPLE MEANS OF ENGAGEMENT:

If students working below grade level have difficulties with the Application Problem, pair them with a partner or put them in a small group to solve the problem. Form a small group, and assign "jobs" so students hold each other accountable.

Decompose Teen Numbers (3 minutes)

Materials: (T) Large Hide Zero cards (Lesson 6 Template 1) (emphasize the breaking apart of numbers by separating the cards as students say numbers the Say Ten way and the regular way.)

Note: Breaking apart teen numbers with the Hide Zero cards prepares students to work with number bonds in today's Concept Development.

T: (Show 12.) Say the number the regular way.

S: 12.

T: (Separate the cards.) Say 12 the Say Ten way.

S: Ten 2.

Continue with the following possible sequence: 13, 14, 19, 11, 10, 15, 17, 16, 18.

NOTES ON MULTIPLE MEANS OF REPRESENTATION:

Scaffold the lesson for English language learners by introducing the terms *total* and *part*. Use pictures of other visual aids of a total and a part. Post the words with the visual on the word wall so that students can continuously refer to them.

Application Problem (5 minutes)

Materials: (S) Hide Zero cards: 1 Hide Zero 10 card (Lesson 6 Template 2) and 5-group cards 1–9 (Lesson 1 Fluency Template 2)

Gregory drew 10 smiley faces and 5 smiley faces. He put them together and had 15 smiley faces. Draw the 15 smiley faces as 10 smiley faces and 5 smiley faces. Then, draw 15 with Hide Zero cards when the zero is hiding and when the zero is not hiding.

Note: Word problems involving quantities above 10 begin in Grade 1. Many of the application problems in Module 5 are simply decomposition and composition experiences (**K.NBT.1**). Note that the problems do not ask, "How many in all?" or "How many?" Also note that there is no unknown in problems of this type.

Concept Development (28 minutes)

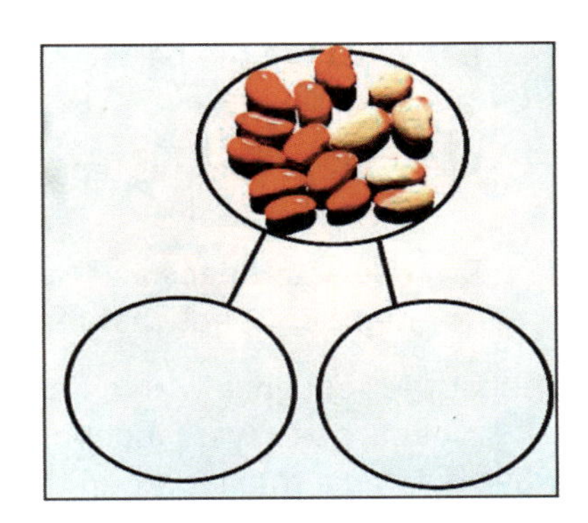

Materials: (T) Large Hide Zero Cards (Lesson 6 Template 1), (S) 20 two-sided counters in a clear plastic bag (white beans spray painted red on one side, commercial two-sided counters, etc.), number bond (Template) within a personal white board, 1 set of Hide Zero cards: 1 Hide Zero 10 card (Lesson 6 Template 2) and 5-group cards 1–9 (Lesson 1 Fluency Template 2) (per pair)

T: Here is Gregory's number with my Hide Zero cards.

T: Show Gregory's number with your 2-sided counters in the "total place" of your number bond. Make 10 ones a different color from the other ones.

S: (Students do so.)

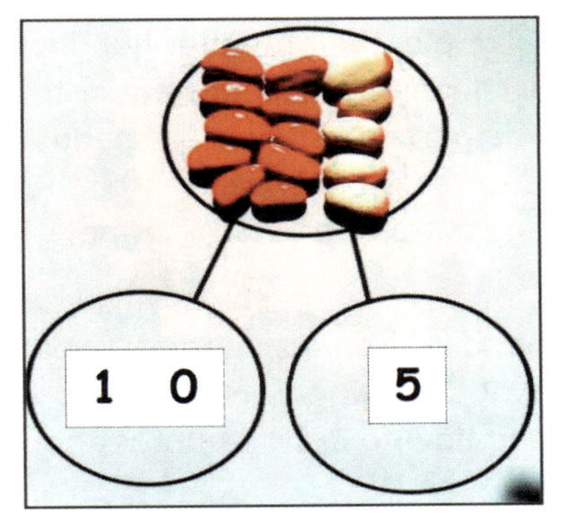

T: Our number bond is not complete! We haven't shown the parts!

T: What number parts are made by the two colors?

S: 10 ones and 5 ones.

T: Show those 2 parts with your own Hide Zero cards.

T: (See the picture to the right.) Is 15 beans the same number as 10 and 5?

S: (Give the students time to recount.) Yes.

T: Now, our number bond is correct!

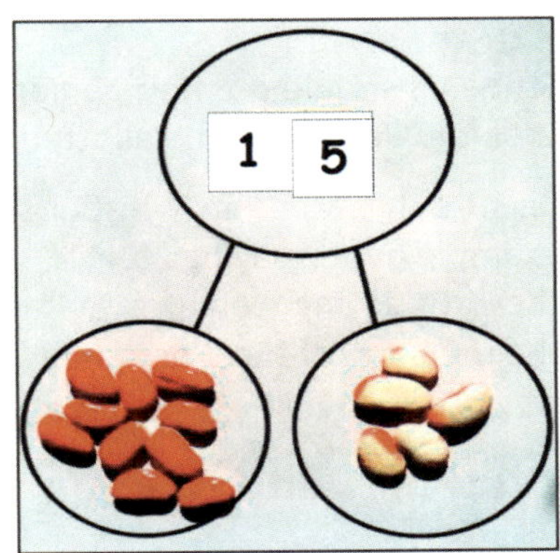

T: Let's switch it. Slide your counters down to be the two parts: 10 ones in a part and 5 ones in a part.

T: Show 15 with your Hide Zero cards in the total place of your number bond.

T: Does 15 tell us the total number of beans in the 2 parts?

S: (Give students time to count.) Yes.

T: Now, our number bond is correct again!

T: Let's replace the Hide Zero cards with a written number. Slide the cards off the total place. What number will you write?

MP.4

S: 15.

T: Slide off your beans from the parts. What numbers will you write to take their place?

S: 10 and 5.

T: Is 15 the same as 10 and 5?

S: Yes.

T: What is the total?

S: 15 (or ten 5).

T: What are the parts?

S: 10 and 5.

T: 15 is the same as ten 5. Our number bond is correct again!

T: Use your beans and Hide Zero cards to make number bonds that are correct.

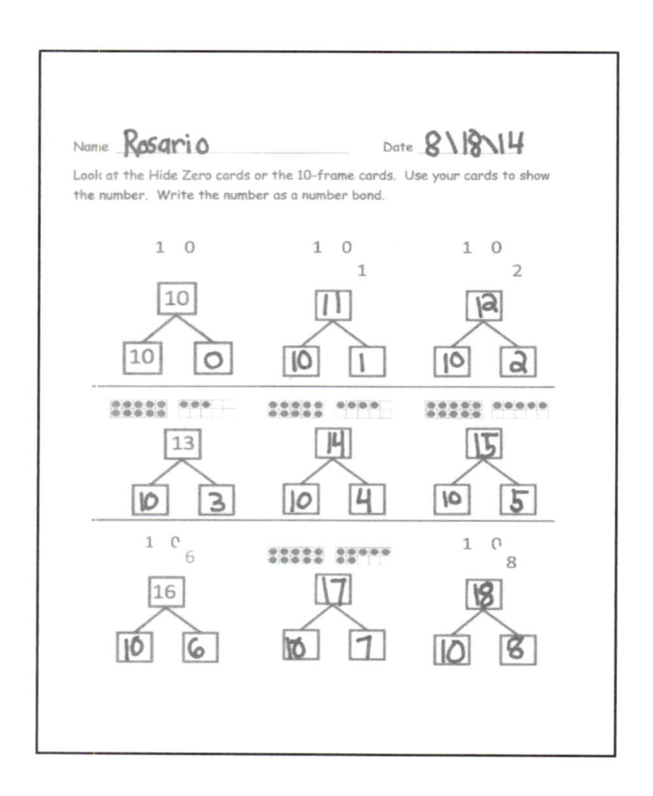

Repeat the sequence with different numbers of beans. Let students go to work independently as they are able while guiding a smaller group that still needs guided practice. Do not let the equality be unresolved. For example, their number bond is not correct if they have 10 beans and 5 beans but nothing in the total place. The parts must always be equal to the total. Students may realize they can switch the order of the 10 ones and extra ones. That is good!

Close the session by having students write a number bond without using the template. This is review from Module 4 where they learned about the "total place" and how to draw a number bond.

Problem Set (8 minutes)

Students should do their personal best to complete the Problem Set within the allotted time.

Be sure that students whisper speak as they work. For example, when saying "ten 2," they write the 1 and then the 2. By saying "ten 2" simultaneously, they internalize the meaning of the 1 as standing for 10 ones.

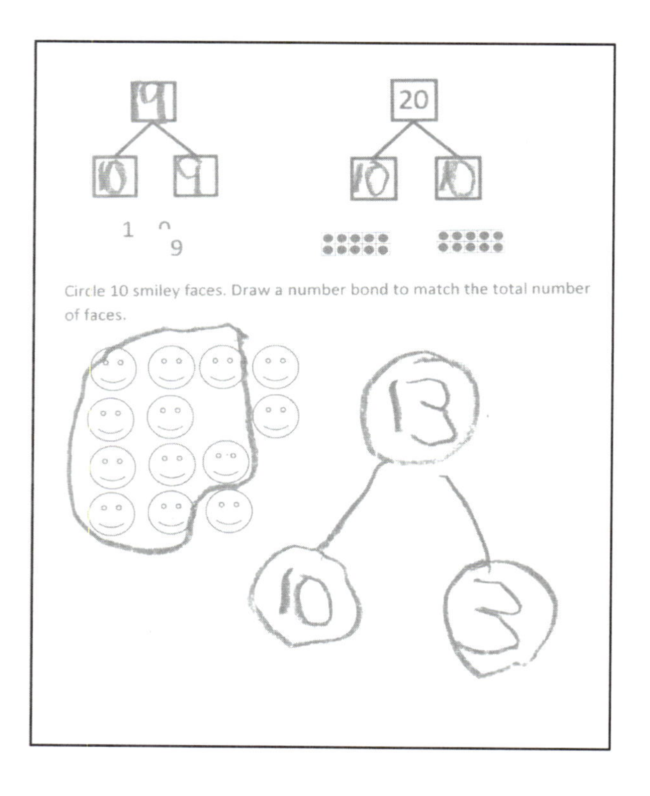

Student Debrief (7 minutes)

Lesson Objective: Model and write numbers 10 to 20 as number bonds.

The Student Debrief is intended to invite reflection and active processing of the total lesson experience. Invite students to review their solutions for the Problem Set. They should check work by comparing answers with a partner before going over answers as a class. Look for misconceptions or misunderstandings that can be addressed in the Debrief. Guide students in a conversation to debrief the Problem Set and process the lesson.

Any combination of the questions below may be used to lead the discussion.

- Tell me about the pattern you see on your Problem Set.
- How are the number bonds and Hide Zero cards helping you to understand the numbers from eleven to twenty?
- How does counting the Say Ten way help you understand?
- How is this 1 in thirteen the same as this 1 in nineteen? When you made your number bonds, what stayed the same and what changed?
- When you see the number eleven, how are those two 1s different?

Exit Ticket (3 minutes)

After the Student Debrief, instruct students to complete the Exit Ticket. A review of their work will help with assessing students' understanding of the concepts that were presented in today's lesson and plan more effectively for future lessons. The questions may be read aloud to the students.

Name ______________________________ Date ______________

Look at the Hide Zero cards or the 5-group cards. Use your cards to show the number. Write the number as a number bond.

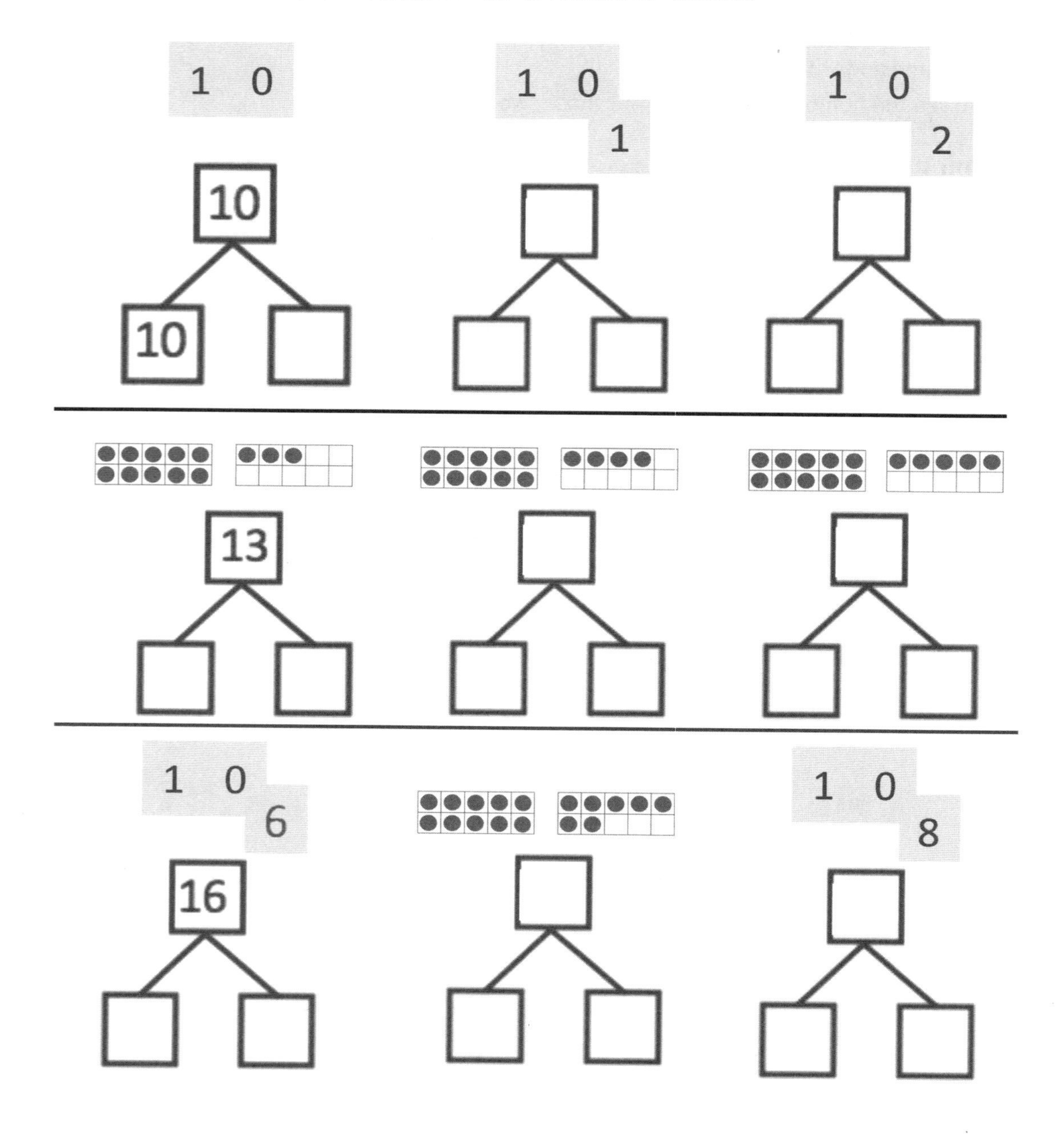

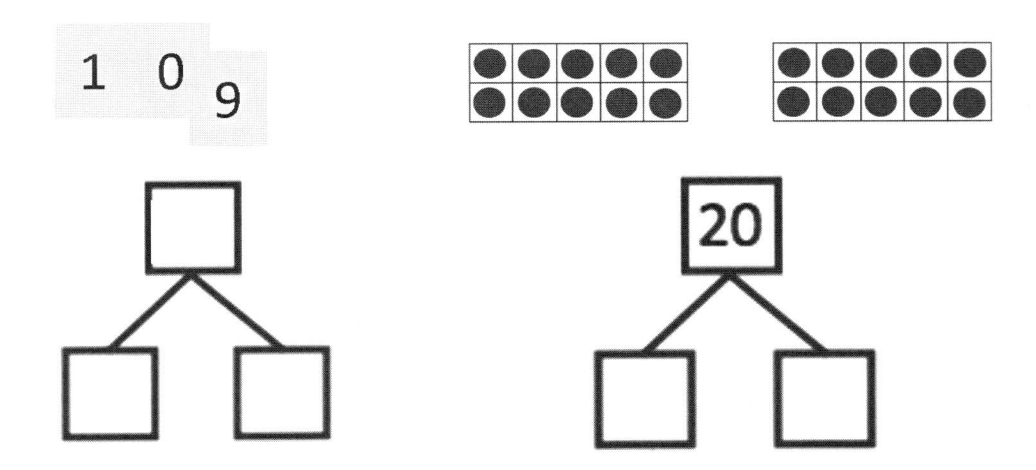

Circle 10 smiley faces. Draw a number bond to match the total number of faces.

Name ______________________________ Date ______________

Look at the Hide Zero cards or the 5-group cards. Use your cards to show the number. Write the number as a number bond.

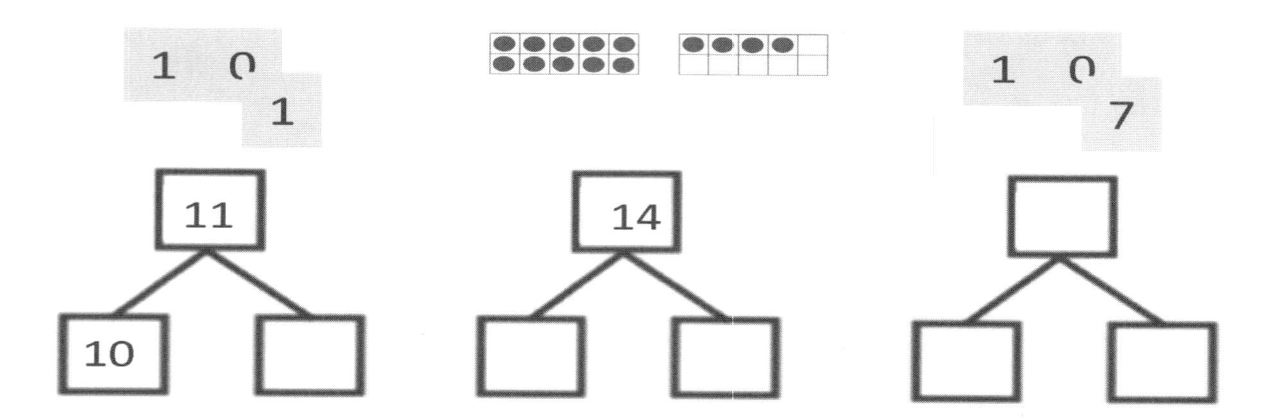

Name ______________________________ Date ______________

Look at the Hide Zero cards or the 5-group cards. Use your cards to show the number. Write the number as a number bond.

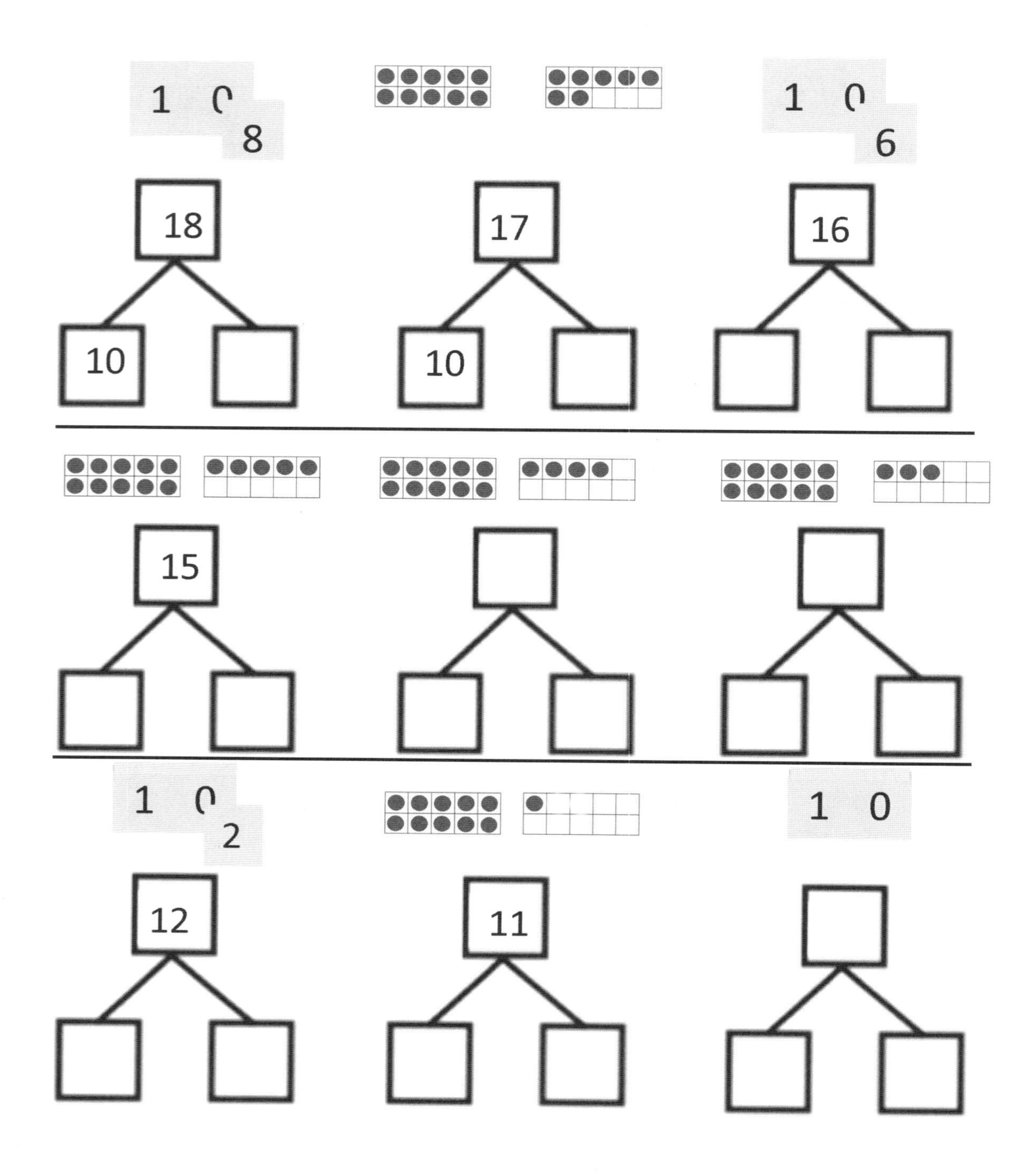
1 0
8
18
10
17
10
16
1 0
6
15
1 0
2
12
11
1 0

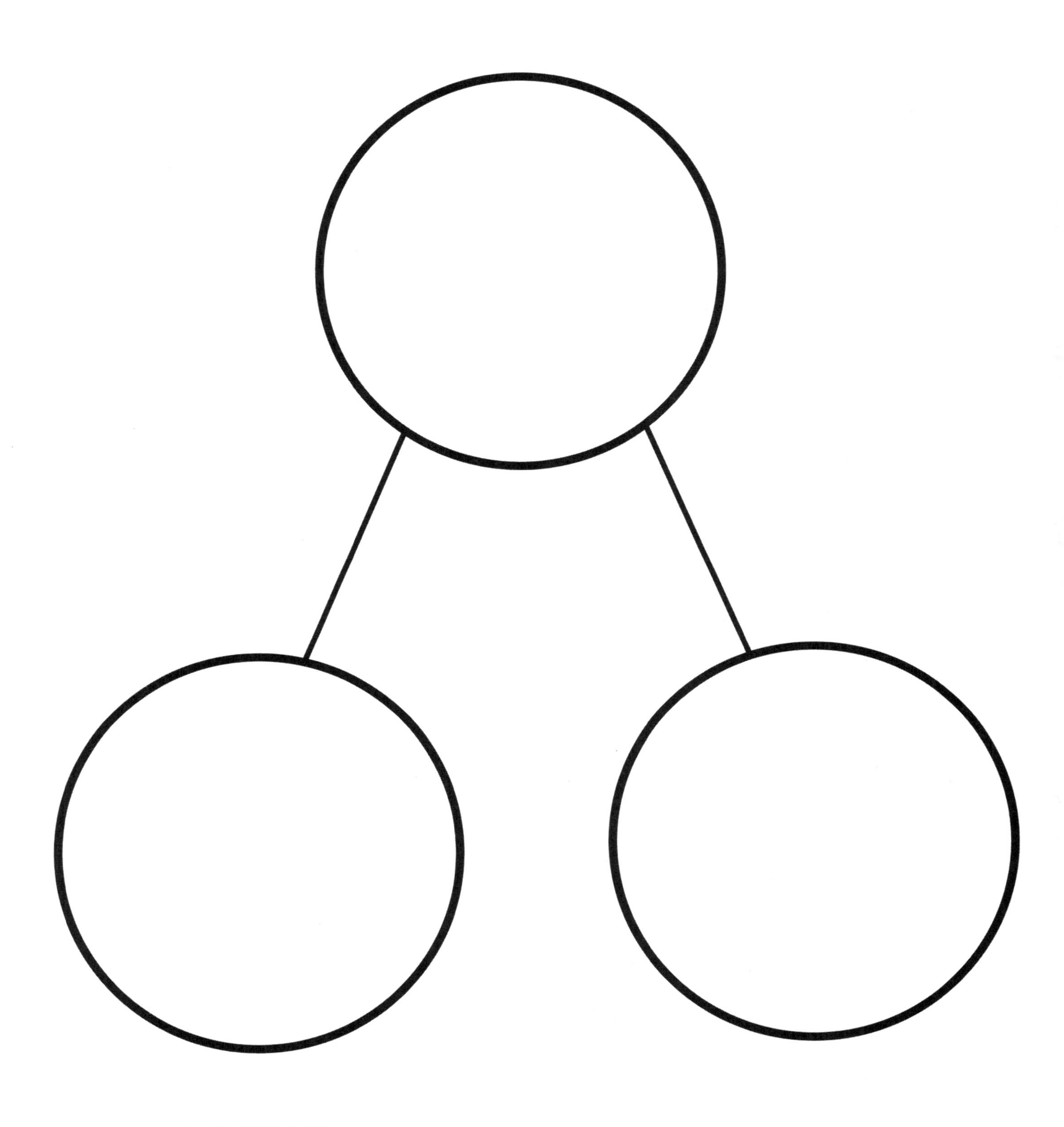

number bond

Lesson 8

Objective: Model teen numbers with materials from abstract to concrete.

Suggested Lesson Structure

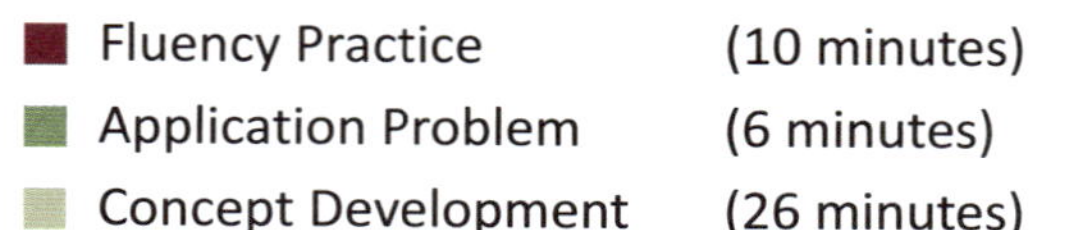

Fluency Practice	(10 minutes)
Application Problem	(6 minutes)
Concept Development	(26 minutes)
Student Debrief	(8 minutes)
Total Time	**(50 minutes)**

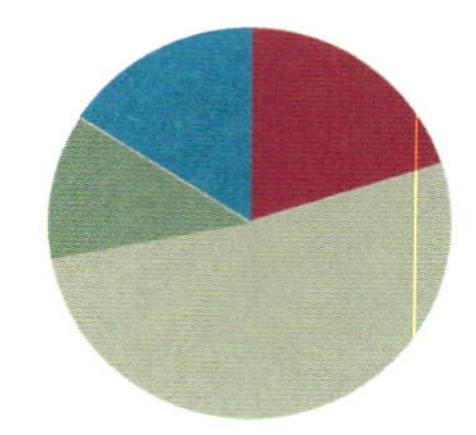

Fluency Practice (10 minutes)

- Number Bonds of Eight **K.CC.2** (4 minutes)
- Separating Ten Ones Inside Teen Numbers **K.NBT.1** (3 minutes)
- Teen Number Bonds **K.NBT.1** (3 minutes)

Number Bonds of Eight (4 minutes)

Materials: (T) Dot cards of 8 (Lesson 6 Fluency Template) (S) Personal white board

Note: This fluency activity gives students an opportunity to develop increased familiarity with compositions of eight and to review number bonds.

T: (Show a dot card, and indicate 7 and 1 as parts.) Say the larger part. (Give students time to count.)
S: 7.
T: Say the smaller part.
S: 1.
T: What is the total number of dots? (Give time to count.)
S: 8.
T: Write your number bond.

Continue with 5 and 3, 4 and 4, 6 and 2, 8 and 0.

Separating Ten Ones Inside Teen Numbers (3 minutes)

Materials: (S) Bag with about 20 small objects

Note: This activity gives continued practice in locating 10 ones embedded in the teen numbers and allows students to experience conservation.

T: Empty your bag. Put all the items on your work mat. Count out 10 ones, and move them together into a bunch.

T: (Wait while students complete the task.) How many things are in your bunch?

S: 10.

T: Are there some outside your bunch?

S: Yes.

T: Push all your things back together. Spread them all out over your work mat.

Repeat this process two or three more times.

Teen Number Bonds (3 minutes)

Materials: (T) Number bond cards (Fluency Template)

Note: This activity advances the work with teen numbers by allowing students to see that the parts of a number bond can be switched around, and the total remains the same.

T: (Show a number bond with 10 and 5 as parts.) Say the number sentence starting with 10.

S: 10 and 5 makes 15.

T: Flip it.

S: 5 and 10 makes 15.

Continue with 10 and 1, 10 and 9, 10 and 4, 10 and 8, 10 and 2, 10 and 6, 10 and 3, 10 and 7.

NOTES ON MULTIPLE MEANS OF ACTION AND EXPRESSION:

To support English language learners in explaining what they see, let them work with a student who speaks their own language. This is the key in illustrating the commutative property in a very student-friendly setting. It is always easier to explain using a familiar language.

Application Problem (6 minutes)

MP.3

Peter drew a number bond of 13 as 10 and 3. Bill drew a number bond, too, but he switched around the 10 and 3. Show both Bill's and Peter's number bonds. Draw a picture of thirteen things as 10 ones and 3 ones. Explain your thinking to your partner about what you notice about the two number bonds.

Note: The students have noticed that the parts of a number bond can be switched around in Module 4. Make it exciting for them to find out that the same rules, or math truths, apply to larger numbers, too!

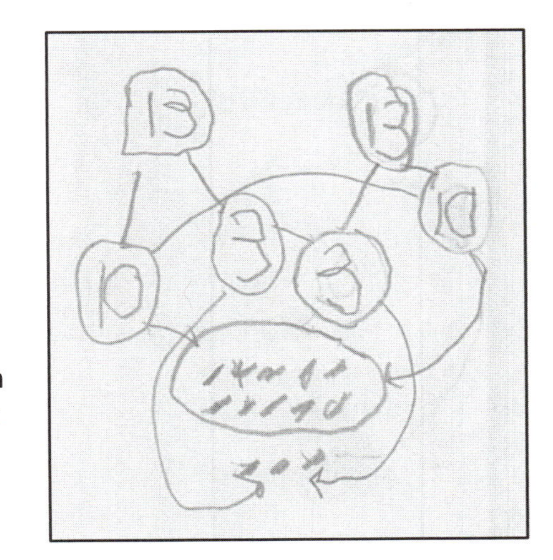

Concept Development (26 minutes)

Materials: (S) personal white board; bag of Hide Zero cards: 1 Hide Zero 10 card (Lesson 6 Template 2) and 5-group cards 1–9 (Lesson 1 Fluency Template 2), bag of 10 linking cubes in one color and 10 linking cubes in another color (per pair)

Part 1: Modeling Teen Numbers 11–20 with Linking Cubes and Hide Zero Cards.

T: Partner A, open the bag with the Hide Zero cards, and put them on your work mat. With your partner, put them in order from 10 to 1. (Wait.)

T: Partner B, open the bag with the linking cubes, and put them on your work mat.

T: (Write 11 on the board.) What number is this?

S: Eleven!

T: How would you say it the Say Ten way?

S: Ten 1.

T: Please write the number 11 on your personal white board. When I ask you to show me your board, show me.

S: (Write 11.)

T: Show me!

S: (Hold up their personal white boards.)

T: Now, I want you to work with your partner to show the number. Partner A, show the number with the Hide Zero cards, and remember to hide the zero!

T: Partner B, show the number with the linking cubes. Use one color to show 10 ones and the other color to show the other ones.

T: Check each other's work. Explain why you're both showing 11.

Repeat the process with the numbers 12–19.

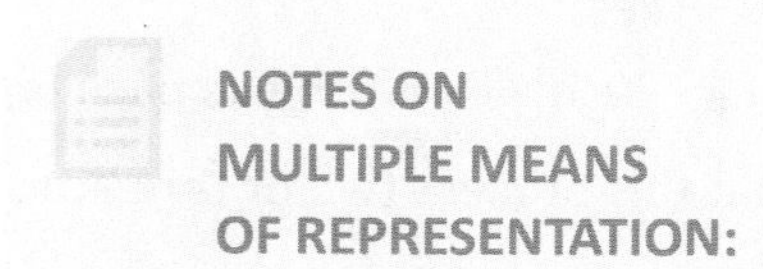

NOTES ON MULTIPLE MEANS OF REPRESENTATION:

Support your English language learners who have difficulty distinguishing between words such as *thirteen* and *thirty* and *fourteen* and *forty* by instructing them to practice saying "thirteen" and "thirty" as you point to both the numeral and the word written under each numeral.

Part 2: Modeling Teen Numbers 11–20 with Hide Zero Cards.

T: (Write 15 on the board.) What is the number?

S: Fifteen!

T: The Say Ten way?

S: Ten 5.

T: Write 15 on your personal white board, and then show me.

T: This time, Partner A is going to show 15 with the dot side of the Hide Zero cards, and Partner B is going to show 15 with the numeral side. After you check each other's work, you'll switch.

Repeat the process above with numbers 11–19.

Problem Set (7 minutes)

Students should do their personal best to complete the Problem Set within the allotted time. Have students use the bag of 20 small objects from today's fluency activity as they complete the Problem Set.

Student Debrief (8 minutes)

Lesson Objective: Model teen numbers with materials from abstract to concrete.

The Student Debrief is intended to invite reflection and active processing of the total lesson experience.

Invite students to review their solutions for the Problem Set. They should check work by comparing answers with a partner before going over answers as a class. Look for misconceptions or misunderstandings that can be addressed in the Debrief. Guide students in a conversation to debrief the Problem Set and process the lesson.

Any combination of the questions below may be used to lead the discussion.

Have a set of 5-group cards, Hide Zero cards, and 20 linking cubes in two different colors ready to display.

- What is the same/different about the 5-group cards and the Hide Zero cards?
- How can you prove 20 is the same as 2 ten?
- When you write the number 18 on your personal white board, how is it the same and different from the number 18 when you show it with Hide Zero cards or 5-group cards?
- Which is your favorite way to show a number—with linking cubes, the Hide Zero cards, the 5-group cards, or just writing the number? Why?
- Count up to 20 in standard form, and count back to 0 the Say Ten way.
- Who can prove that the 1 in 14 is 10 ones, not 1 one?

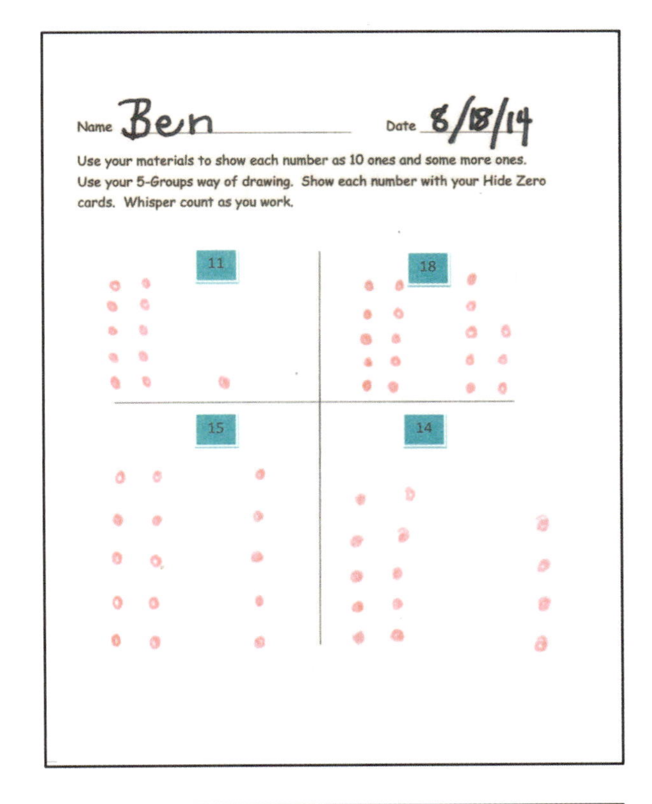

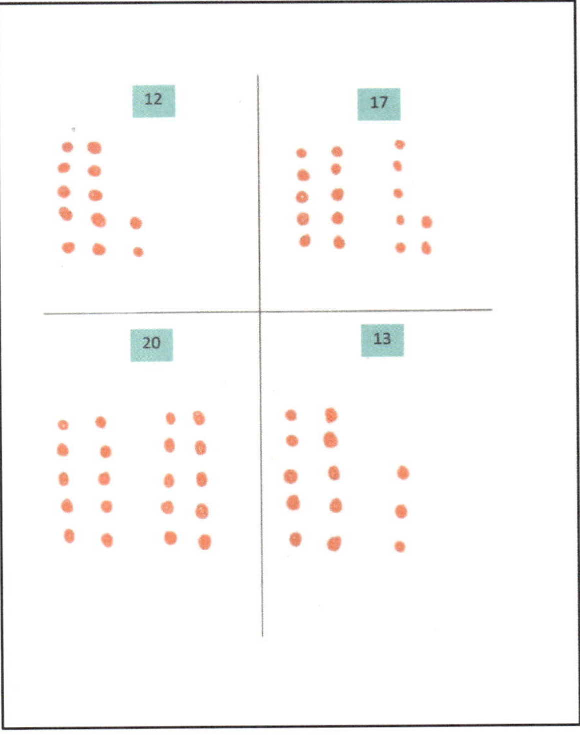

Exit Ticket (3 minutes)

After the Student Debrief, instruct students to complete the Exit Ticket. A review of their work will help with assessing students' understanding of the concepts that were presented in today's lesson and planning more effectively for future lessons. The questions may be read aloud to the students.

Name ______________________________ Date ______________

Use your materials to show each number as 10 ones and some more ones. Use your 5-groups way of drawing. Show each number with your Hide Zero cards. Whisper count as you work.

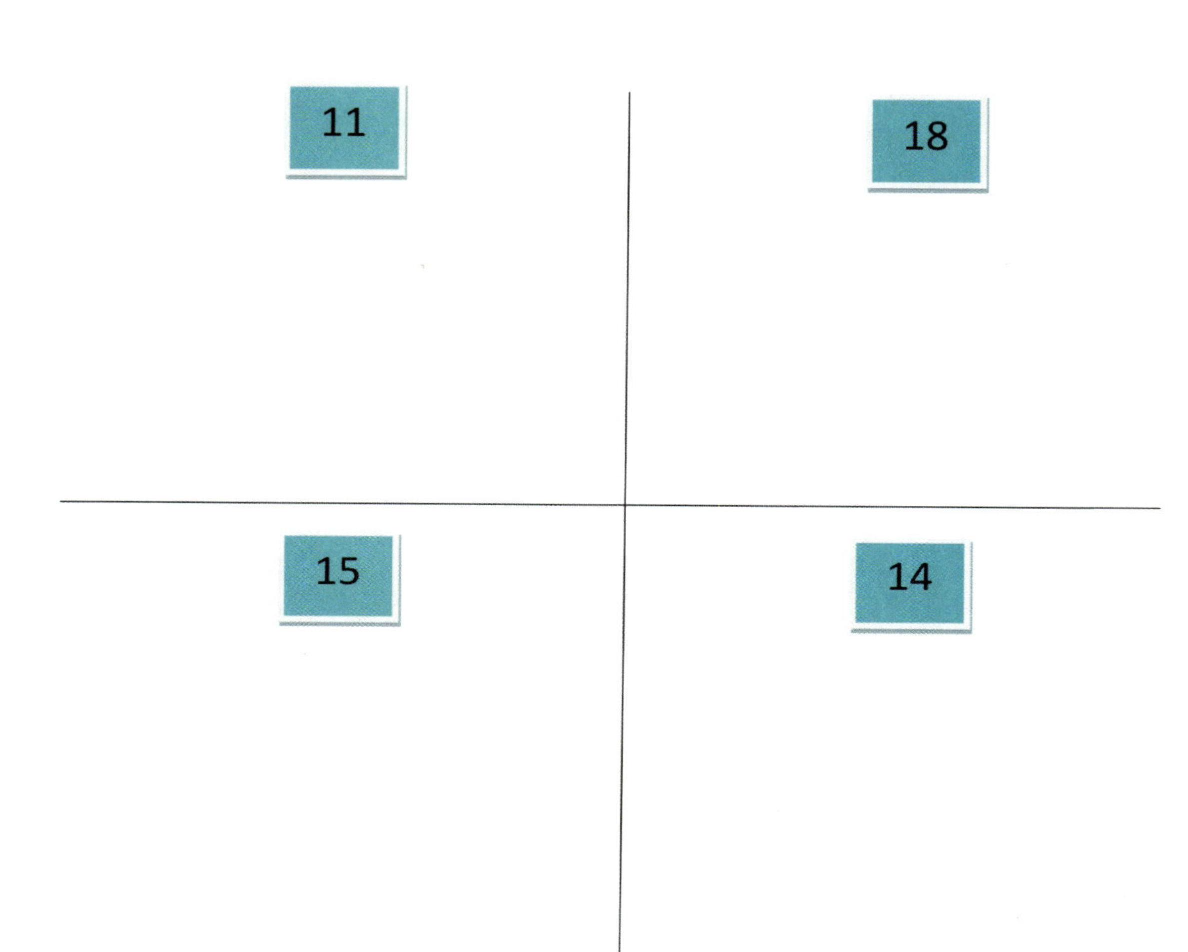

12

17

20

13

Name ______________________________ Date ____________

Use your materials to show the number as 10 ones and some more ones. Use your 5-groups way of drawing.

1 6

Use your cubes to show the number. Then, color in the cubes to match the number.

1 2

Name ______________________________ Date ______________

Use your materials to show each number as 10 ones and some more ones. Use your 5-groups way of drawing.

1 5	1 3
Ten seven	Ten one

1 2	1 6
2 ten	Ten four

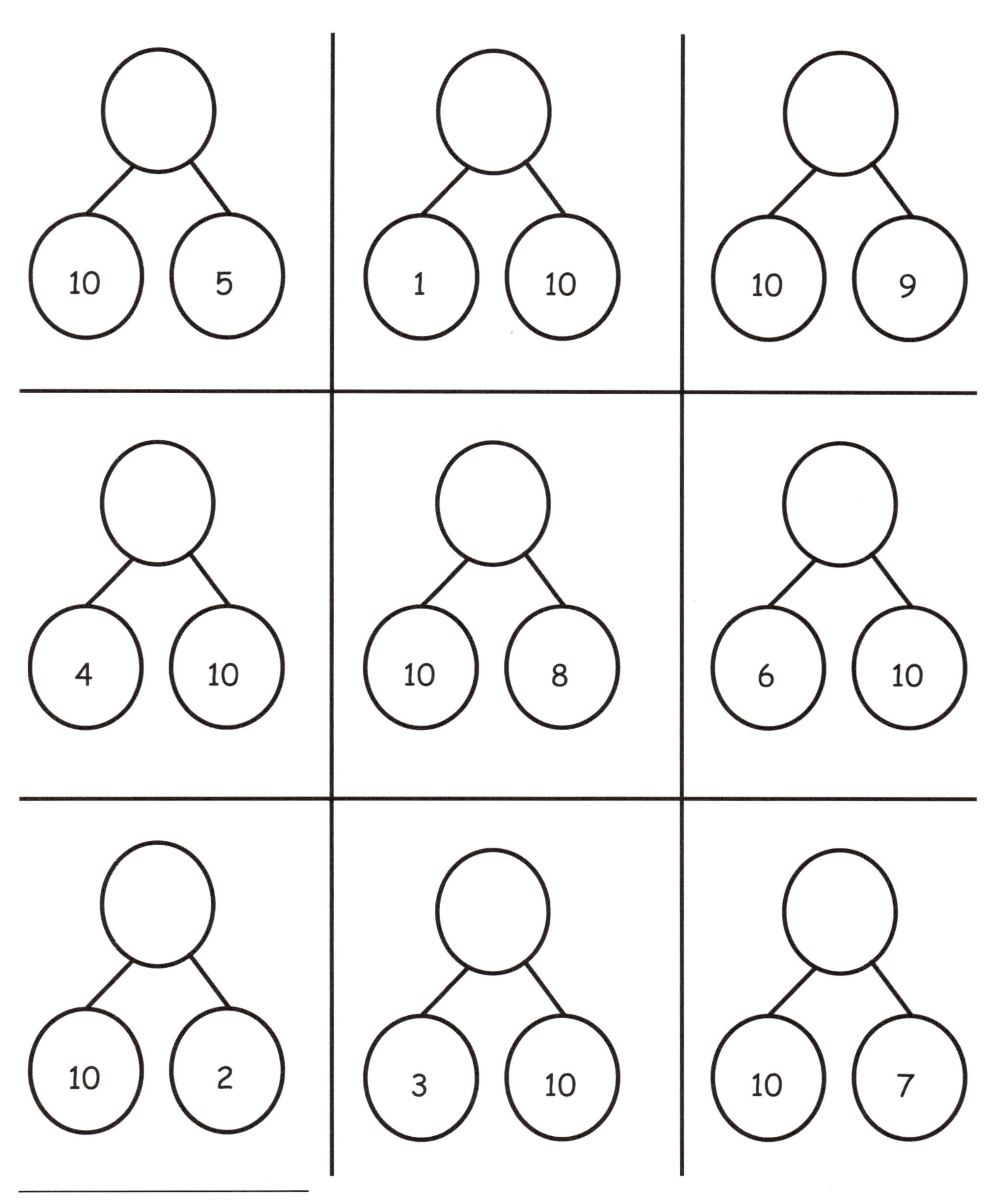

number bond cards

Lesson 9

Objective: Draw teen numbers from abstract to pictorial.

Suggested Lesson Structure

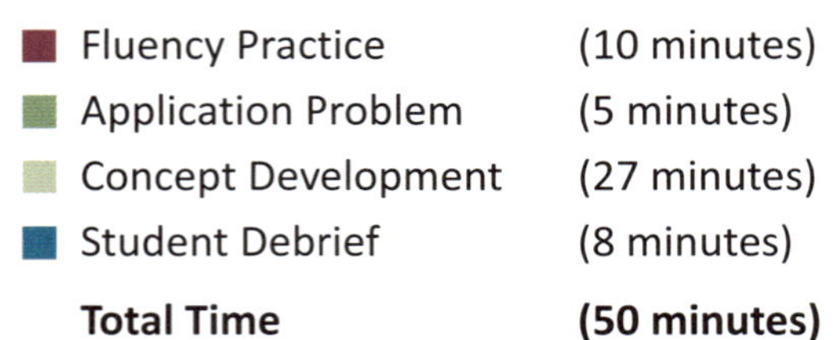

- Fluency Practice (10 minutes)
- Application Problem (5 minutes)
- Concept Development (27 minutes)
- Student Debrief (8 minutes)

Total Time (50 minutes)

Fluency Practice (10 minutes)

- Dot Cards of Nine **K.CC.5, K.CC.2** (4 minutes)
- How Many Is One More? **K.CC.2** (2 minutes)
- Grouping Teen Numbers into 10 Ones **K.NBT.1** (4 minutes)

Dot Cards of Nine (4 minutes)

Materials: (T/S) Dot cards of 9 (Fluency Template)

Note: This fluency activity gives students an opportunity to develop increased familiarity with decompositions of nine and practice seeing part–whole relationships.

T: (Show a card with 9 dots.) How many dots do you count? Wait for the signal to tell me. Get ready (snap).

S: 9.

T: How can you see them in two parts?

S: (Students come up to the card.) I saw 5 here and 4 here. → I saw 3 here and 6 here. → I saw 2 here and 7 here.

Repeat with other cards. Pass out the cards for students to work with a partner.

How Many Is One More? (2 minutes)

Materials: (T) Large 5-group cards (Lesson 1 Fluency Template 1)

Note: This fluency activity advances the familiar work with the pattern of *1 more* as it requires students to visualize an additional dot on the 5-groups.

T: (Show 3.) How many dots do you see?

S: 3.

T: What's one more than 3?

S: 4.

Repeat with all the numbers through 10.

Grouping Teen Numbers into 10 Ones (4 minutes)

Materials: (S) Bag with about 20 small objects and work mat

Note: The bags should have a variety of objects between 11 and 20.

Note: Practice separating and counting objects as ten ones and some ones solidifies students' understanding of teen numbers.

T: Empty your bag. Put all the items on your work mat. Count out 10 ones, and move them together into a bunch.

T: (Wait while they work.) How many ones are in your bunch?

S: 10 ones.

T: How many are not in your bunch?

S: 3 ones.

T: Say the number sentence.

S: 10 ones and 3 ones equals 13 ones.

T: Push all your things back together. Spread them all out over your work mat.

Repeat process 2 or 3 more times. Ask students if the same 10 things are in the bunch each time.

NOTES ON MULTIPLE MEANS OF REPRESENTATION:

Students working below grade level may need to model Jenny's mistake and count the quantity so that they can compare it to the fifteen chips. Provide students with counters so that they can show the correct solution to the problem and explain her mistake.

Application Problem (5 minutes)

A Pre-Kindergarten friend named Jenny drew 15 things with 1 chip and 5 more chips. Draw 15 things as 10 ones and 5 ones, and explain to your partner why you think Jenny made her mistake.

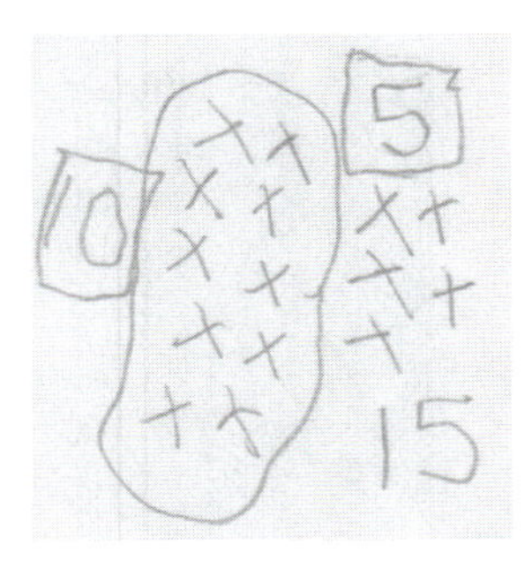

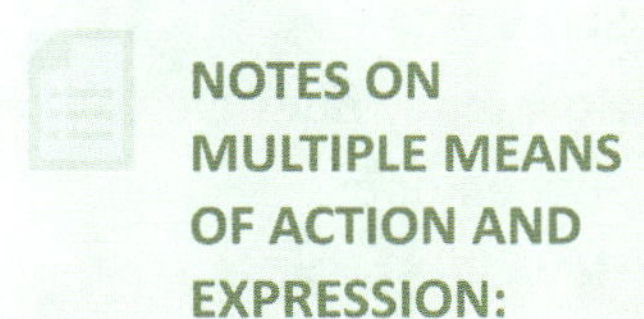

NOTES ON MULTIPLE MEANS OF ACTION AND EXPRESSION:

Challenge students working above grade level by extending the Application Problem and asking, "If Jenny made the same mistake representing 18, how might she show it?" and "How many more chips does Jenny need to correct her mistake?"

Concept Development (27 minutes)

Materials: (S) Double 10-frame (Template) within a personal white board

T: I'm going to write a number on the board. I want you to show that number by putting circles or dots inside the 10-frames.
T: (Write 10 on the board.) Say the number.
S: Ten!
T: Draw circles or dots to show ten. When I say show me, hold up your white board.
T: Show me. How many ones did you draw?
S: Ten ones.
T: Very good. Erase your boards. (Write 14.) Say the number.
S: Fourteen!
T: Whisper the number the Say Ten way as you fill in your 10-frames to show it.
S: Ten 4 (whispering while filling in 10-frames).
T: Talk with a partner to explain your drawing and how you grouped the dots.
T: (Write 18.) Say the number the Say Ten way.
S: Ten 8.
T: Whisper the number the regular way as you fill in your 10-frames.
S: Eighteen (whispering while filling in 10-frames).
T: Talk with your partner. Explain why your picture shows ten 8.

Continue this way with 15 and 19.

T: Now, let's try something different. Turn your board over to the blank side. I'm going to show a number. I want you to make a drawing that shows that many circles. Then, I want you to circle 10 ones so we can see the parts that make up the number.
T: (Show 16. Wait.)
T: Show me.
T: How many ones did you draw?
S: Sixteen ones.
T: How did you group the sixteen ones?
S: Ten ones and 6 ones.
T: Yes! Let's do another.

Continue this way through the other teen numbers.

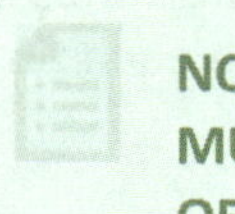

NOTES ON MULTIPLE MEANS OF REPRESENTATION:

Support your English language learners in comparing the 10-frame drawing and circle drawings by referring to the images. For the teen numbers, be sure to post the numerals along with the written word. Students have a difficult time hearing that *thirteen* is a different number from *thirty* because they sound alike. Having these clearly differentiated on the word wall will help them keep them apart.

Problem Set

Students should do their personal best to complete the Problem Set within the allotted time. Direct students to count as they represent the numbers. Have them whisper count as they work and fill one complete 10-frame before moving on to the next. Have them show their numbers with Hide Zero cards.

Name Dan Date

Whisper count as you draw the number. Fill one ten-frame first. Show your numbers with your Hide Zero cards.

12

17

16

13

Student Debrief (8 minutes)

Lesson Objective: Draw teen numbers from abstract to pictorial.

The Student Debrief is intended to invite reflection and active processing of the total lesson experience.

Invite students to review their solutions for the Problem Set. They should check work by comparing answers with a partner before going over answers as a class. Look for misconceptions or misunderstandings that can be addressed in the Debrief. Guide students in a conversation to debrief the Problem Set and process the lesson.

Any combination of the questions below may be used to lead the discussion.

- How are your 10-frame drawings and your circle drawings the same and different?
- Look at your 10-frame drawings with your partner. Did you draw the number 17 the same way? If not, explain why both drawings show 17. Do the same for the number 16.
- Compare your 10-frame drawings with your circle drawings. Is one drawing easier to read and understand than the other? Explain your thinking.
- (Do a finger flash in mixed order from 10 to 20, and have students say the numbers the Say Ten way.)

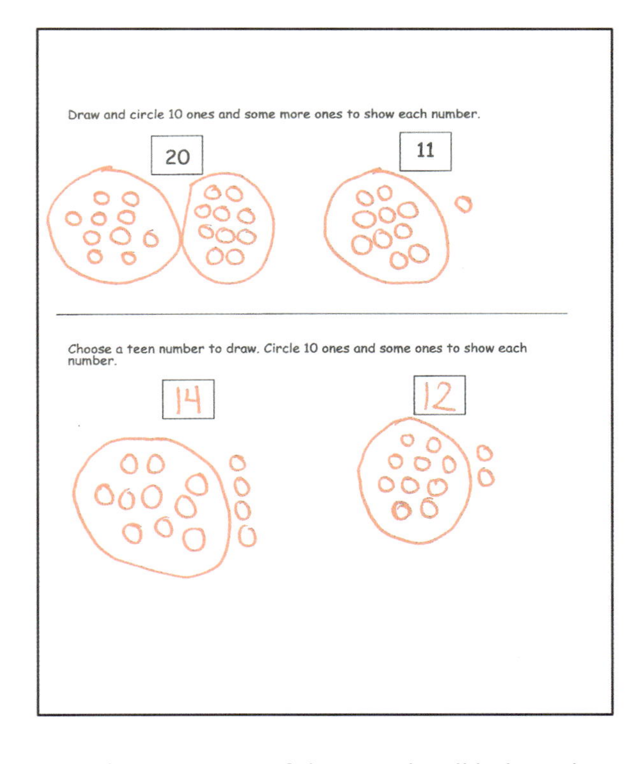
Draw and circle 10 ones and some more ones to show each number.

20

11

Choose a teen number to draw. Circle 10 ones and some ones to show each number.

14

12

Exit Ticket (3 minutes)

After the Student Debrief, instruct students to complete the Exit Ticket. A review of their work will help with assessing students' understanding of the concepts that were presented in today's lesson and planning more effectively for future lessons. The questions may be read aloud to the students.

Name ______________________________ Date ____________

Whisper count as you draw the number. Fill one 10-frame first. Show your numbers with your Hide Zero cards.

12

17

16

13

Draw and circle 10 ones and some more ones to show each number.

20

11

Choose a teen number to draw. Circle 10 ones and some ones to show each number.

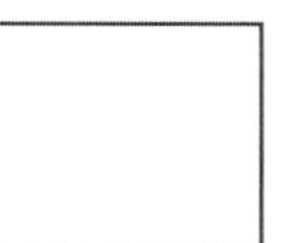

Name ______________________________ Date ______________

Show the number by filling in the 10-frames with circles.

15

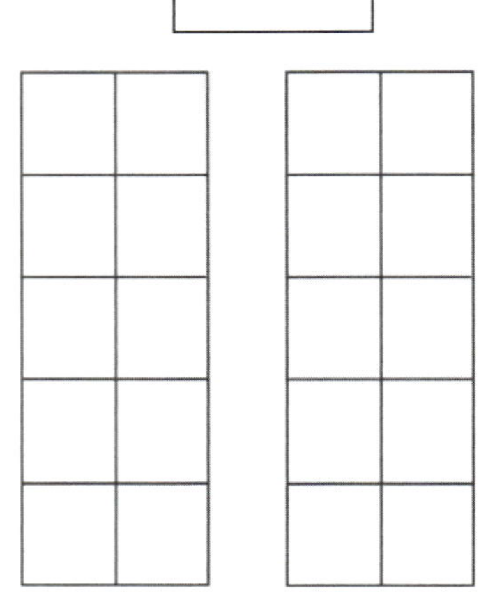

19

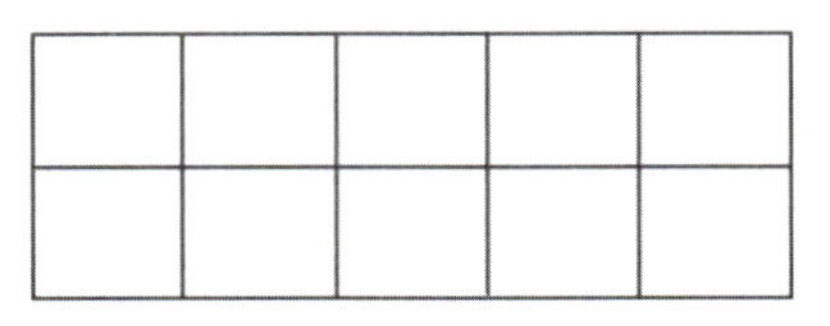

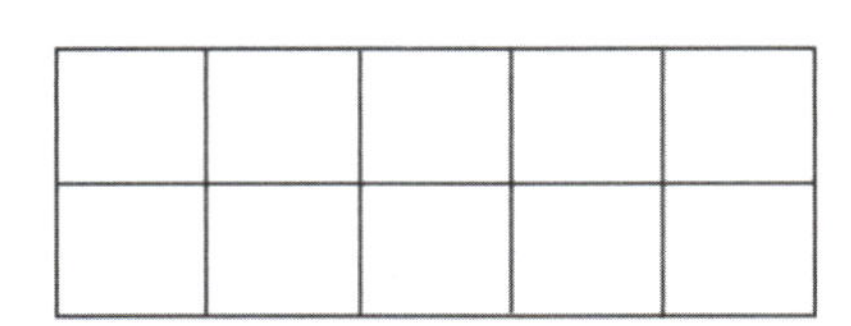

Draw circles to show the number. Circle 10 ones.

18

14

Name ______________________________ Date ______________

For each number, make a drawing that shows that many objects.
Circle 10 ones.

11

16

20

19

14

12

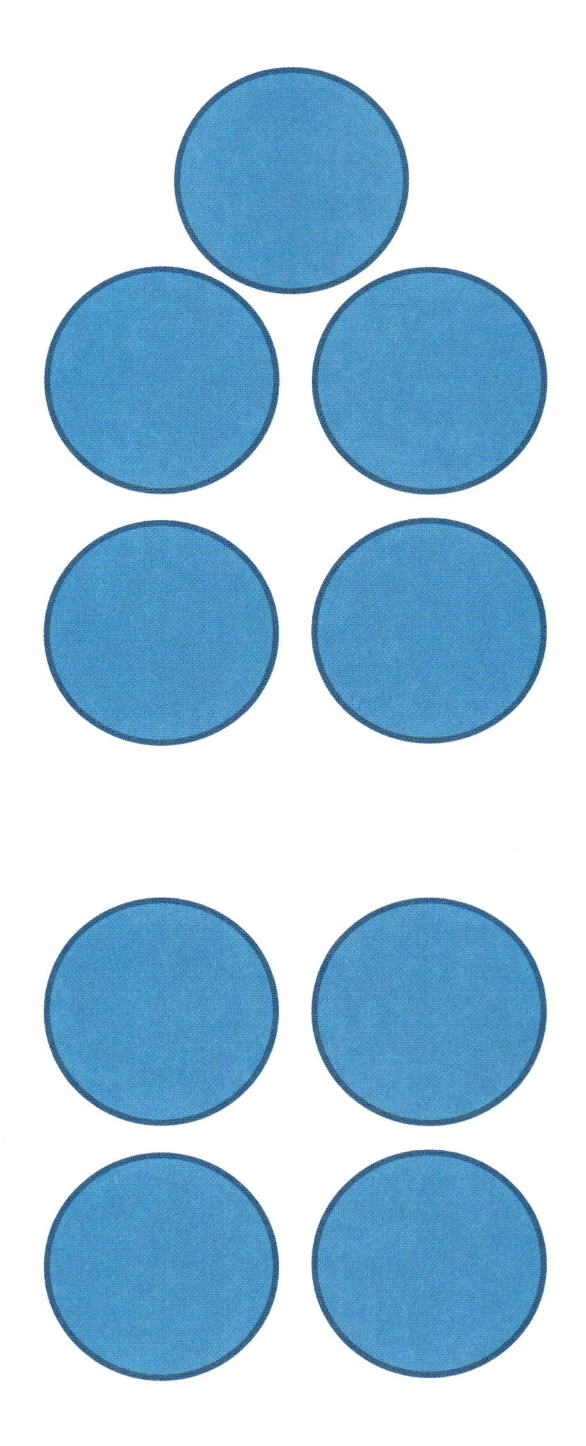

dot cards of 9

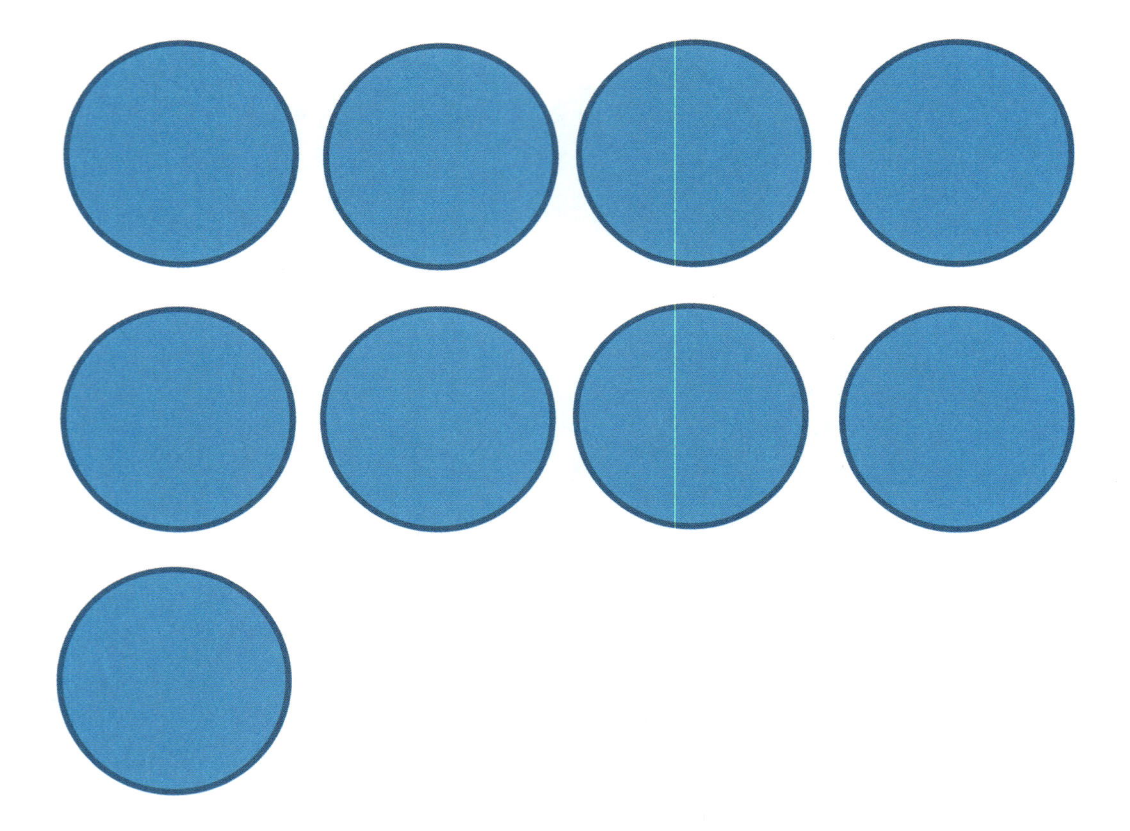

dot cards of 9

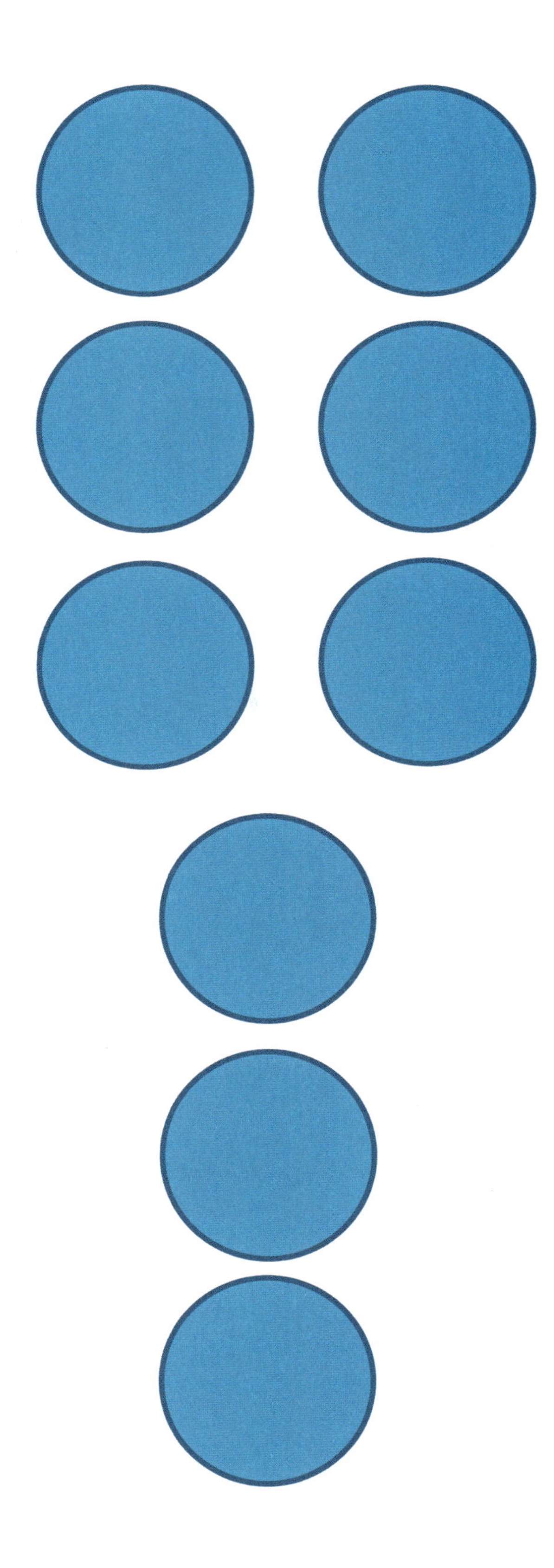

dot cards of 9

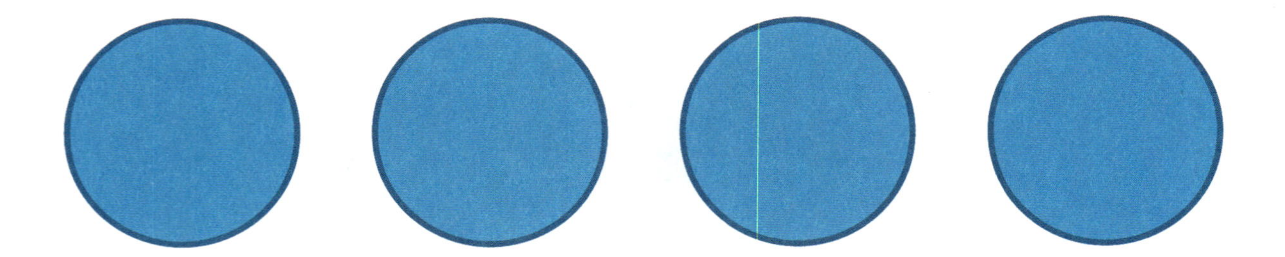

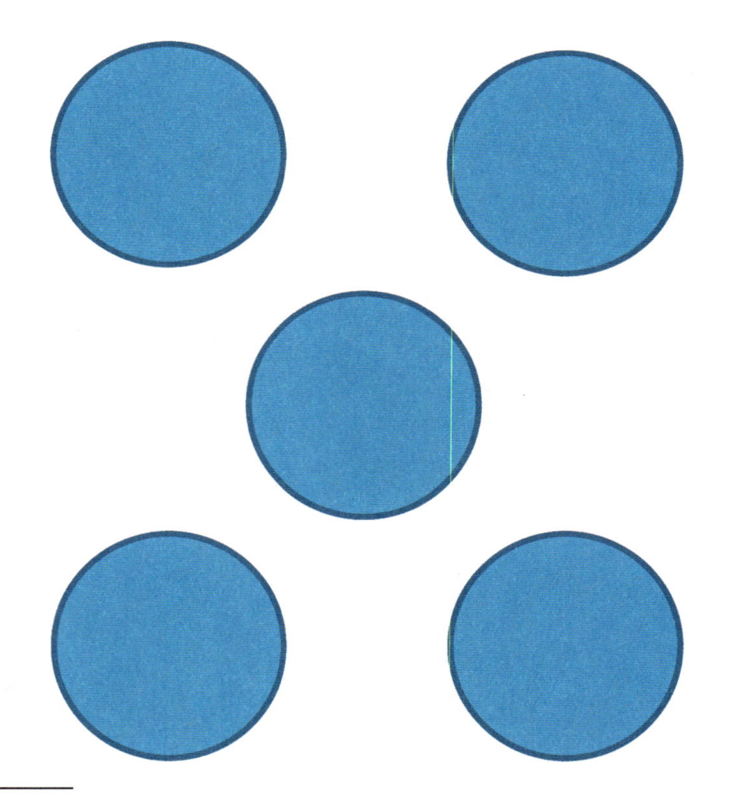

dot cards of 9

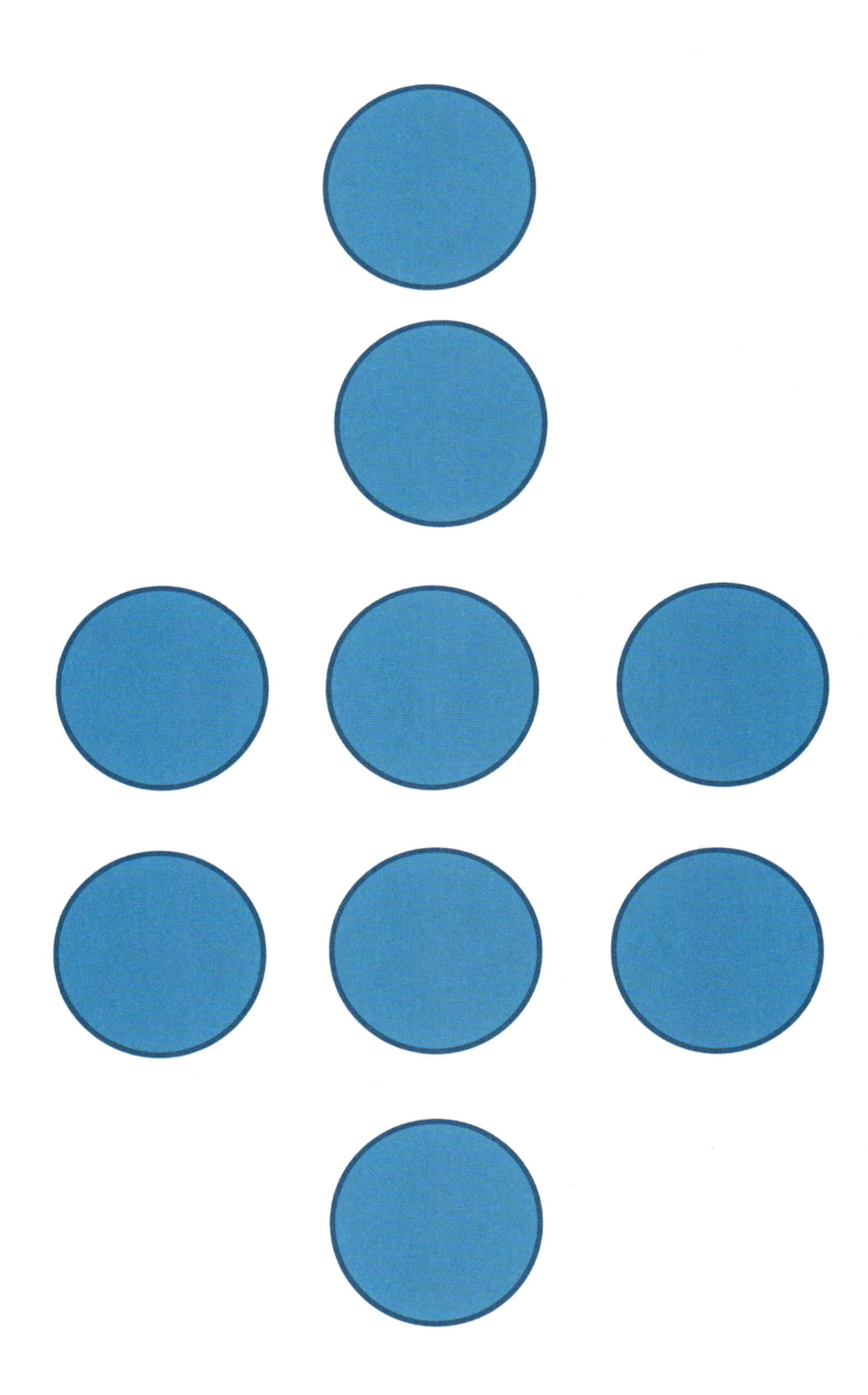

dot cards of 9

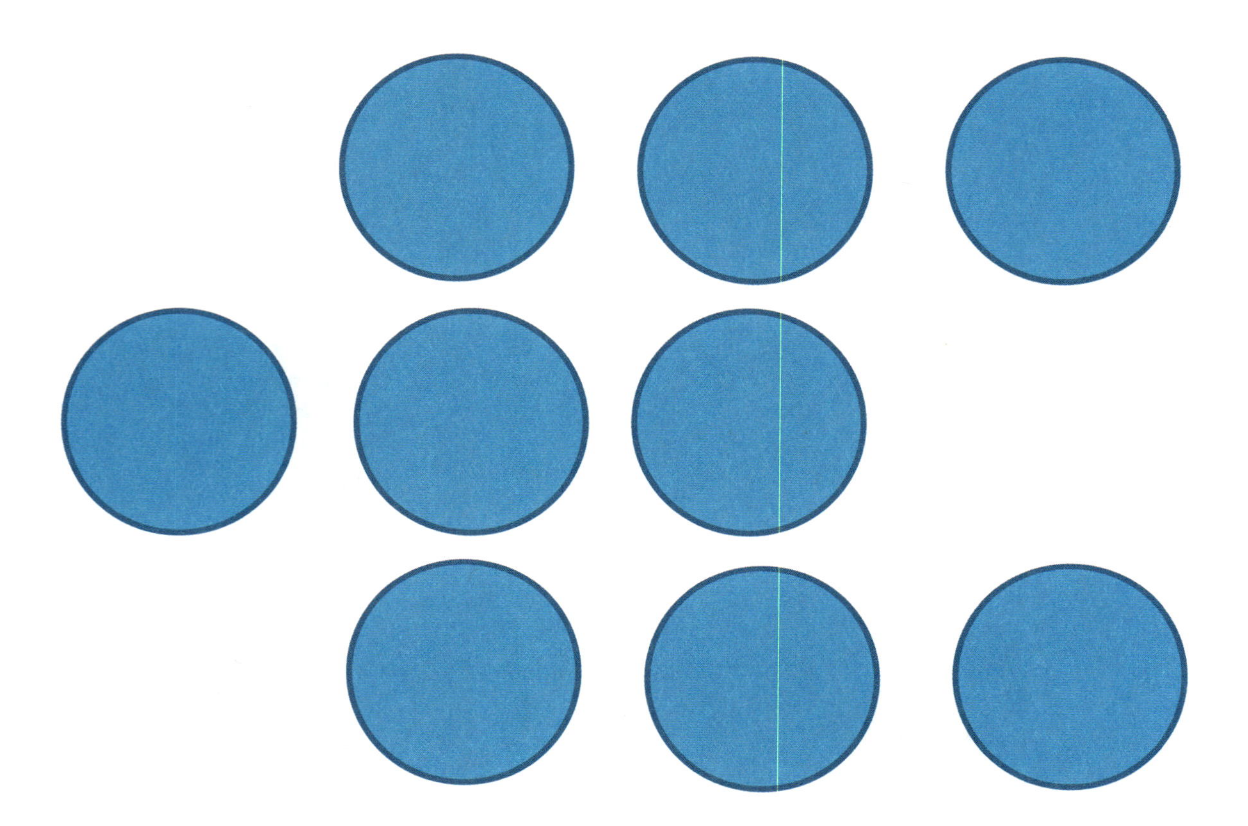

dot cards of 9

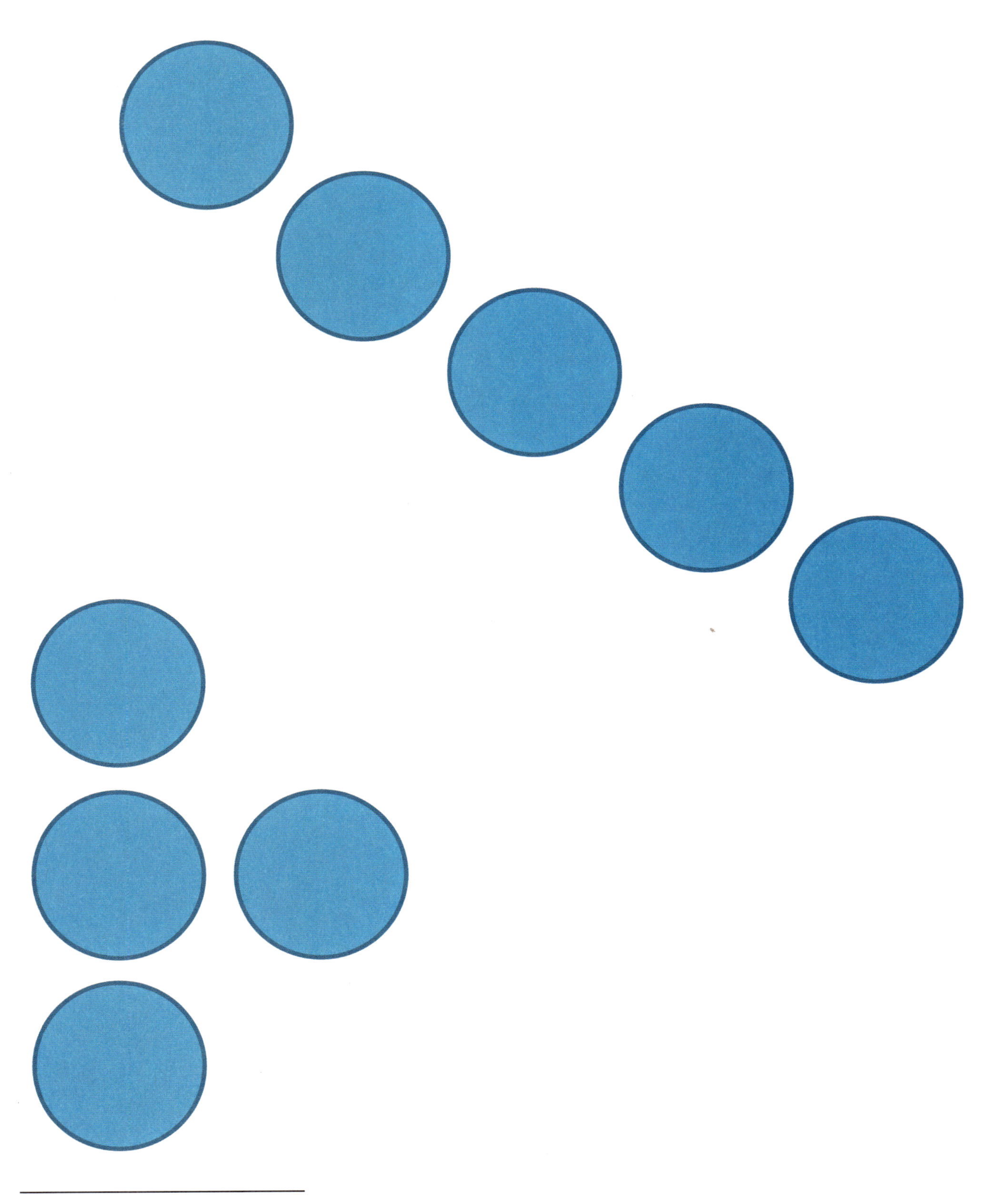

dot cards of 9

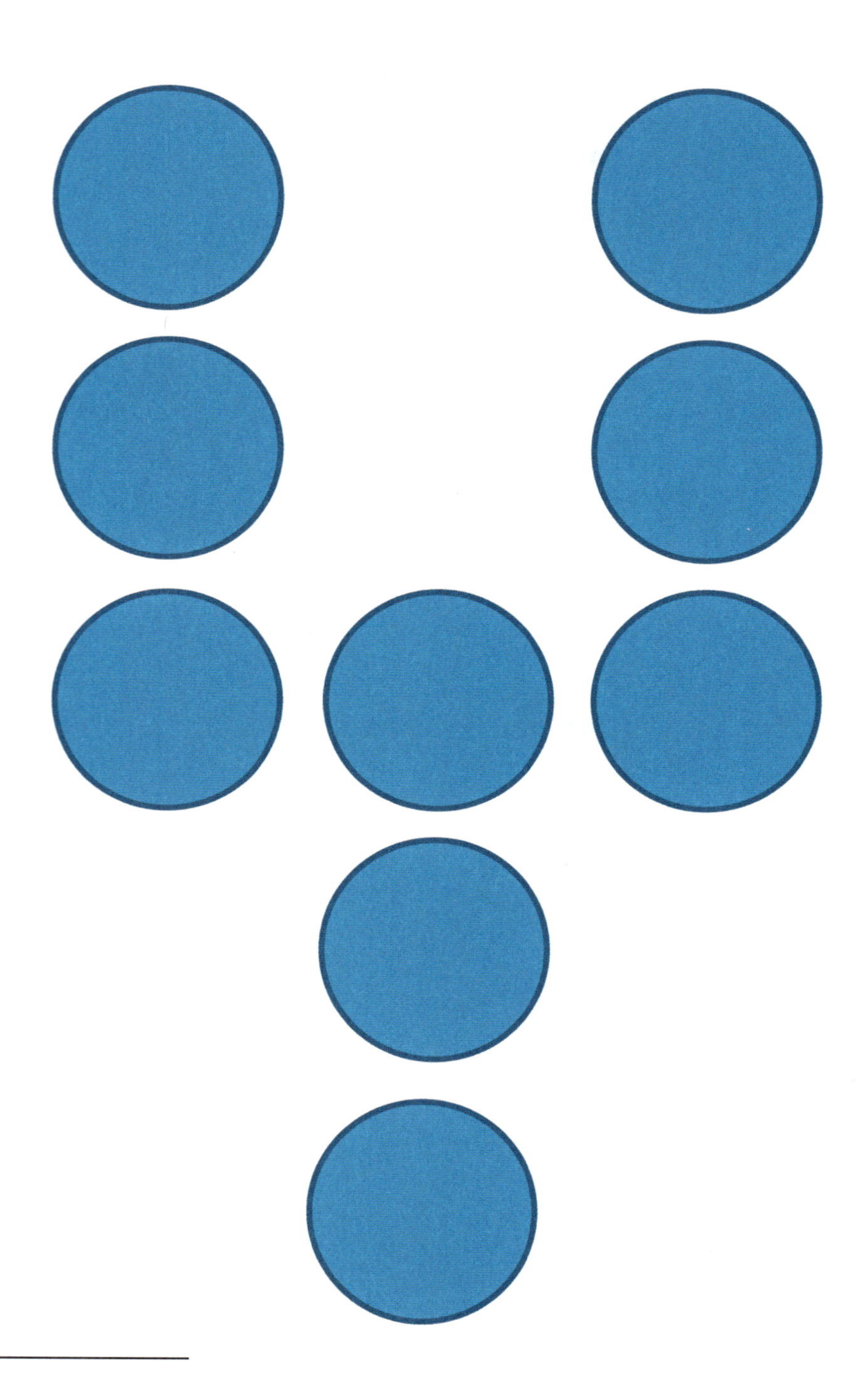

dot cards of 9

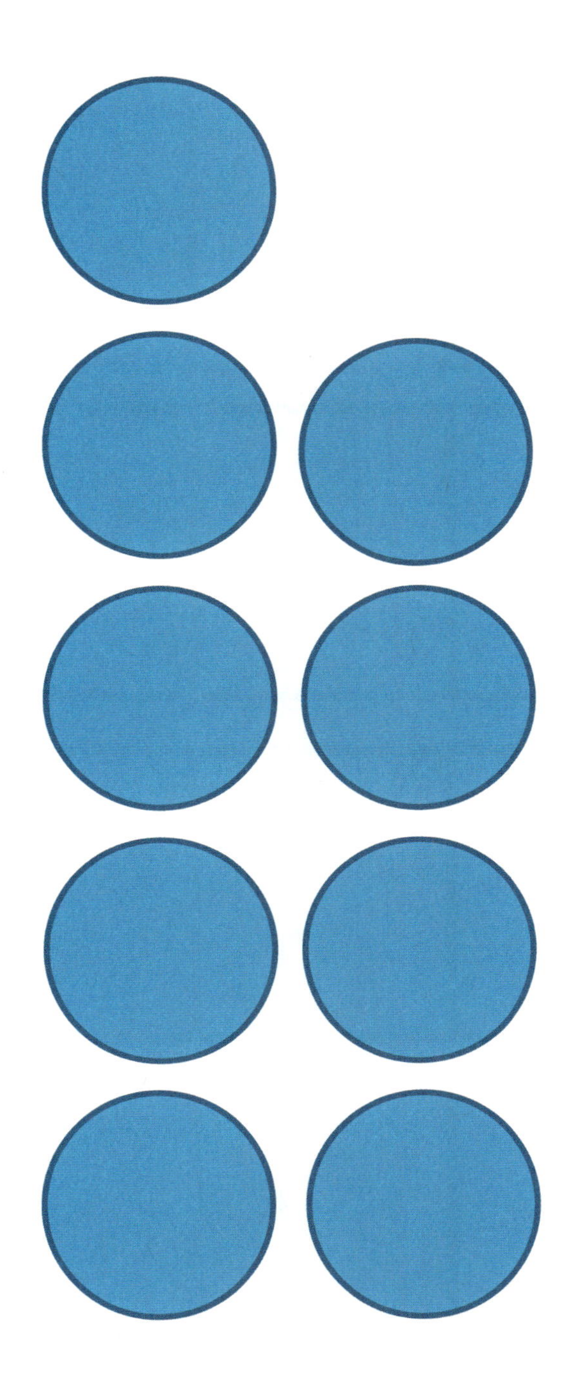

dot cards of 9

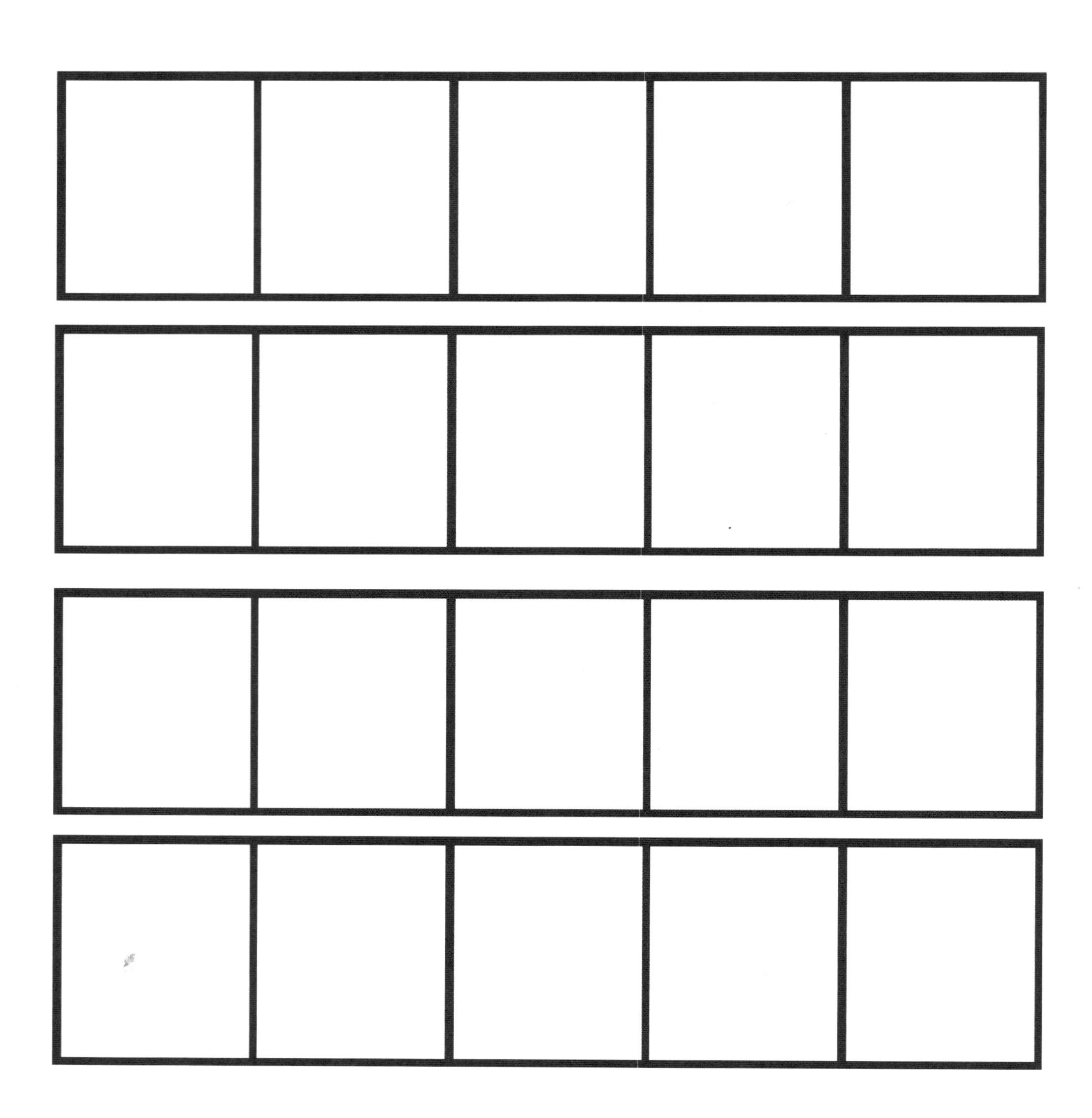

double 10-frame

Topic C

Decompose Numbers 11–20, and Count to Answer "How Many?" Questions in Varied Configurations

K.CC.4bc, K.CC.5, K.NBT.1, K.CC.3, K.CC.4a

Focus Standards:	K.CC.4	Understand the relationship between numbers and quantities; connect counting to cardinality. b. Understand that the last number name said tells the number of objects counted. The number of objects is the same regardless of their arrangement or the order in which they were counted. c. Understand that each successive number name refers to a quantity that is one larger.
	K.CC.5	Count to answer "how many?" questions about as many as 20 things arranged in a line, a rectangular array, or a circle, or as many as 10 things in a scattered configuration; given a number from 1–20, count out that many objects.
	K.NBT.1	Compose and decompose numbers from 11 to 19 into ten ones and some further ones, e.g., by using objects or drawings, and record each composition or decomposition by a drawing or equation (e.g., 18 = 10 + 8); understand that these numbers are composed of ten ones and one, two, three, four, five, six, seven, eight, or nine ones.
	5	
Instructional Days:	GPK–M5	Addition and Subtraction Stories and Counting to 20
Coherence -Links from:	G1–M2	Introduction to Place Value Through Addition and Subtraction Within 20
-Links to:		

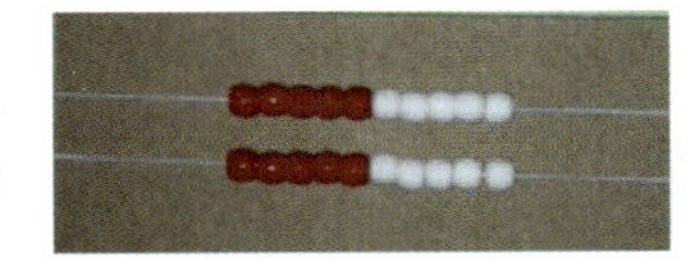

Topic C opens in Lesson 10 with students building a Rekenrek to 20, which they use to count and model numbers for the balance of the year. They deepen their understanding of the composition and decomposition of teen numbers as 10 ones and some more ones (**K.NBT.1**) by showing, counting, and writing (**K.CC.3**) the numbers 11 to 20 using a variety of configurations: vertical towers, linear, array, and circular configurations. In each configuration, students count to answer "how many?" questions (**K.CC.5**) and realize that whatever the configuration, a teen number can be decomposed into 10 ones and some ones.

Lessons 11 and 12 represent each teen number as a part of a set of number stairs to 20. Each vertical tower is set within the ordered continuum. This configuration allows students to see each teen number in relationship to the others, as one larger than the number before it (**K.CC.4c**), in relationship to 10, and in relationship to numbers 1–9 since the lesson's Problem Set has a color change after 10 ones. Next, in Lesson 13, students move teen quantities back and forth between linear and array configurations, practice counting strategies, and recognize that when they answer "how many?" the total has not changed. Finally, the topic culminates with the most challenging configuration, the circle. Students circle 10 and see that, yes, the circle is composed of 10 ones and some ones, too. They become proficient at counting in all configurations to answer "how many?" questions (**K.CC.5**).

A Teaching Sequence Toward Mastery of Decomposing Numbers 11–20, and Counting to Answer "How Many?" Questions in Varied Configurations	
Objective 1:	**Build a Rekenrek to 20.** **(Lesson 10)**
Objective 2:	**Show, count, and write numbers 11 to 20 in tower configurations increasing by 1—a pattern of *1 larger*.** **(Lesson 11)**
Objective 3:	**Represent numbers 20 to 11 in tower configurations decreasing by 1—a pattern of *1 smaller*.** **(Lesson 12)**
Objective 4:	**Show, count, and write to answer *how many* questions in linear and array configurations.** **(Lesson 13)**
Objective 5:	**Show, count, and write to answer *how many* questions with up to 20 objects in circular configurations.** **(Lesson 14)**

Lesson 10

Objective: Build a Rekenrek to 20.

Suggested Lesson Structure

Fluency Practice	(10 minutes)
Application Problem	(7 minutes)
Concept Development	(13 minutes)
Student Debrief	(20 minutes)
Total Time	**(50 minutes)**

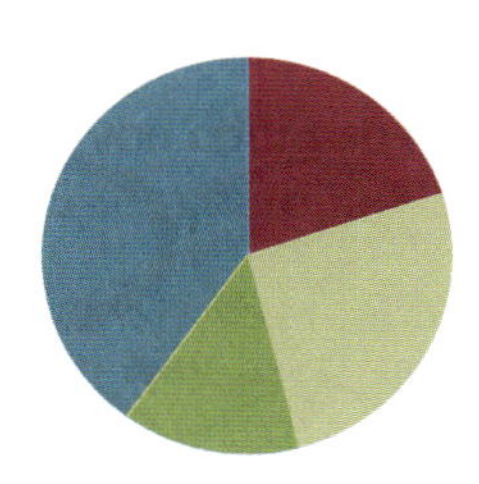

Fluency Practice (10 minutes)

- Writing Teen Numbers **K.CC.3** (4 minutes)
- Showing Numbers with Hands **K.CC.4, K.NBT.1** (3 minutes)
- Counting **K.CC.2** (3 minutes)

Writing Teen Numbers (4 minutes)

Materials: (T) Linking cubes (S) Personal white board

Note: By writing the corresponding numeral for each part, and then the whole, students are continually reminded that the *1* in teen numbers refers to 10 ones.

T: (Show 3 cubes.) Write the number.
S: (Students write the numeral 3.)
T: (Show 10 cubes.) Write the number.
S: (Students write the numeral 10.)
T: (Show 13 cubes.) Write the number.
S: (Students write 13.)

Repeat the process for the following possible sequence: 10, 13, 19, 5, 17, 8, 18, 15, 12, 14, 16.

Showing Numbers with Hands (3 minutes)

Materials: (T) 20-bead Rekenrek

Note: Relating the group of 10 on the Rekenrek to students' own hands helps them internalize the structure of teen numbers.

T: (Show 12 on the Rekenrek.)
T: Show the two parts of the number on your fingers. Say the parts at the same time.
S: 10 (flashing ten fingers) and 2 (showing two fingers).

Continue with the following possible sequence: 13, 14, 19, 16, 18, 15, 11, 17, 20.

Counting (3 minutes)

Materials: (T) 20-bead Rekenrek

Note: Students relate Say Ten counting to conventional teen number names in this activity. Counting both ways, and in both directions, ensures that students remain alert to the sequence and do not simply extend a pattern of number words. If students struggle, return to a more manageable range (such as within 13 or 15), and later build up to work within 20.

Count by ones from 11–20, changing directions both the Say Ten way and the regular way.

Application Problem (7 minutes)

Ms. Garcia is painting her fingernails. She has painted all the nails on her left hand except her thumb. How many more nails does she need to paint? How many does she have left to paint after she paints her left thumb? Draw a picture to help you.

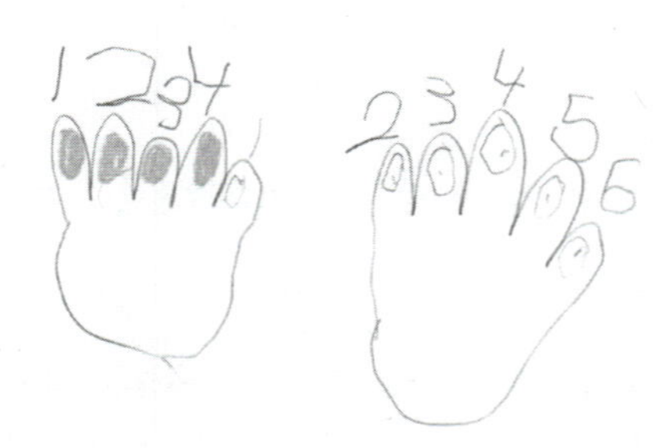

Note: This problem is an application of **K.OA.4**, wherein students learn the number that makes 10 from any number less than 10. As a word problem, this is a *change unknown*, which is a Grade 1 problem type. Therefore, the number sentence is not asked for since missing addends are introduced in the fall of Grade 1.

Concept Development (13 minutes)

Materials: (S) Problem Set, 10 red pony beads, 10 white pony beads, a red crayon, a black crayon

T: (Distribute the Problem Set. Have students put the beads on the circles below the first pair of hands, 5 red on the left, and 5 white on the right.)
T: Imagine these red beads are Ms. Garcia's painted fingernails. Show me how many she painted at first (in the Application Problem). Put them on her fingernails.
S: (Move 4 beads from the circles to the fingernails, starting with the left pinky finger.)

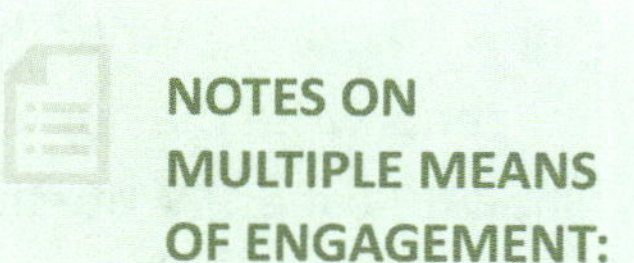

NOTES ON MULTIPLE MEANS OF ENGAGEMENT:

Push the thinking of students working above grade level by asking, "What would happen if Ms. Garcia also paints her toenails? How many nails will be painted when she is completely done?" Consider extending their thinking further by asking, "If Ms. Garcia draws two green polka dots on each finger, how many polka dots does she paint altogether?"

T: How many fingernails did she paint, and how many does she need to paint? Use these words to help. Listen.

T: "She painted ____ fingernails. She needs to paint ____ fingernails."

S: She painted 4. She needs to paint 6.

T: Paint one more nail on her left hand. (Pause.) Tell me what she's painted and what she needs to paint.

S: She painted 5. She needs to paint 5.

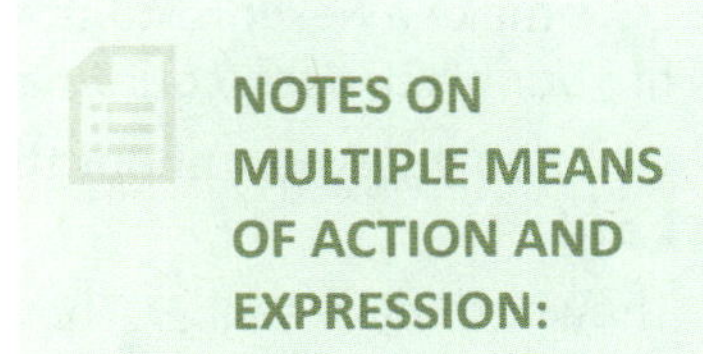

NOTES ON MULTIPLE MEANS OF ACTION AND EXPRESSION:

Scaffold the lesson for English language learners by pointing to the painted hand and asking, "How many did she paint?" Then, point to the hand that is not painted and ask, "How many does she need to paint?"

MP.7

Continue the pattern of painting one more fingernail and making the statements that describe how many have been painted and need to be painted. Have the students work independently as soon as they can. Once they have finished the first pair of hands, have them use the second pair of hands for Ms. Garcia's daughter's unpainted nails. Have them put the beads on her fingers, counting and making statements as they go. Engage them in counting all the beads, analyzing how many are red and how many are white, how many are on the left hands, and how many on the right hands.

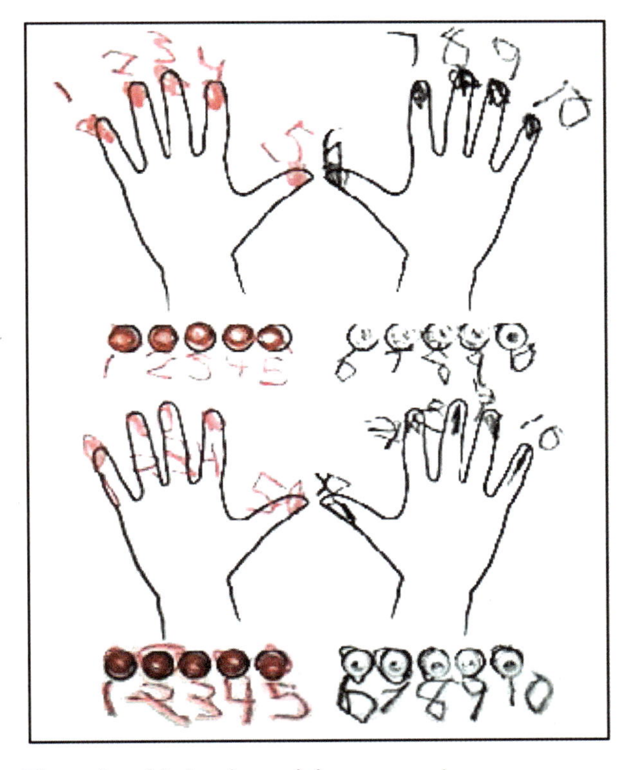

Problem Set (5 minutes)

Students color the left-hand fingernails red and color the right-hand fingernails black (in lieu of white), counting as they go. They color the corresponding *beads* below to match the hands, counting as they go. They can write their numbers 1 to 10, too.

Student Debrief (20 minutes)

Lesson Objective: Build a Rekenrek to 20.

The Student Debrief is intended to invite reflection and active processing of the total lesson experience.

Invite students to review their solutions for the Problem Set. They should check work by comparing answers with a partner before going over answers as a class. They can count on or count all, as needed. Look for misconceptions or misunderstandings that can be addressed in the Debrief. Guide students in a conversation to debrief the Problem Set and process the lesson. Any combination of the questions below may be used to lead the discussion.

Materials: (S) 10 red and 10 white pony beads from the Concept Development, two 12-inch lengths of elastic, one 2.75-inch by 5.5-inch piece of chipboard (or cardboard strip) with an indentation (note that each 8 ½-inch by 11-inch chipboard makes 4 Rekenreks.)

T: Let's make a Rekenrek. Put your red beads on top of your red dots and your white beads on top of your black dots, counting as you go.

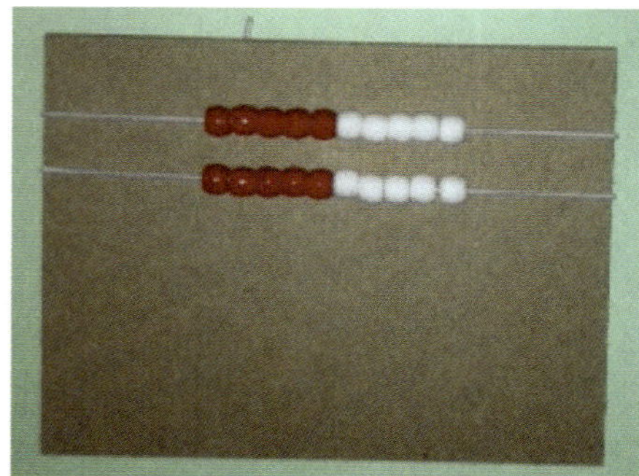

T: What do you know about the number of your red and white beads?

S: They both have ten. → They are the same number. → They are an equal number.

T: How do you say the total number of beads the Say Ten way?

S: 2 tens.

T: How many beads is that the regular way?

S: Twenty.

After showing students how to thread the elastic through from left to right, red beads first, give each student a 12-inch elastic. Once they have finished one row, have them do the other row. Show them how to pinch the elastics at either end to pick up the row and place it on their chipboard (or cardboard strip), one row under the other. The teacher can circulate and tie the elastics, or have helpers tie the elastics after class for use in future lessons.

The discussion should establish a correlation between students' fingernails and the beads of the Rekenrek.

- Talk to your partner about what is the same and what is different about the number of your fingernails and the number of beads.
- How many people do we need to have the same number of fingernails as on your Rekenrek?
- If the beads were purple and green, how many nails and beads would be purple, and how many would be green?
- What if you hide two hands? How many beads would you see?

Exit Ticket (3 minutes)

After the Student Debrief, instruct students to complete the Exit Ticket. A review of their work will help with assessing students' understanding of the concepts that were presented in today's lesson and planning more effectively for future lessons. The questions may be read aloud to the students.

Name ______________________________ Date ______________

Name ______________________________ Date ______________

Use your red crayon and yellow crayon to draw the beads from your Rekenrek in two lines.

How many beads did you draw?

Trace your hands. Draw your fingernails. How many fingernails do you have on your two hands?

Name ______________________________ Date ______________

Color the number of fingernails and beads to match the number bond. Show by coloring 10 ones above and extra ones below. Fill in the number bonds.

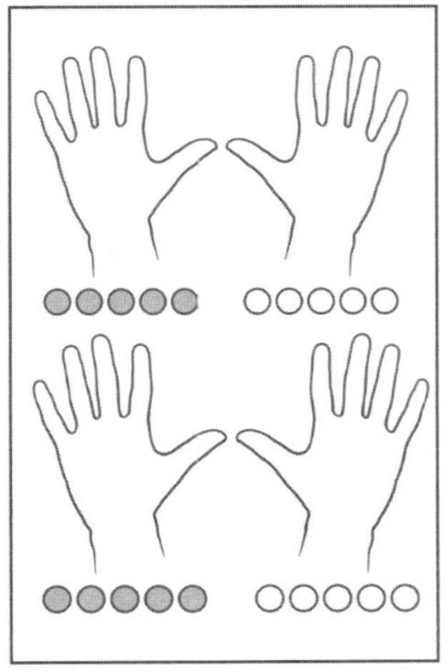

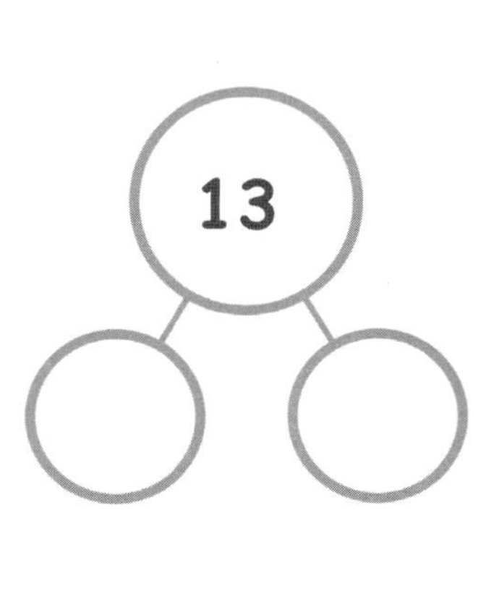

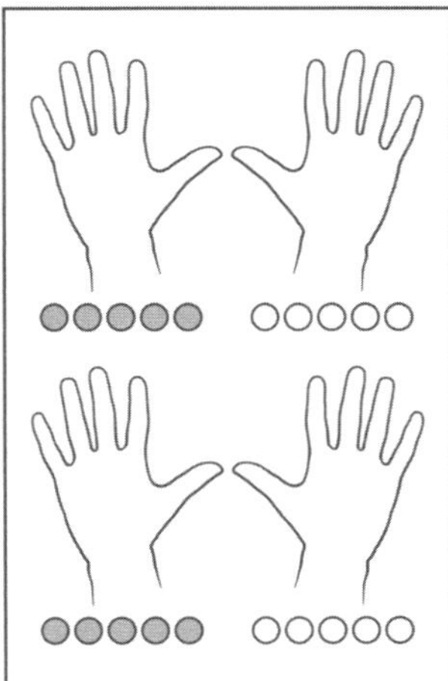

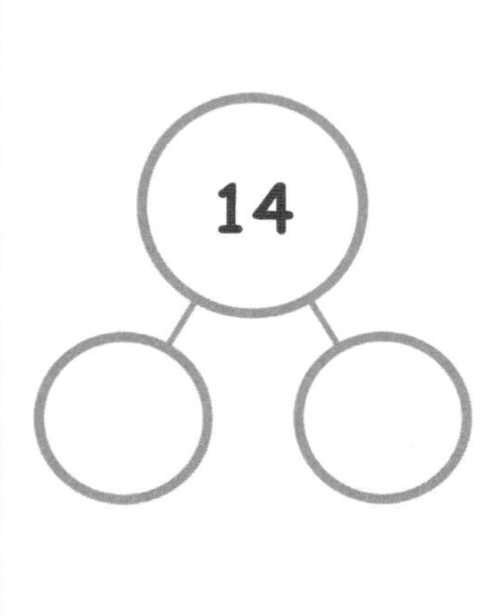

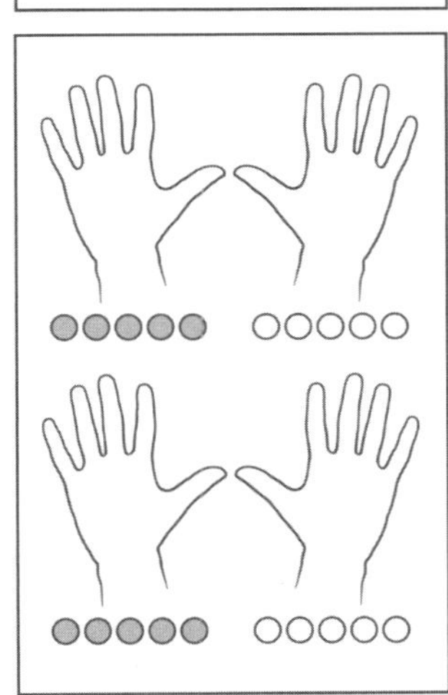

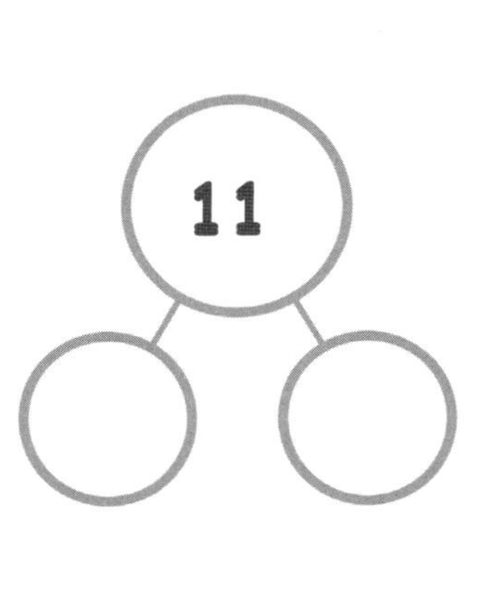

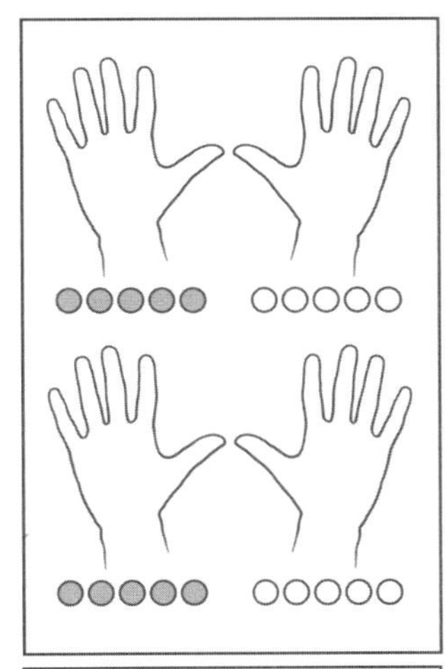

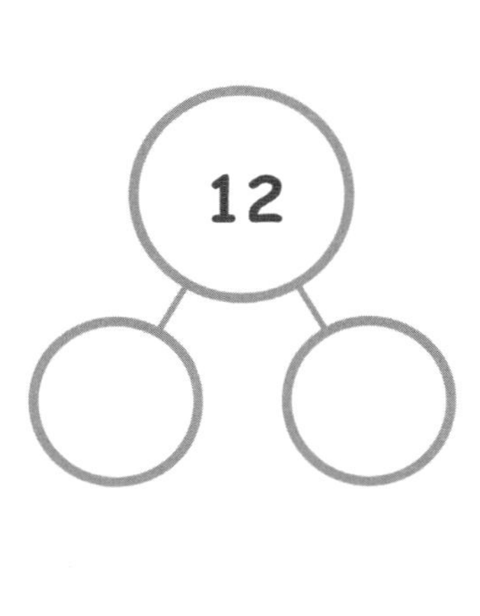

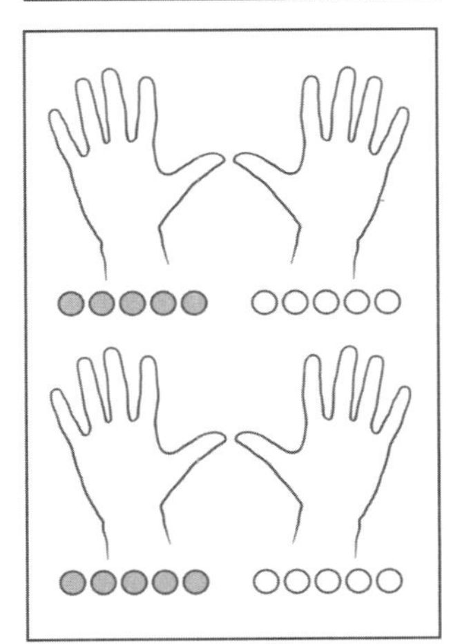

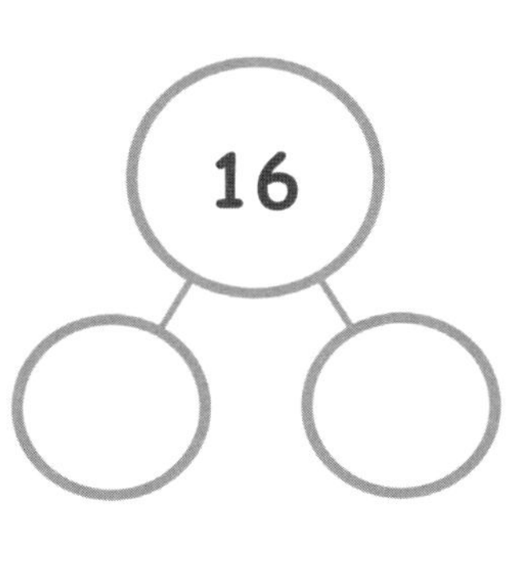

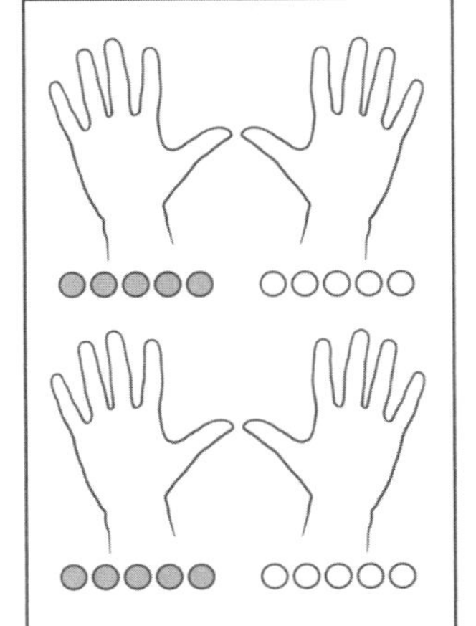

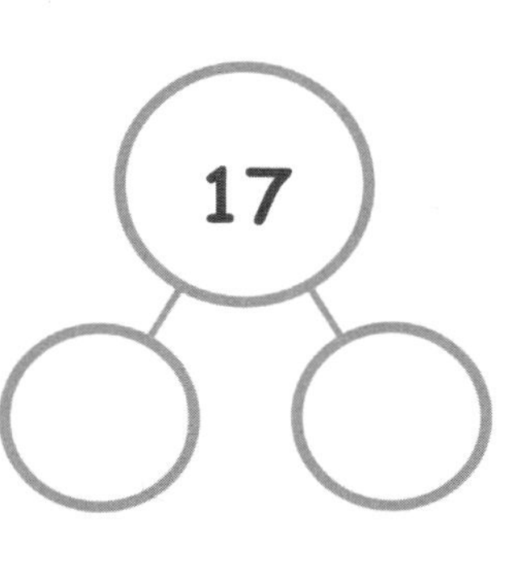

Lesson 11

Objective: Show, count, and write numbers 11 to 20 in tower configurations increasing by 1—a pattern of *1 larger*.

Suggested Lesson Structure

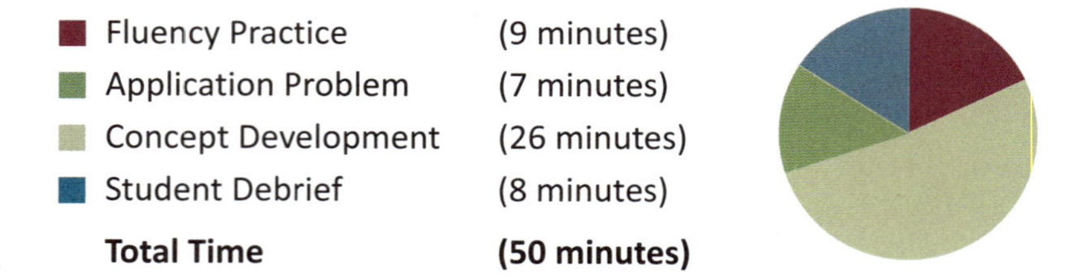

- Fluency Practice (9 minutes)
- Application Problem (7 minutes)
- Concept Development (26 minutes)
- Student Debrief (8 minutes)

Total Time **(50 minutes)**

Fluency Practice (9 minutes)

- Counting on a Rekenrek **K.CC.5** (4 minutes)
- One More **K.CC.2** (3 minutes)
- Saying Teen Numbers the Say Ten Way **K.NBT.1** (2 minutes)

Counting on a Rekenrek (4 minutes)

Materials: (S) Personal Rekenrek (Built in Lesson 10)

Note: Encourage students to show teen numbers in both horizontal (e.g., 13 as 10 on the top row and 3 on the bottom) and vertical (e.g., 13 as 10 red and 3 white) orientations. Students might also show numbers in 2 parts (e.g., 5 as 3 and 2).

T: Take out the Rekenrek that you made yesterday. I'm going to call out a number, and I want you to show it on your Rekenrek. (Wait while students prepare their Rekenreks.)

Possible sequence: 1, 2, 5, 6, 10, 11, 12, 13, 14, 15, 16, 15, 16, 17, 18, 19, 20, 19, 18, 17, 16, 15, 10, 5, 4, 3, 2, 1.

One More (3 minutes)

Materials: (T) 20-bead Rekenrek

Note: Students make use of the pattern of 1 more in numbers 1–9, to determine 1 more with teen numbers. Knowing that 4 ones are part of 14, for example, allows them to determine that 1 more is 15, just as 1 more than 4 is 5.

T: I want you to say one more than the number that you see on the Rekenrek. (Show 3.)
S: 4.
T: (Show 13.)
S: 14.

Continue with the following possible sequence: 5, 15, 1, 11, 4, 14, 7, 17, 8, 18, 9, 19, 6, 16.

Saying Teen Numbers the Say Ten Way (2 minutes)

Note: Now that students have had ample experience with counting the Say Ten way, the goal is to build speed and accuracy.

T: I'm going to say a number. You say it the Say Ten way. Eleven.
S: Ten 1.
T: Twelve.
S: Ten 2.

Repeat process for possible sequence: 13, 17, 19, 14, 16, 18, 15, 20.

Application Problem (7 minutes)

Mary has 10 toy trucks. She told her mom she likes to spread them out on the floor. She said she doesn't like to put them away neatly in the little toy box because then there are fewer toys. Draw a picture to prove to Mary that the number of toy trucks is the same when they are all spread out as when they are in the little toy box.

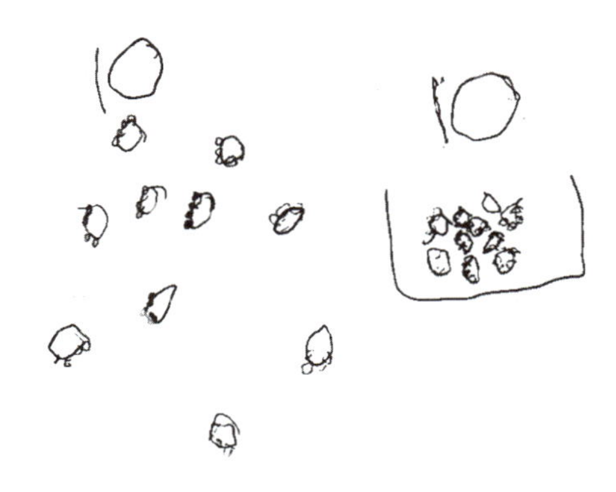

Note: This Application Problem provides an opportunity for students to model conservation. Students draw to prove that the number of objects remains the same, despite the perceptual change.

Concept Development (26 minutes)

Materials: (S) Two sets of 10 linking cubes (10 in one color and 10 in another color), sentence frame (Template)

Note: Notice that we are not saying "20 is 1 more *than* 19." This is very complex linguistically for many kindergarten students who can say "19 is more than 18" without quantifying the difference. They simply are seeing and analyzing that each successive number is one larger (**K.CC.4c**).

NOTES ON MULTIPLE MEANS OF REPRESENTATION:

Focus on academic vocabulary to help English language learners with the Application Problem. Provide students with a template for their work. Adapt the template so that one side has a graphic or a picture to represent the floor and one side has a graphic to represent the toy box.

T: Show me a tower of 10 cubes using one color.
T: (Students show a tower of 10.) How many cubes are you holding?
S: Ten.
T: How many ones is that?
S: 10 ones.
T: How many more cubes do you need to put on you tower to make 11?
S: 1 more!
T: Show me 11. (Point to the first sentence frame.) While you do that, say, "10. 1 more is 11."
S: 10. 1 more is 11.
T: And how do we say 11 the Say Ten way?
S: Ten 1.
T: Good! Put one more cube on your tower.
S: (Show 12.)
T: How many cubes do you have now?
S: 12.
T: Repeat with me, "11. 1 more is 12."
S: 11. 1 more is 12.

Use the sentence frames to help students express the relationship of each number to the preceding number. Continue adding one more cube for each number up to 20. Release as many students as possible to continue the pattern with a partner: "13. 1 more is 14." Continue releasing students as they demonstrate skill and understanding.

Problem Set (7 minutes)

Students should do their personal best to complete the Problem Set within the allotted time. As students color the squares and write the numbers to complete the pattern, have them continue to say the relationship of each number to its preceding number. Example: Fourteen. 1 more is 15. Fifteen. 1 more is 16, etc.

Note: Have students use a different color crayon after they color 10 ones.

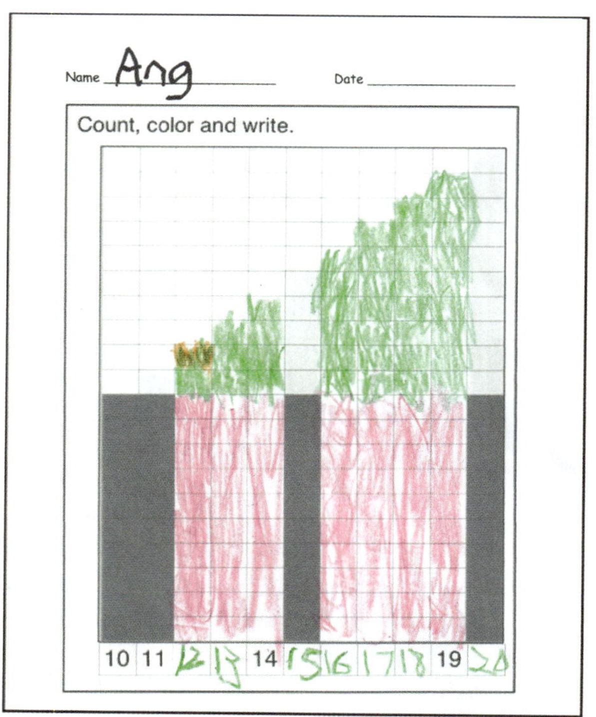

Student Debrief (8 minutes)

Lesson Objective: Show, count, and write numbers 11 to 20 in tower configurations increasing by 1—a pattern of *1 larger*.

The Student Debrief is intended to invite reflection and active processing of the total lesson experience.

Invite students to review their solutions for the Problem Set. They should check work by comparing answers with a partner before going over answers as a class. They can count on or count all, as needed. Look for misconceptions or misunderstandings that can be addressed in the Debrief. Guide students in a conversation to debrief the Problem Set and process the lesson.

Any combination of the questions below may be used to lead the discussion.

- What do you notice when you look at your paper?
- How is your drawing like the towers you made?
- How many cubes did you put on your tower each time?
- Did the number get larger or smaller when you put on one more?
- How is the number tower you made the same as the Rekenrek you made? How is it different?
- Fold your paper in half, and look just at the green stairs. How are they the same and different from the stairs for the larger numbers?

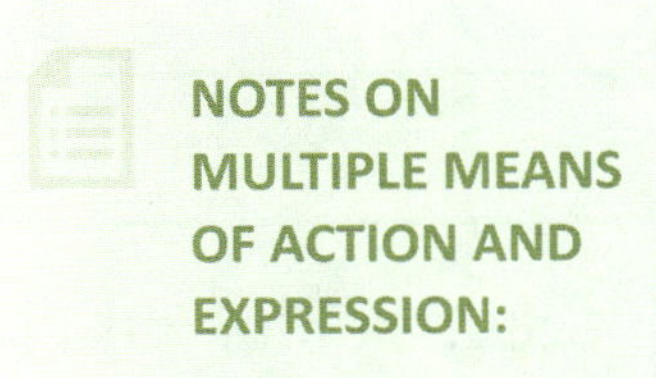

NOTES ON MULTIPLE MEANS OF ACTION AND EXPRESSION:

For students working below grade level, have them regularly work with you when they come to the carpet rather than with a partner. This provides them with much-needed extra time with the teacher.

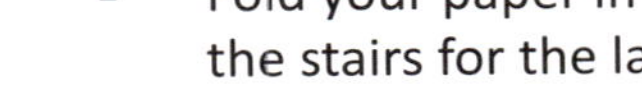

Exit Ticket (3 minutes)

After the Student Debrief, instruct students to complete the Exit Ticket. A review of their work will help with assessing students' understanding of the concepts that were presented in today's lesson and planning more effectively for future lessons. The questions may be read aloud to the students.

Name ______________________________ Date ______________

Count, color and write.

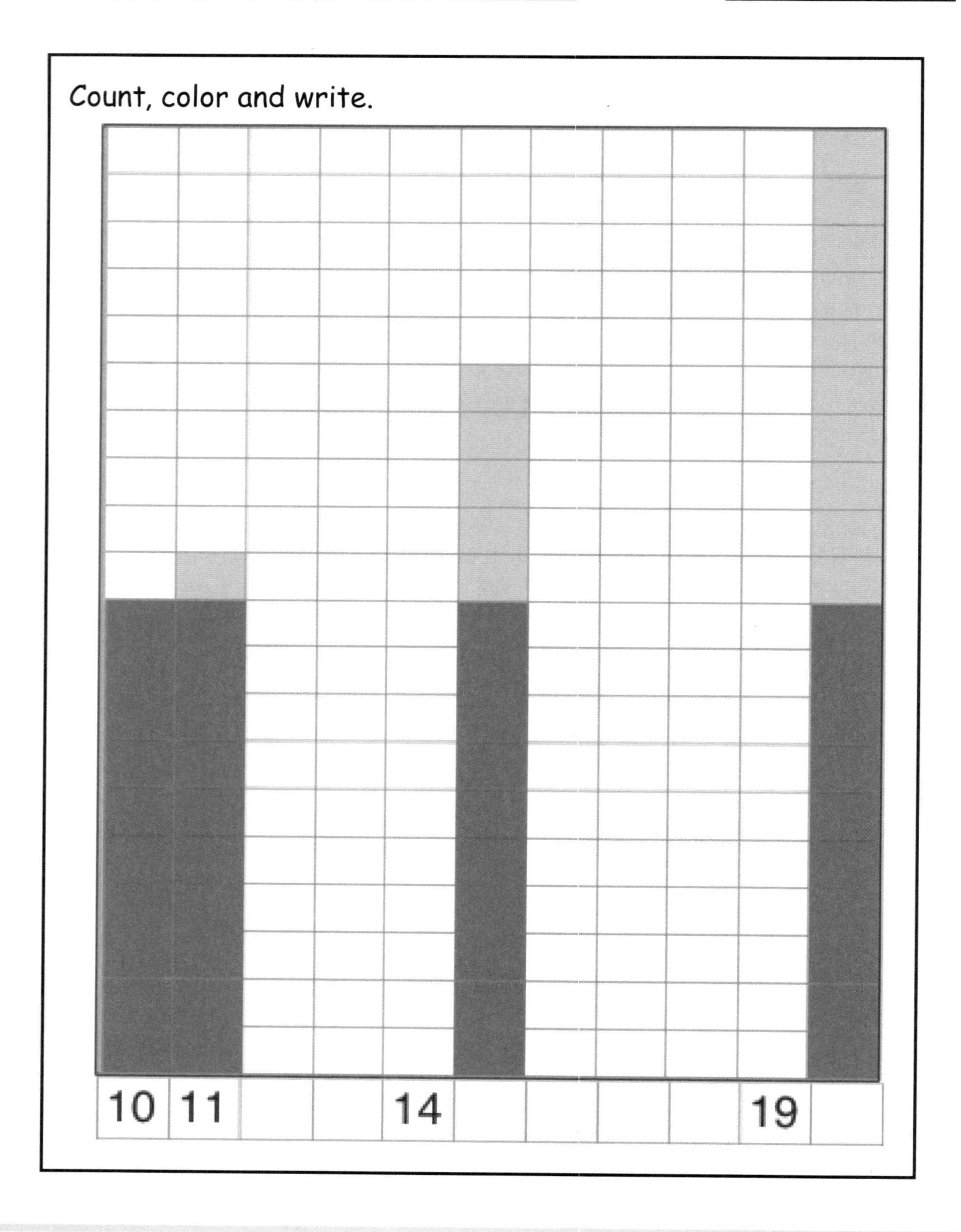

Name ________________________ Date ____________

Start at the bottom. Draw lines to put the numbers in order on the tower. Then, write the numbers in the tower. Say each number the regular way and the Say Ten way as you work.

12 ●

19 ●

16 ●

14 ●

17 ●

20
18
15
13
11
10

Name ______________________________ Date ______________

Write the missing numbers. Then, count and draw X's and O's to complete the pattern.

O O O O O O O O O O	X O O O O O O O O O O		X X X O O O O O O O O O O		X X X X X O O O O O O O O O O				X X X X X X X X X O O O O O O O O O O	
10		12		14		16	17	18		20

____. 1 more is ____.

sentence frame

Lesson 12

Objective: Represent numbers 20 to 11 in tower configurations decreasing by 1—a pattern of *1 smaller*.

Suggested Lesson Structure

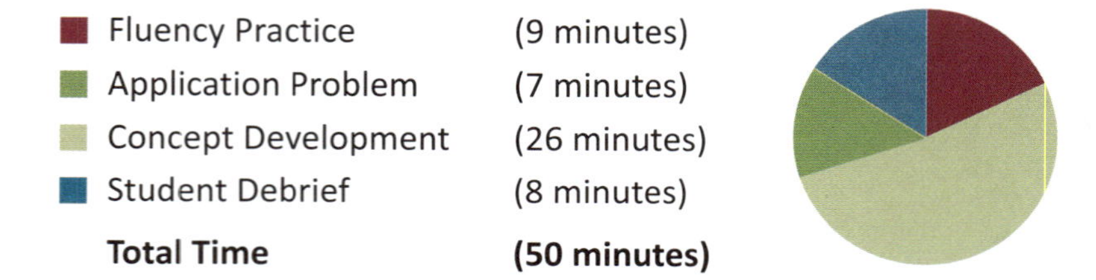

Fluency Practice (9 minutes)

- Write Teen Numbers **K.CC.3** (3 minutes)
- Show Teen Numbers **K.NBT.1** (3 minutes)
- Count the Say Ten Way **K.NBT.1** (3 minutes)

Write Teen Numbers (3 minutes)

Materials: (S) One stick of 10 linking cubes that are the same color, 10 loose cubes of a different color, personal white board

Note: By writing the corresponding numeral for each part and then the whole, students are continually reminded that the 1 in teen numbers refers to 10 ones.

T: Place your stick of ten cubes on your personal white board.
T: Place 3 cubes next to your 10 cubes.
T: Write the number of cubes that you placed on your board.
T: (Students write 13.) Say the number.
S: Ten 3. → Thirteen!

Repeat process for several other teen numbers.

Show Teen Numbers (3 minutes)

Materials: (S) One stick of 10 linking cubes that are the same color, 10 loose cubes of a different color

Note: A color change at 10 makes the two parts stand out visually, allowing students to compose teen numbers with efficiency.

T: Hold up your stick of 10 cubes.
T: Show me 11 cubes. Say the number the Say Ten way.
S: Ten 1.
T: Take off the extra one, and put it back in the pile of 10 ones.

Repeat process for several other teen numbers.

NOTES ON MULTIPLE MEANS OF ACTION AND EXPRESSION:

Give your English language learners extra time to allow them to process the meanings of the essential terms in your lesson before calling for responses. Review and post key vocabulary (cube, more, less, remove), and allow extra conversation time while students are working.

Count the Say Ten Way (3 minutes)

Note: Counting up and down prepares students to work with the pattern of 1 less in the Concept Development.

T: Let's count the Say Ten way.

Guide students to count forward and backward between 10 and 20.

Application Problem (7 minutes)

MP.2

Peter was sitting at lunch eating his french fries. He counted 8 left on his plate. He ate 1 french fry. He ate another french fry. Then, he ate another french fry. How many french fries did Peter have then?

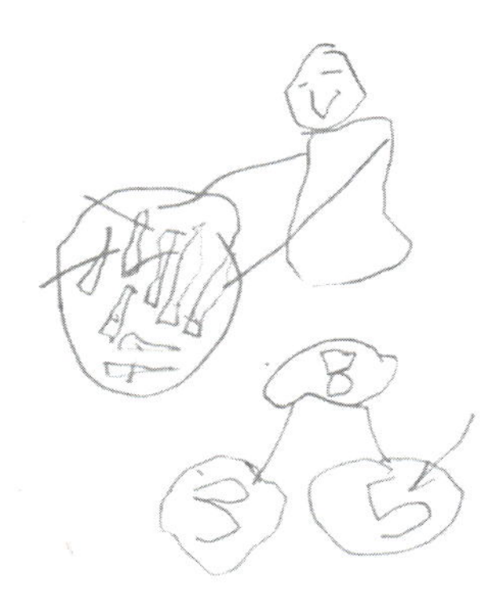

Note: The purpose of this Application Problem is to simply prepare students for thinking about 1 less. Eight. 1 less is 7. Seven. 1 less is 6.

Concept Development (26 minutes)

NOTES ON MULTIPLE MEANS OF ACTION AND EXPRESSION:

Challenge students who are working above grade level by providing them with extensions of the Application Problem to solve. Ask, "If Peter ate two fries at a time, how many would he have then? If Peter started with 18 fries and ate one at a time, how many would he have left? And, if Peter had 50 fries and he ate 1 and then another and then another, how many would he have then?"

Materials: (S) 2 sets of 10 linking cubes (10 in one color and 10 in another color), sentence frame (Template)

Note: Notice that we are not saying "19 is one less *than* 20." This is very complex linguistically for many kindergarten students who can say "19 is less than 20" without quantifying the difference. We simply are extending the "one more" lesson to "one less" as an opportunity for the students to do counting of teen numbers in a linear configuration: the tower (**K.CC.5**).

T: Build a tower with all the cubes of one color.
T: How many cubes are in your tower?
S: Ten!
T: How many ones is that?

S: 10 ones!

T: Now, build a tower using the other cubes.

T: How many cubes are in this tower?

S: Ten!

T: Join the two towers. What is 10 ones and 10 ones?

S: Twenty! → 2 tens!

T: How can we show 19?

S: Take off 1 cube. (Students remove one cube.)

T: Say this with me: "20. 1 less is 19." (Use sentence frame for support.)

S: 20. 1 less is 19.

T: Take off one cube. Be sure to take the same color cube as before. Talk to your partner. How many cubes are in your tower now?

S: (Allow time for students to figure it out.) 18.

Students continue in this manner, taking off one cube each time, down to 10. As they remove each cube, have them express the relationship of each number to the preceding number, for example, 18. 1 less is 17. As in the preceding lesson, release students to work independently as soon as possible.

Problem Set (7 minutes)

Students should do their personal best to complete the Problem Set within the allotted time. As students color the squares and write the numbers to complete the pattern, have them continue to say the relationship of each number to its preceding number, for example, 13. 1 less is 12. 12. 1 less is 11.

Student Debrief (8 minutes)

Lesson Objective: Represent numbers 20 to 11 in tower configurations decreasing by 1—a pattern of *1 smaller*.

The Student Debrief is intended to invite reflection and active processing of the total lesson experience.

Invite students to review their solutions for the Problem Set. They should check work by comparing answers with a partner before going over answers as a class. Look for misconceptions or misunderstandings that can be addressed in the Debrief. Guide students in a conversation to debrief the Problem Set and process the lesson.

Any combination of the questions below may be used to lead the discussion.

- What do you notice when you look at your work?
- How is your drawing like the towers you made?

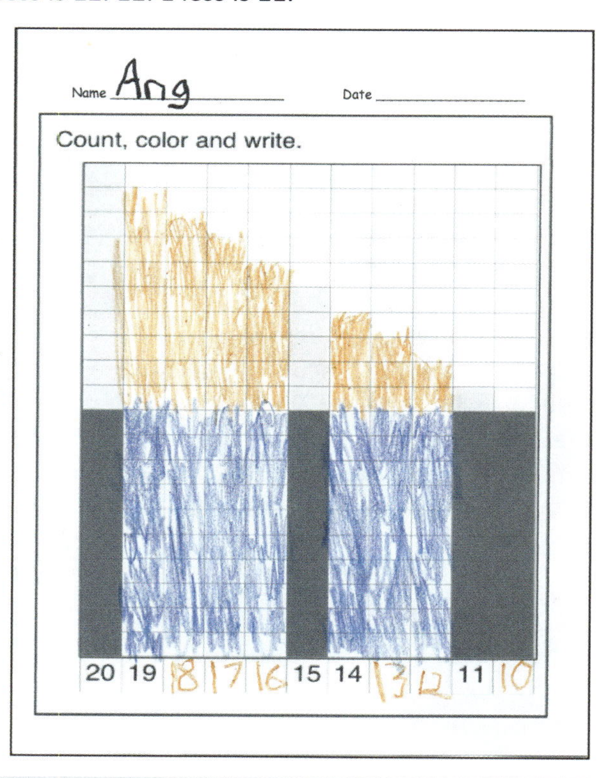

- How many cubes did you remove from your tower each time?
- When you take one cube off, does the number get larger or smaller?
- How is this work similar to the story problem of the french fries?
- How is what we did today alike and different from what we did yesterday?

Exit Ticket (3 minutes)

After the Student Debrief, instruct students to complete the Exit Ticket. A review of their work will help with assessing students' understanding of the concepts that were presented in today's lesson and planning more effectively for future lessons. The questions may be read aloud to the students.

Name ____________________ Date ____________

Count, color and write.

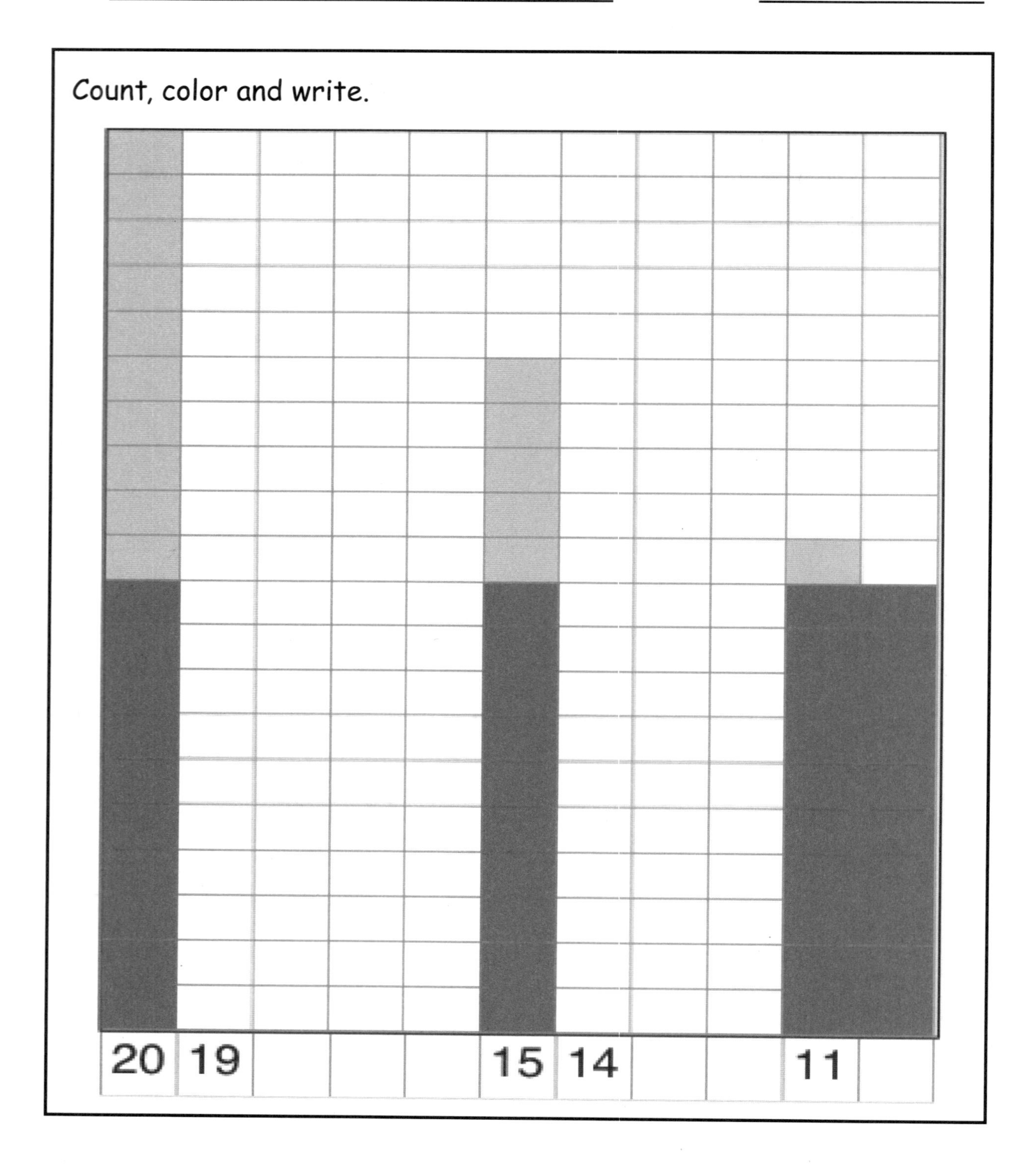

Name ______________________________ Date ______________

Write the missing numbers, counting down.

14, 13, 12, 11, ________
15, 14, ________, 12, ________, ________
13, 12, ________, ________, ________

Name ______________________________ Date ______________

Write the missing numbers. Then, draw X's and O's to complete the pattern.

X										
X	X									
X	X									
X	X		X							
X	X		X							
X	X		X		X					
X	X		X		X					
X	X		X		X					
X	X		X		X					
X	X		X		X				X	
O	O		O		O				O	
O	O		O		O				O	
O	O		O		O				O	
O	O		O		O				O	
O	O		O		O				O	
O	O		O		O				O	
O	O		O		O				O	
O	O		O		O				O	
O	O		O		O				O	
O	0		O		O				O	
20		18		16		14	13	12		10

______. 1 less is ______.

sentence frame

Lesson 13

Objective: Show, count, and write to answer *how many* questions in linear and array configurations.

Suggested Lesson Structure

Fluency Practice	(9 minutes)
Application Problem	(5 minutes)
Concept Development	(28 minutes)
Student Debrief	(8 minutes)
Total Time	**(50 minutes)**

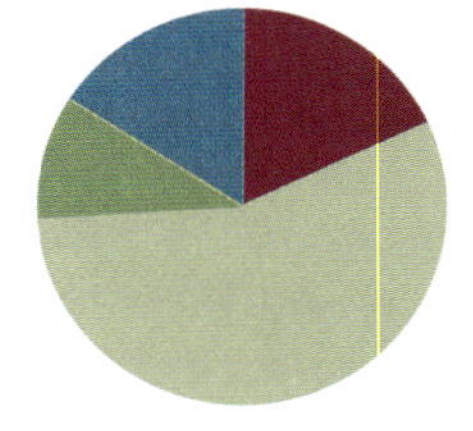

Fluency Practice (9 minutes)

- Count the Say Ten Way **K.NBT.1** (3 minutes)
- Show Teen Numbers **K.NBT.1** (3 minutes)
- Write Teen Numbers with Tower Configurations **K.CC.3** (3 minutes)

Count the Say Ten Way (3 minutes)

Note: Counting up and down prepares students to count and answer *how many* questions accurately in the Concept Development.

T: Let's count the Say Ten way.

Guide students to count forward and backward between 10 and 20.

Show Teen Numbers (3 minutes)

Materials: (S) 2 sticks of 10 linking cubes that are different colors

Note: This activity gives students continued practice with counting in linear configurations and guides students to efficiency with the color change at 10.

T: There are 10 cubes on each of your sticks. Connect your 2 cube sticks.
S: (Students connect cube sticks.)
T: Say the number the Say Ten way.
S: 2 tens.
T: Take away 1 cube, and put it on the carpet space in front of you.
S: (Students do so.)

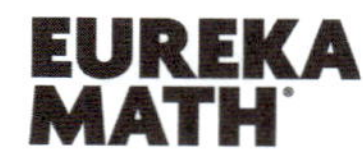

T: Say how many you have now the Say Ten way.

S: Ten 9.

T: Say how many you have the regular way.

S: 19.

Repeat the process for three or four other teen numbers.

Write Teen Numbers with Tower Configurations (3 minutes)

Materials: (T) 1 stick of 10 linking cubes that are the same color, 10 loose cubes of a different color
(S) Personal white board

Note: The color change, along with the Say Ten way, supports students in accurately writing teen numbers. Guide students to recognize groups of cubes as ten ones and some ones, rather than count all.

T: (Hold a tower of 12 connected linking cubes, with the bottom 10 a different color than the top 2.) Write the number on your personal white board.

S: (Students write 12.)

T: Say the number the Say Ten way.

S: Ten 2.

T: Say the number the regular way.

S: 12.

Repeat the process for several other teen numbers.

> **NOTES ON MULTIPLE MEANS OF ENGAGEMENT:**
>
> Provide students with disabilities learning aids in the form of counters to model the Application Problem. Give them a number bond template to complete the task. Scaffolding the Application Problem allows students with disabilities to complete the task and focus on the lesson.

Application Problem (5 minutes)

Vincent's father made 15 tacos for the family. Show the 15 tacos as 10 tacos and 5 tacos. Draw a number bond to match.

Note: This Application Problem is a simple experience of decomposition (**K.NBT.1**). We can ask students to draw the decomposition in 5-groups, another name for a ten-frame configuration, but which has the advantage of emphasizing the five.

Concept Development (28 minutes)

Materials: (S) 2 sticks of 10 linking cubes with a color change at five, personal white board, personal Rekenrek (from Lesson 10); set of Hide Zero cards: 1 Hide Zero 10 card (Lesson 6 Template 2) and 5-group cards 1–9 (Lesson 1 Fluency Template 2) (per pair)

T: Count in order from 1 to 20.

S: 1, 2, 3, ..., 20.

T: Count from 10 to 20 the Say Ten way.

S: Ten 1, ten 2, ten 3, ten 4, ten 5, ten 6, ten 7, ten 8, ten 9, 2 tens.

T: Partner A, show the number that is one more than 13 on the Rekenrek.

T: Partner B, show the number that is one more than 13 with the Hide Zero cards.

T: Check that you are each showing the same number. What is the number?

S: 14.

T: Count from 14 up to 20.

S: 14, 15, 16, 17, 18, 19, 20.

T: Partner B, show the number that is one more than 7 on the Rekenrek.

T: Partner A, show the number that is one more than 7 with the Hide Zero cards.

T: What is the number?

S: 8.

T: Count from 8 up to 20.

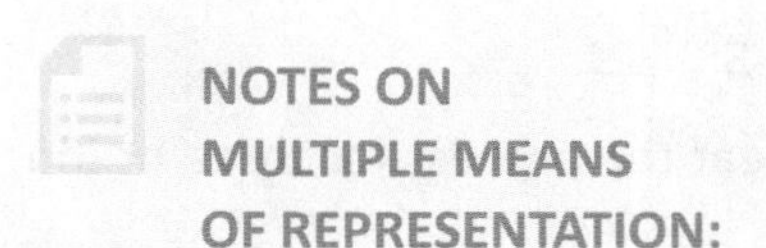

NOTES ON MULTIPLE MEANS OF REPRESENTATION:

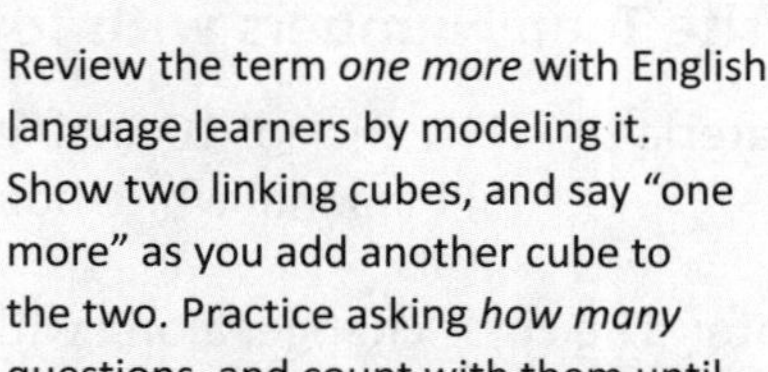

Review the term *one more* with English language learners by modeling it. Show two linking cubes, and say "one more" as you add another cube to the two. Practice asking *how many* questions, and count with them until they get accustomed to answering the question.

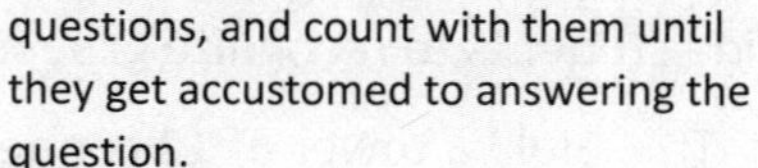

Repeat with two more numbers so that each partner uses both representation tools a second time.

T: (Pass out the linking cubes.)

Have students connect the linking cubes to create a continuous number train to 20. Have them count to see they have 2 sticks of 10 ones.

T: Show me ten 7 cubes.

T: (Allow students time to finish.) How many cubes is that?

S: Ten 7. → Seventeen!

T: Make your long number train of 2 sticks of 10 again. Break it, and put 1 stick below the other. How many cubes do you have now?

S: (Count again, as needed.) 10 here and 10 here. → 2 tens. → Twenty!

Have students break the linking cube sticks at the color change. Have them place the shorter sticks one below the other. Guide students to place the sticks in four rows and recount the cubes from left to right starting from the top with number 1 and continuing this way to the fourth row of 16 to 20. Have them recount to get better at it. They will enjoy the chance to recount.

T: (Allow students time to finish.) How many cubes did you count?

S: 20.

T: (Revisit the process.) Put the sticks back into one train from 1 to 20. Count. Break the stick into 2 sticks of 10 cubes. Count. Break the sticks to make 4 sticks of 5. Count.

T: (Allow students time to finish.) How many cubes do you have now? Count to check.

S: 20.

Before doing the Problem Set, give students a personal white board or blank paper, and have them use their 10-sticks to draw what they just did in the lesson.

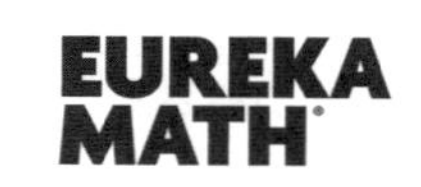

Problem Set (7 minutes)

Distribute Problem Sets to students. Students should do their personal best to complete the Problem Set within the allotted time.

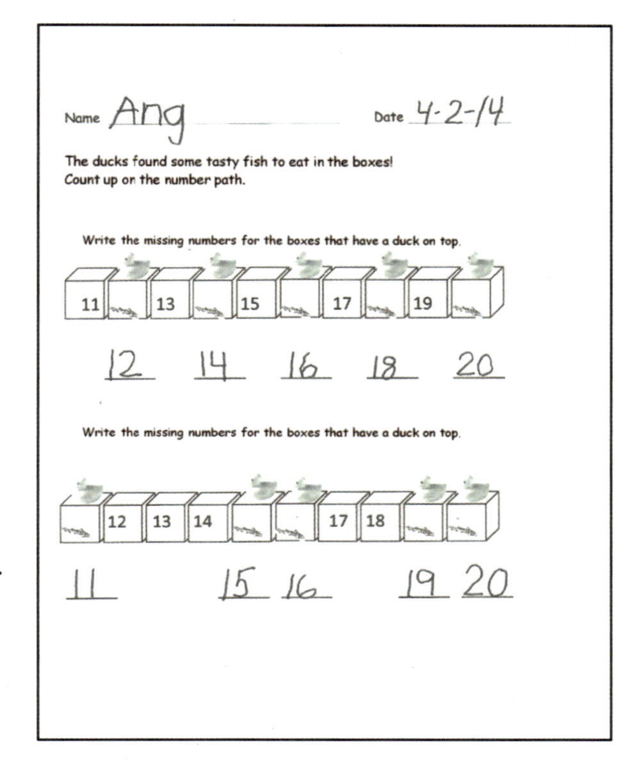

Student Debrief (8 minutes)

Lesson Objective: Show, count, and write to answer *how many* questions in linear and array configurations.

The Student Debrief is intended to invite reflection and active processing of the total lesson experience.

Have students always check their work with a partner once they bring it to the carpet. Encourage them to notice, if they don't, that the number of ducks is the same. Ask: "How do they look different?" "Is there another way we can put the 16 ducks?"

Be sure they compare how they showed 15 and 12 in rows in the last two problems. Then, possibly discuss:

T: Count the cubes as I lay them down. (Place 10 ones in a horizontal line.)

S: 1, 2, 3, 4, 5, 6, 7, 8, 9, 10.

T: What is one more than 10? (Add a cube.)

S: 11.

T: One more than 11? (Add a cube.)

S: 12.

T: How many cubes do you see?

S: 12.

T: (Slide the cubes into a vertical line.) Do I still have 12 cubes? How do you know?

T: (Slide the cubes into different rectangular array configurations, asking after each change, "How many do I have now?")

Guide students to see that the number of objects is the same regardless of how they are arranged. Let them close the lesson by showing 12 cubes in different rows to a partner. (Rows do not have to be complete.)

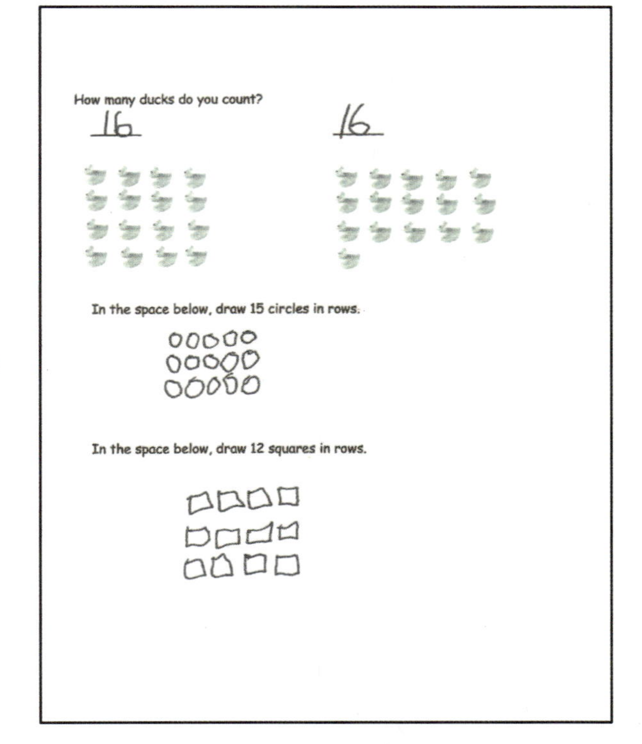

Exit Ticket (3 minutes)

After the Student Debrief, instruct students to complete the Exit Ticket. A review of their work will help with assessing students' understanding of the concepts that were presented in today's lesson and planning more effectively for future lessons. The questions may be read aloud to the students.

Name ______________________________ Date ______________

The ducks found some tasty fish to eat in the boxes!
Count up on the number path.

Write the missing numbers for the boxes that have a duck on top.

______ ______ ______ ______ ______

Write the missing numbers for the boxes that have a duck on top.

______ ______ ______ ______ ______

How many ducks do you count?

______ ______

In the space below, draw 15 circles in rows.

In the space below, draw 12 squares in rows.

Name ______________________________ Date ______________

Count and write how many.

☆ ☆ ☆ ☆
☆ ☆ ☆ ☆
☆ ☆ ☆ ☆

____

Look at the 3 sets of blocks below. Count the shaded blocks in each set. Circle the set that has the same number of shaded blocks as stars.

Early finishers: Which was easier to count, stars or blocks? Why?

Name ______________________________ Date ______________

Count the objects. Draw dots to show the same number on the double 10-frames.

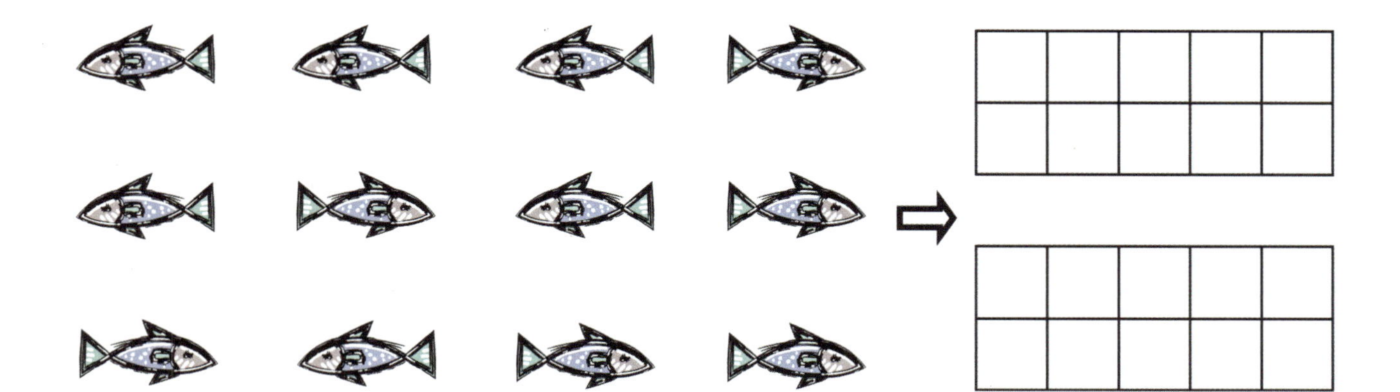

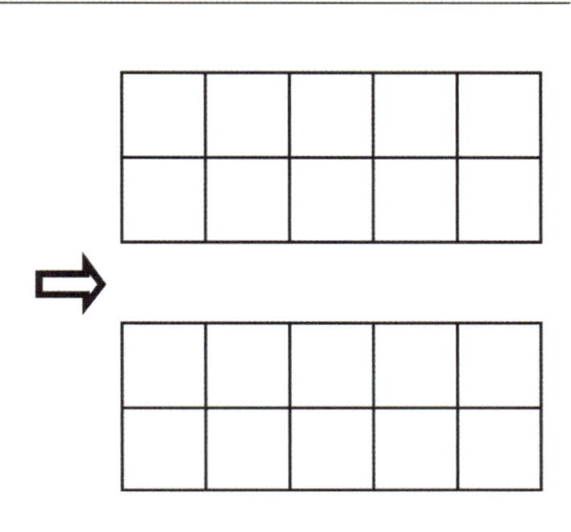

Lesson 14

Objective: Show, count, and write to answer *how many* questions with up to 20 objects in circular configurations.

Suggested Lesson Structure

Fluency Practice	(9 minutes)
Application Problem	(7 minutes)
Concept Development	(26 minutes)
Student Debrief	(8 minutes)
Total Time	**(50 minutes)**

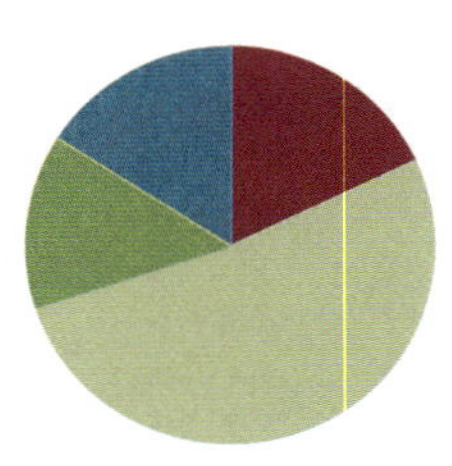

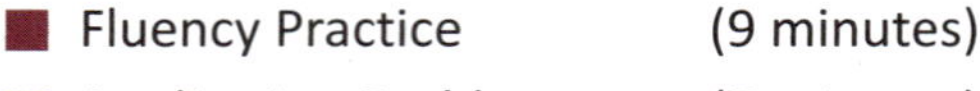

Fluency Practice (9 minutes)

- Write Teen Numbers with Arrays **K.CC.3** (3 minutes)
- Hide Zero for Teen Numbers **K.NBT.1** (3 minutes)
- Teen Counting Array Template **K.CC.5** (3 minutes)

Write Teen Numbers with Arrays (3 minutes)

Materials: (T) Pre-drawn arrays (S) Personal white board

Note: Now that counting in arrays with teen numbers has been introduced, the goal is to develop speed and accuracy. Encourage students to locate 2 fives, or a group of 10, within each array to facilitate counting.

T: (Project a 5 by 3 array of stars.) On your personal white board, write the number of stars you see.
S: (Students write 15.)
T: Say the number the Say Ten way.
S: Ten 5.
T: Say the number the regular way.
S: 15.

Repeat the process for three or four other teen numbers.

Hide Zero for Teen Numbers (3 minutes)

Materials: (T) Large Hide Zero cards (Lesson 6 Template 1)

Note: This activity reminds students that the *1* in teen numbers refers to 10 ones, preparing them for answering *how many* questions in writing.

T: (Hold the 10 card and 5 card so that it appears as 15.) Say the number.
S: 15.
T: Say the number the Say Ten way.
S: Ten 5.

Break apart the cards into 10 and 5. Repeat the process for other teen numbers.

Teen Counting Array Template (3 minutes)

Materials: (S) Teen counting array (Fluency Template)

Note: Repeated experiences with counting in arrays lead students to efficiency over time. Guide students to see 10 as 2 fives to determine the total skillfully.

Distribute the teen counting array. Have students count how many are in each array.

Application Problem (7 minutes)

Eva put her 12 cookies on her cookie sheet in 2 rows of 6. Draw Eva's cookies. Show her 12 cookies as a number bond of 10 ones and 2 ones using your Hide Zero cards. Then, find and circle the 10 cookies that are inside the 12 cookies.

Have students explain how the parts of the number bond match the parts of their drawing and the Hide Zero cards with a partner.

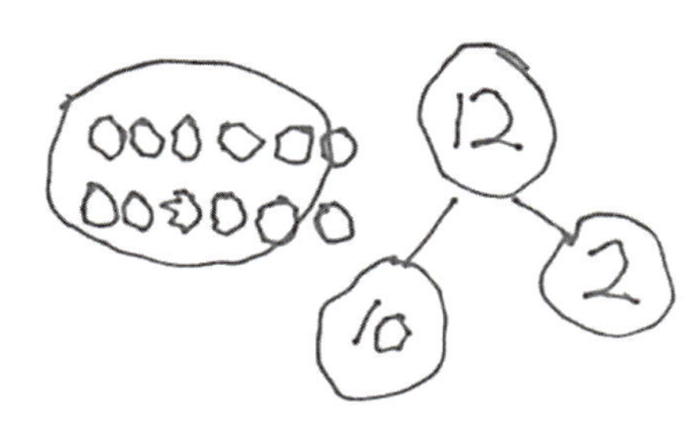

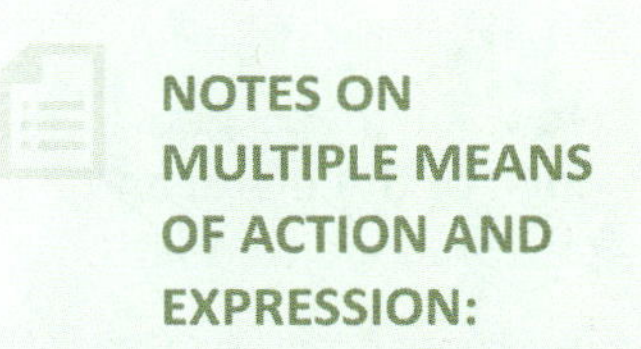

NOTES ON MULTIPLE MEANS OF ACTION AND EXPRESSION:

Scaffold the Application Problem for English language learners by adding gestures when reading the Application Problem. Hold both arms straight out when reading "rows," and make a large circle with both arms while reading the direction "circle the 10."

Note: This Application Problem serves as a bridge from the previous lesson's focus on organizing and counting objects in an array configuration. It also reviews the grade-level standard of understanding teen numbers as ten ones and some more ones.

Concept Development (26 minutes)

Materials: (S) double 10-frame mat (Lesson 9 Template) within a personal white board; Teen numeral and dot cards (only numeral cards from 10–20) (Template), paper plate or round mat, bag of 20 counting objects (per pair)

T: Let's see how well you can show, count, and write numbers!

T: Partner A, draw a card and tell your partner the number. You can say the number the regular way or the Say Ten way.

T: Partner B, put that number of objects around the outside edge of your plate. (Guide them to use the edge of the plate to make a circular configuration.)

T: Now, take turns counting the objects. How many are there?

T: Partner B, now you get to draw the card, and Partner A will show it.

T: Count the objects. How many are there?

Repeat the process two or three times.

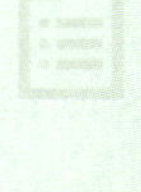

NOTES ON MULTIPLE MEANS OF ACTION AND EXPRESSION:

For students working below grade level, scaffold Concept Development work. Provide a plate with 20 empty circles drawn around the edge. This will serve as a visual container for students when they are showing numbers up to 20. For further support, label the circles with numbers 1–20 to help students with sorting.

T: Let's try something different. We won't use the number cards for this.

T: Partner A, put any number of objects you want in a circle around the edge of your plate.

T: Partner B, count the objects and write the number on your personal white board.

T: Now, Partner B gets to put any number of objects in a circle around the edge of the plate, and Partner A counts them and writes the number on her personal white board.

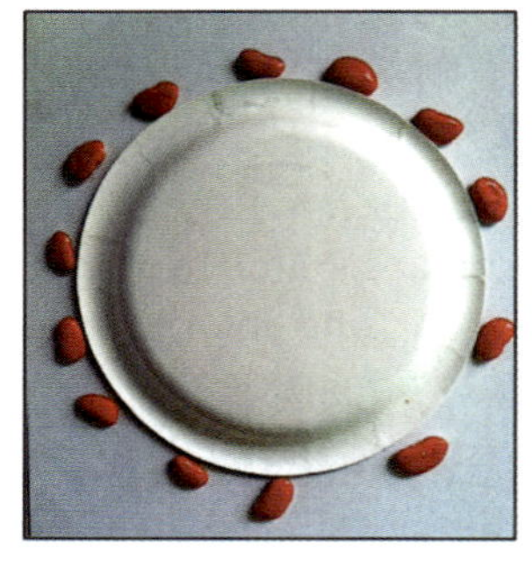

Repeat the process two or three times.

T: This time, Partner A, write any number between 11 and 20 on your personal white board. Partner B, count out that many objects as you place them in a circle around the edge of the plate. How many objects are there?

T: Partner A, count each object as you move it from the circle to the 10-frame to check that the count is correct. How many objects are there?

T: Now, Partner B, you get to write any number between 11 and 20 on your personal white board. Partner A, count out that many objects as you place them in a circle around the edge of the plate. How many objects are there?

T: Partner B, count each object as you move it from the circle to the 10-frame to check that the count is correct. How many objects are there?

Repeat the process two or three times.

Before using the Problem Set, have students use the plate to draw dots in a circular shape and count each other's dots. Have them circle 10 dots to prove that they counted correctly (as pictured below).

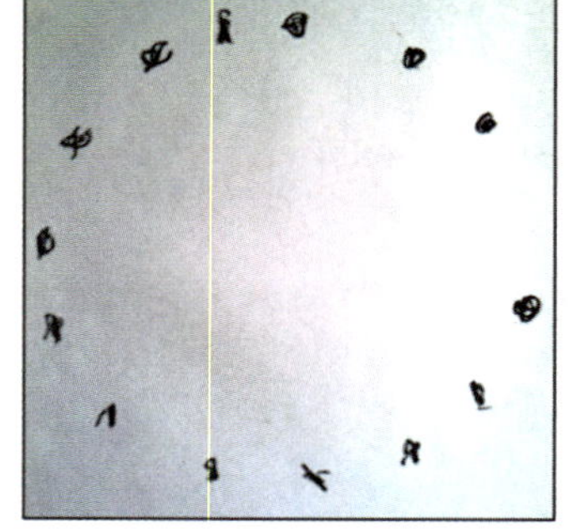

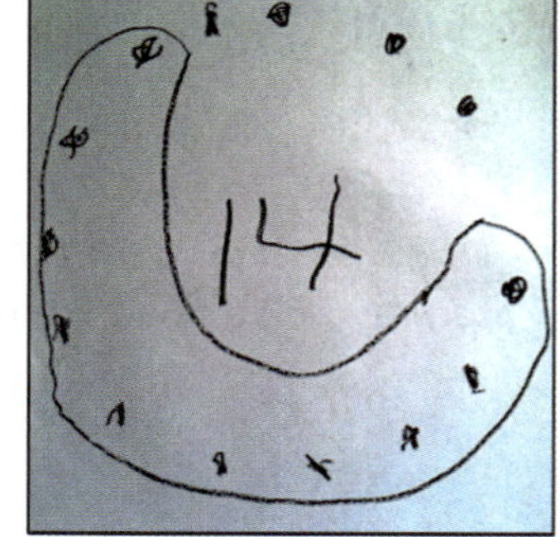

Problem Set (7 minutes)

Students should do their personal best to complete the Problem Set within the allotted time.

Student Debrief (8 minutes)

NOTES ON MULTIPLE MEANS OF ENGAGEMENT:

For students working above grade level, provide an opportunity for deeper understanding.

- Ask students how many different ways they can count the objects.
- Possible answers can be as follows: by ones, by twos, by threes, and by counting a ten and then counting the remaining objects.

Lesson Objective: Show, count, and write to answer *how many* questions with up to 20 objects in circular configurations.

The Student Debrief is intended to invite reflection and active processing of the total lesson experience.

Invite students to review their solutions for the Problem Set. They should check work by comparing answers with a partner before going over answers as a class. Look for misconceptions or misunderstandings that can be addressed in the Debrief. Guide students in a conversation to debrief the Problem Set and process the lesson.

Any combination of the questions below may be used to lead the discussion.

- What do you notice about all of the pictures?
- Is it easier or harder for you to count objects when they are in circles like these pictures? Why?
- Which way is easier for you to count—when we show the number in a circle or when we show it as a tower? Why?
- Did the number change when you moved the objects from the circle to the 10-frame? Why not?

MP.3

- (Show objects in a circle configuration, and have students count how many. Then, slide the objects to change the circle into a line.) How can you prove that the number is still the same? Tell your partner. Did he prove it to you? What are some ways you proved it? Which ways were the most convincing?

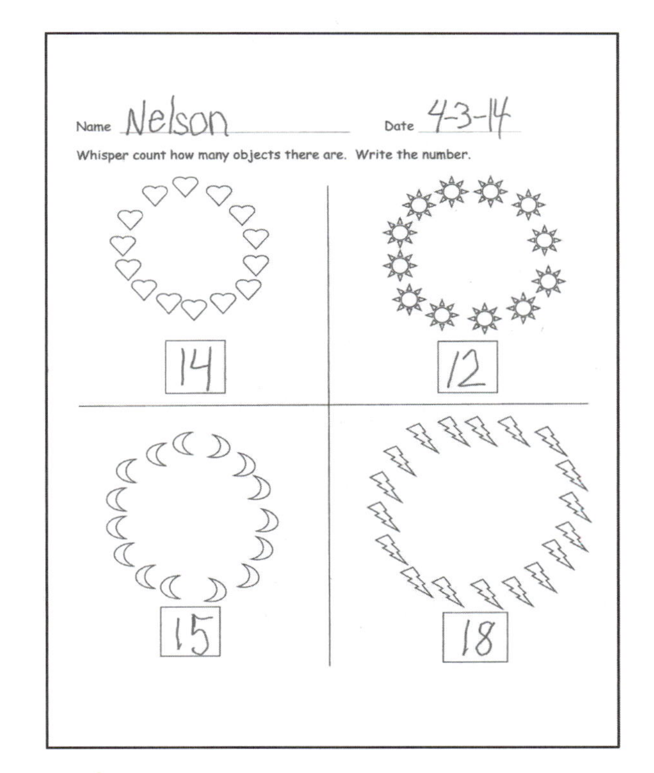

Exit Ticket (3 minutes)

After the Student Debrief, instruct students to complete the Exit Ticket. A review of their work will help with assessing students' understanding of the concepts that were presented in today's lesson and planning more effectively for future lessons. The questions may be read aloud to the students.

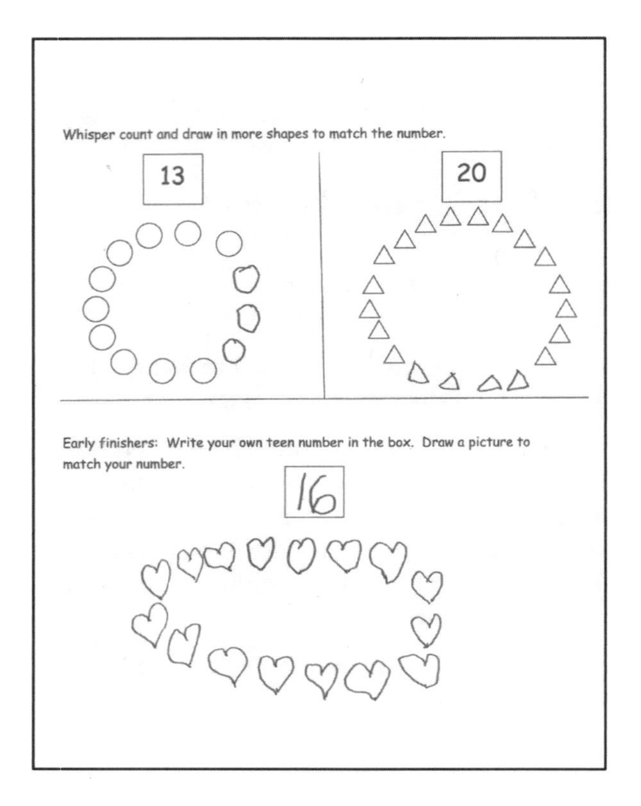

Name ______________________________ Date ____________

Whisper count how many objects there are. Write the number.

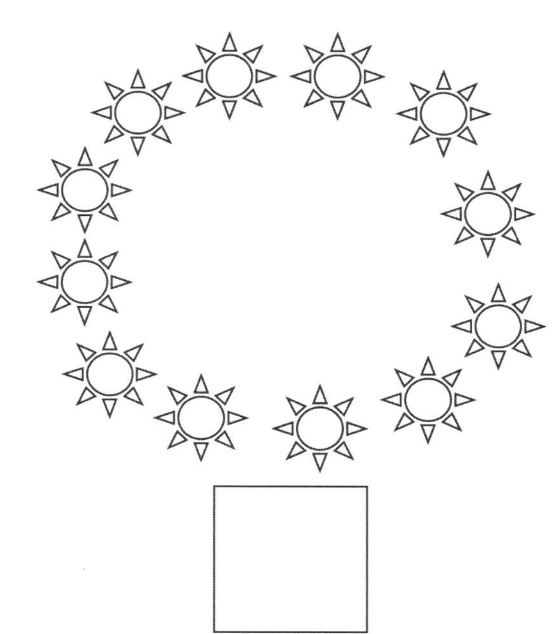

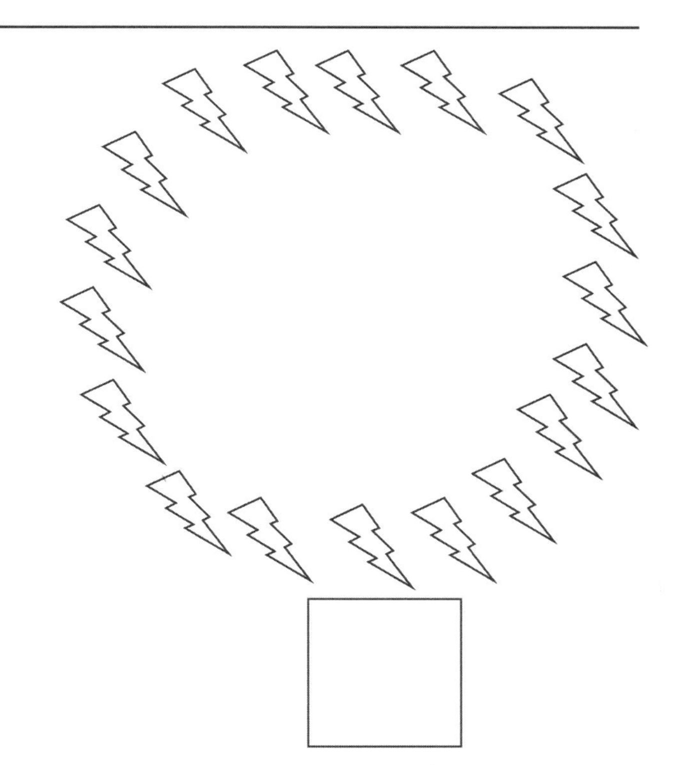

Whisper count and draw in more shapes to match the number.

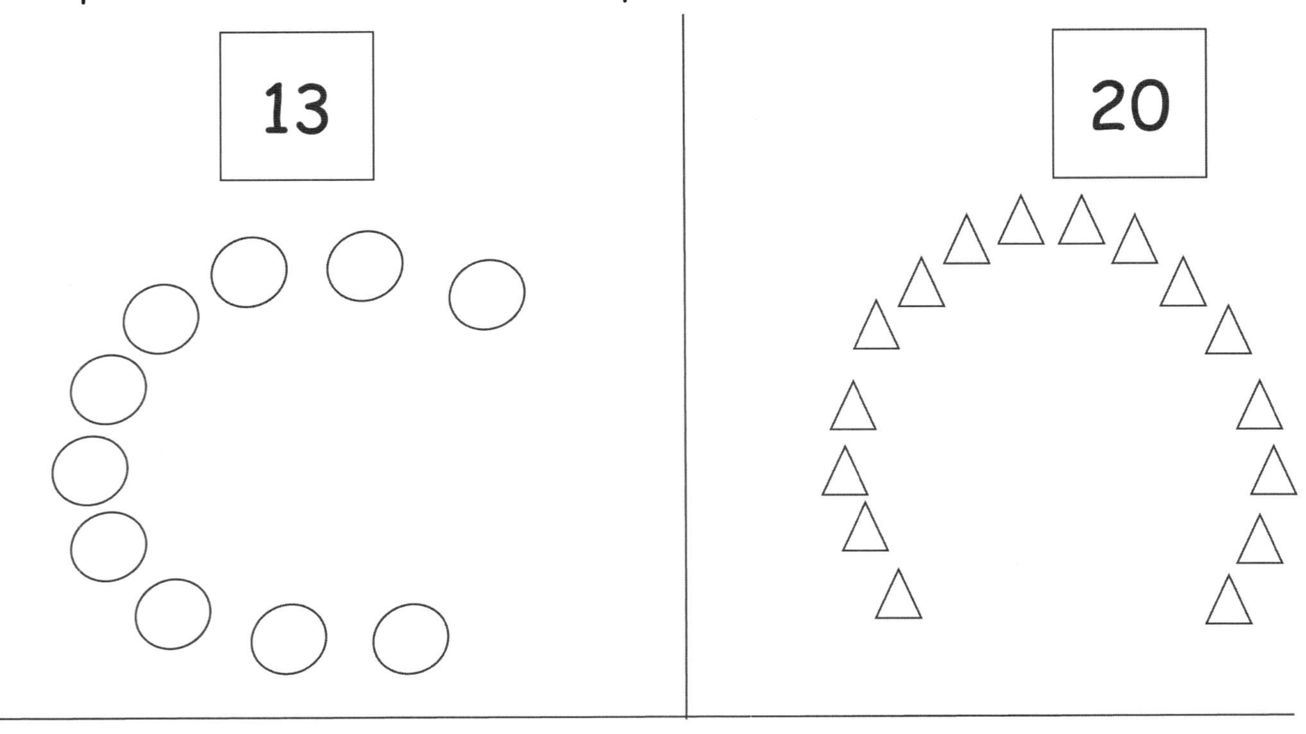

Early finishers: Write your own teen number in the box. Draw a picture to match your number.

Name ______________________________ Date ____________

Count the stars. Write the number in the box.

Whisper count and draw in more dots to match the number.

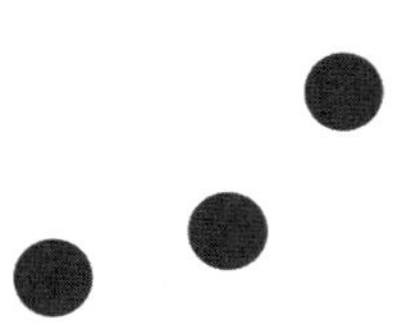

15

Name ______________________ Date ____________

Count the objects in each group. Write the number in the boxes below the pictures.

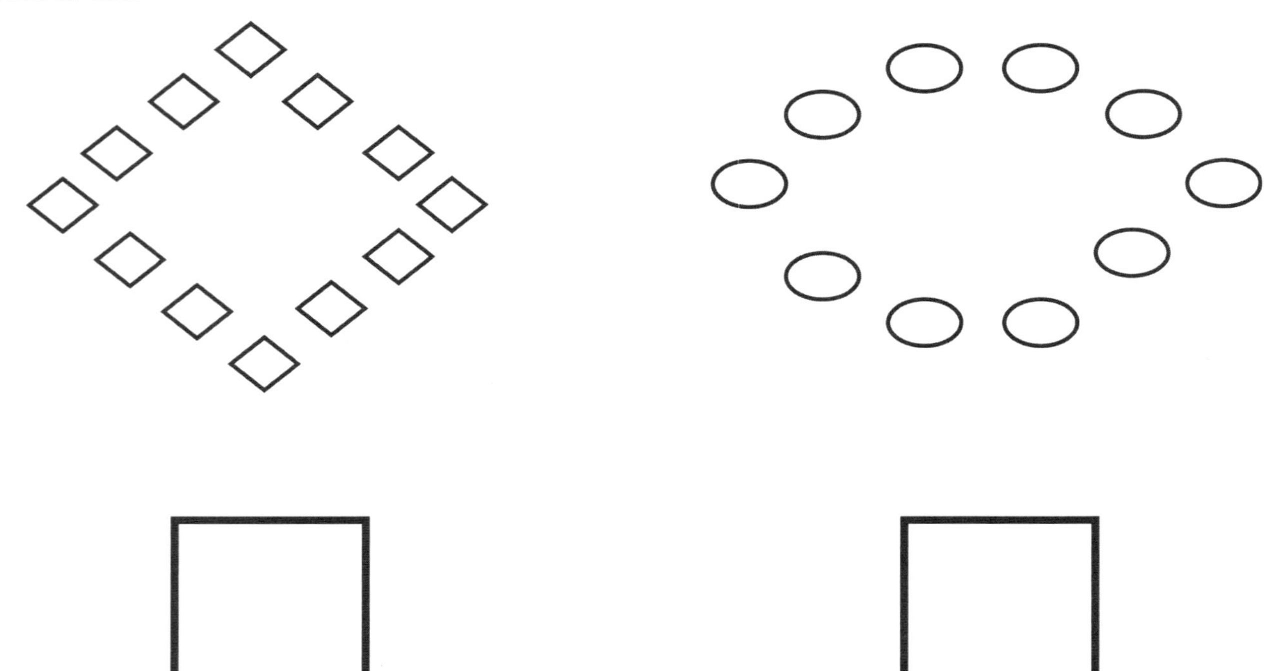

Count and draw in more shapes to match the number.

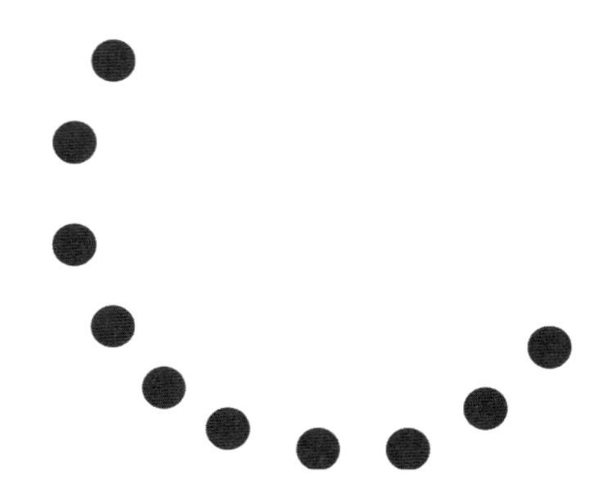

19

Count the dots. Draw each dot in the 10-frame. Write the number in the box below the 10-frames.

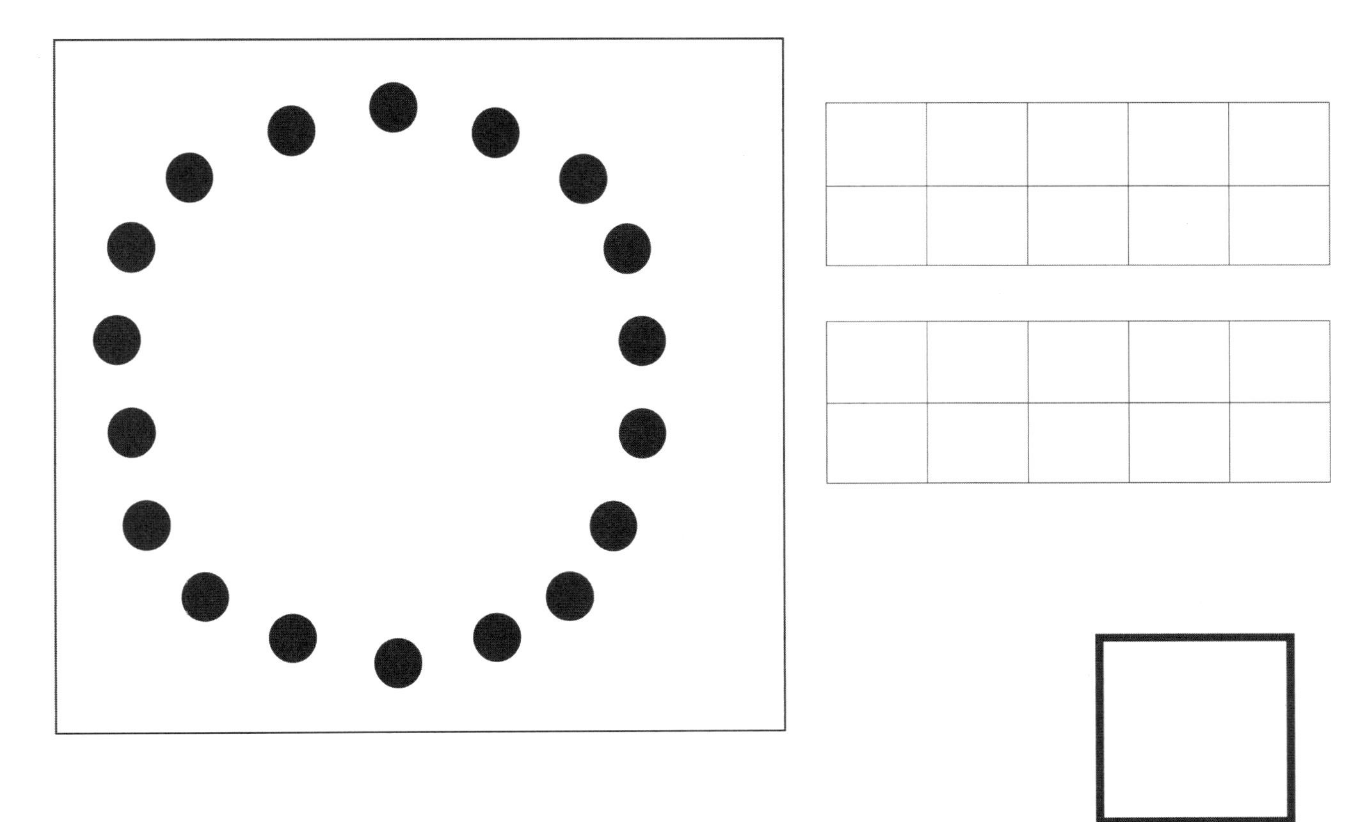

Write a teen number in the box below. Draw a picture to match your number.

Name ______________________________ Date ______________

Count the objects in each group and write the number.

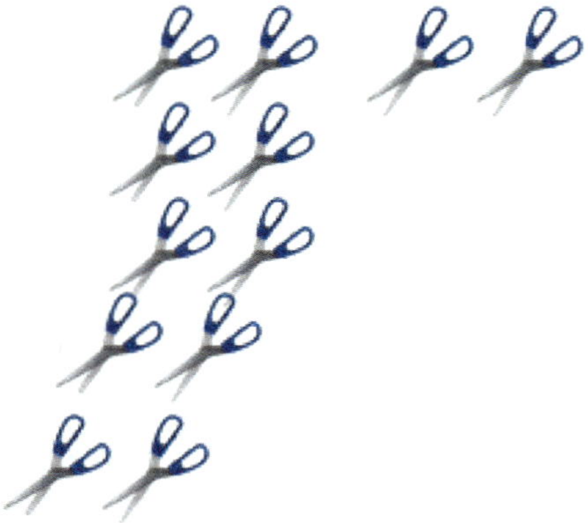

teen counting array

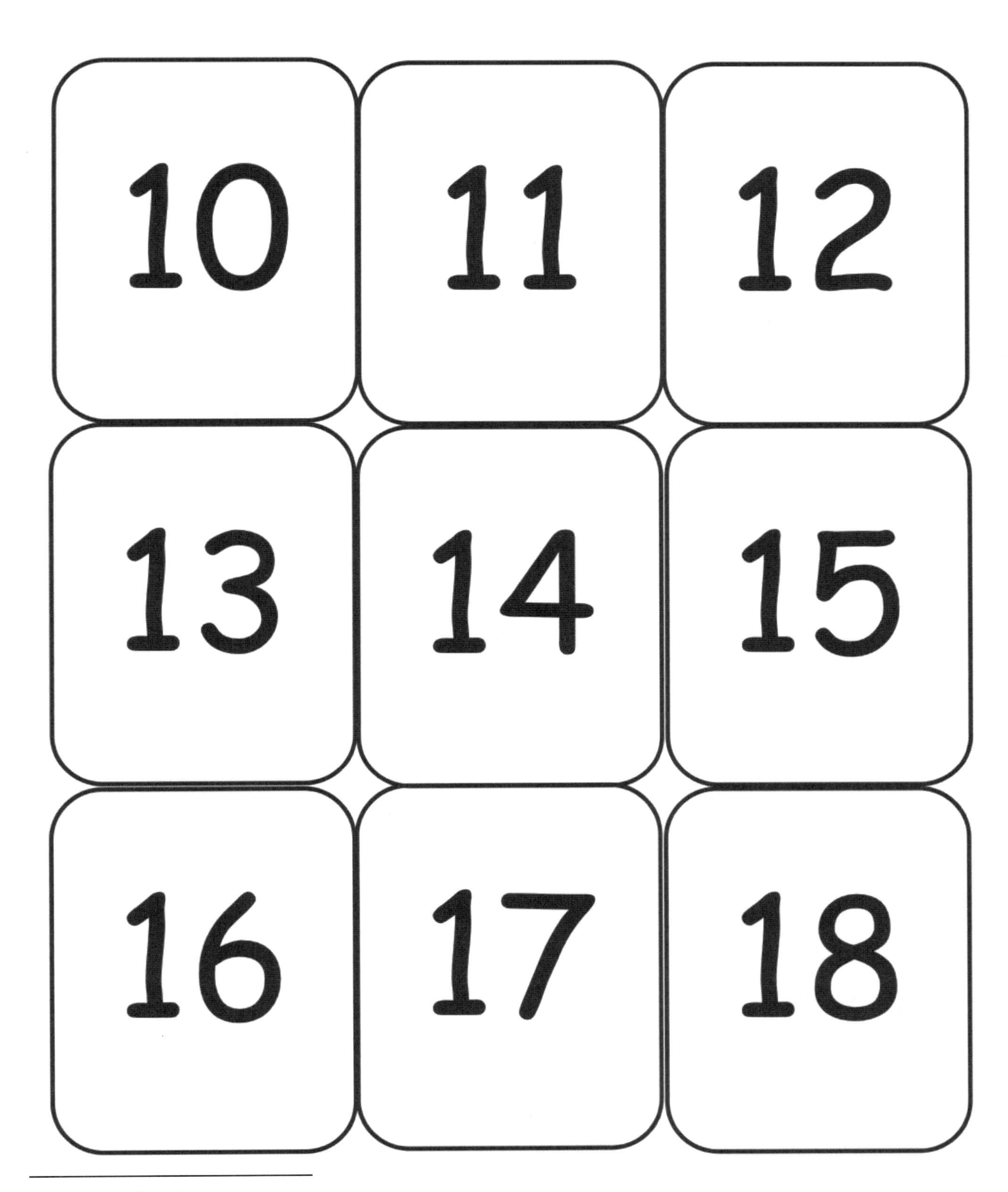

teen numeral cards

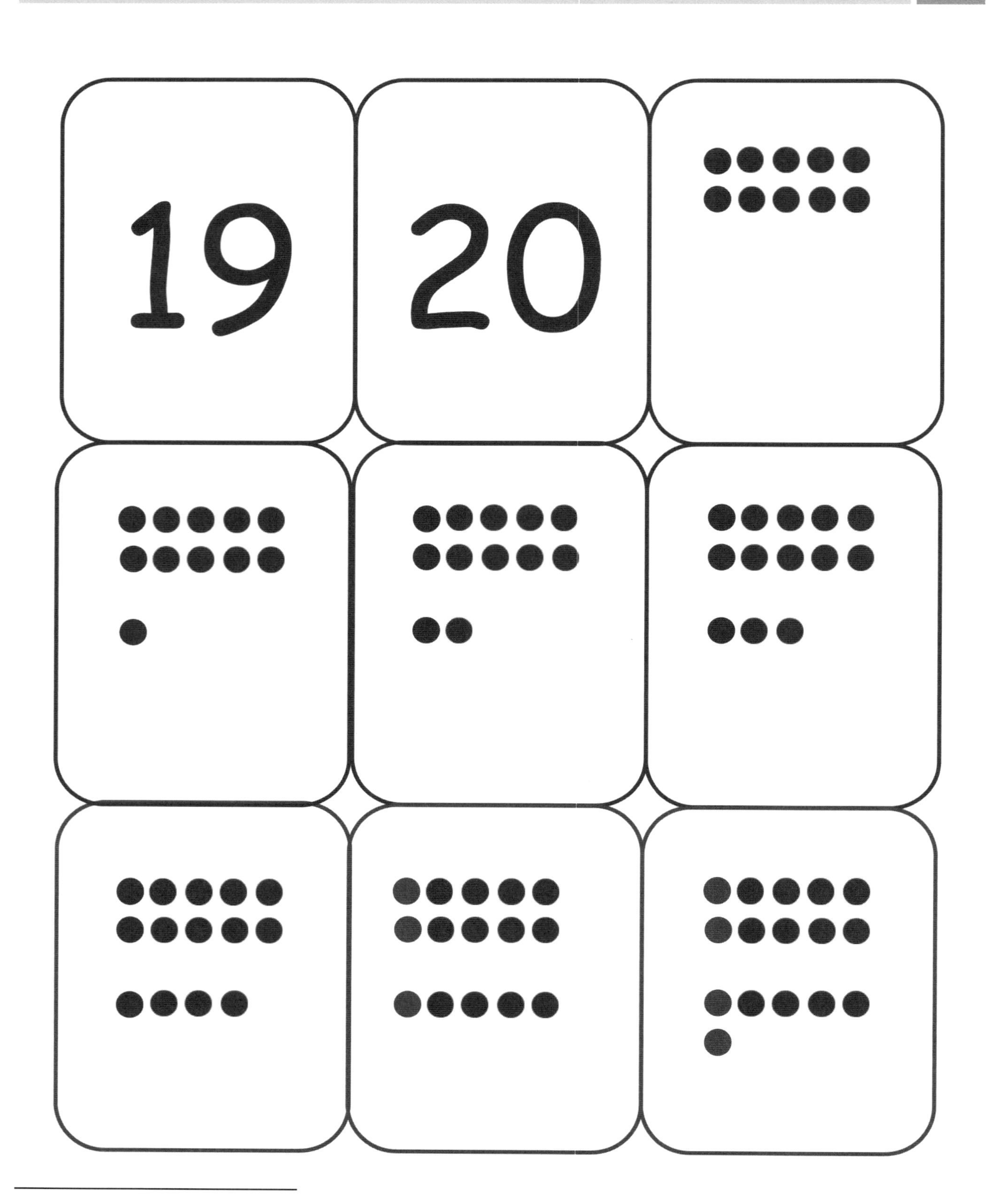

teen numeral and dot cards

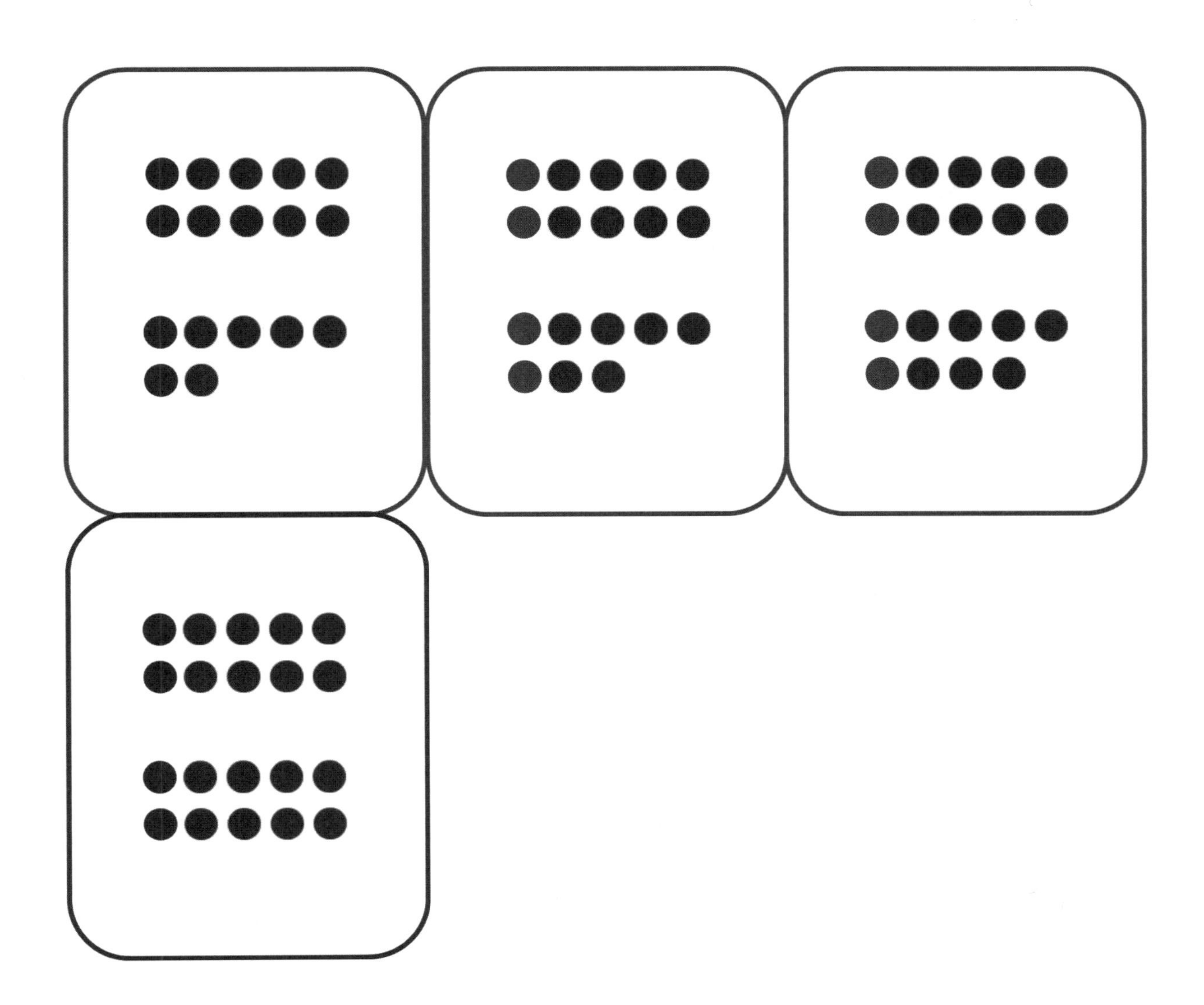

Note: Only numeral cards are used in this lesson. Set aside full set for later use. Consider copying on card stock for durability.

teen numeral and dot cards

Kindergarten Mid-Module 5 Assessment (Administer after Topic C)

Kindergarten End-of-Module 5 Assessment (Administer after Topic E)

Assessment time is a critically important component of the student-teacher relationship. It is especially important in the early grades to establish a positive and collaborative attitude when analyzing progress. Sit next to the student rather than opposite, and support the student in understanding the benefits of sharing and examining her level of mastery.

Please use the specific language of the assessment and, when possible, translate for English language learners. (This is a math rather than a language assessment.) If a student is unresponsive, wait about 15 seconds for a response. Record the student's results in two ways: (1) the narrative documentation after each topic set, and (2) the overall score per topic using A Progression Toward Mastery. Use a stopwatch to document the elapsed time for each response.

Within each assessment, there is a set of problems targeting each topic. Each set comprises three or four related questions. Document what the student did and said in the narrative, and use the rubric for the overall score for each set.

If the student is unable to perform any part of the set, her score cannot exceed Step 3. However, if the student is unable to use her words to tell what she did, do not count that against her quantitatively. Be aware of the difference between a non-native English speaker's and a native English speaker's ability to articulate something. If the student asks for or needs a hint or significant support, provide either, but the score is automatically lowered. This ensures that the assessment provides a true picture of what a student can do independently.

If a student scores at Step 1 or 2, repeat that topic set again at two-week intervals, noting the date of the reassessment in the space at the top of the student's record sheet. Document progress on this one form. If the student is very delayed in her response but completes it, reassess to see if there is a change in the time elapsed.

House the assessments in a three-ring binder or student portfolio. By the end of the year, there will be 10 assessments for each student. Modules 1, 3, 4, and 5 have two assessments each, whereas Modules 2 and 6 only have one. Use the Class Record Sheet for an easy reference look at students' strengths and weaknesses.

These assessments can be valuable for daily planning, parent conferences, and Grade 1 teachers preparing to receive these students.

Student Name ____________________________

	Date 1	Date 2	Date 3
Topic A			
Topic B			
Topic C			

Topic A: Count 10 Ones and Some Ones

Rubric Score ____________ Time Elapsed ____________

Materials: (S) 19 loose straws (or another set of objects in the classroom)

T: Count 10 straws into a pile. Whisper while you count so I can hear you.
T: Count 6 more straws into a different pile.
T: Count 10 straws and 6 more straws the Say Ten way. (Pause.) How many straws do you have? (If the student says the number the Say Ten way, ask the student to also say it the regular way.)

What did the student do?	What did the student say?

Topic B: Compose Numbers 11–20 from 10 Ones and Some Ones; Represent and Write Teen Numbers

Rubric Score ____________ Time Elapsed ____________

Materials: (S) 19 cubes, work mat, marker, Hide Zero cards: 1 Hide Zero 10 card (Lesson 6 Template 2) and 5-group cards 1–9 (Lesson 1 Fluency Template 2)

T: (Show the numeral 13.) Move this many cubes onto your work mat.
T: Use the Hide Zero cards to show the number of cubes on your work mat.
T: Hand me the cubes that the 1 is telling us about. (Point to the 1 of 13 on the numeral 13.)
T: (Put 3 more cubes.) This is 16 cubes. Please write the number 16 on your work mat.

What did the student do?	What did the student say?

Topic C: Decompose Numbers 11–20, and Count to Answer "How Many?" Questions in Varied Configurations

Rubric Score ___________ Time Elapsed ___________

Materials: (S) 19 cubes

T: (Set out 15 cubes in a scattered configuration.) Count 12 cubes into a straight line. (Pause.) How many cubes are there counting the regular way? The Say Ten way?

T: Move the cubes into 2 rows.

a. How many cubes are there? (Assessing for conservation.)

b. Please show me how you count these cubes that are now in rows.

T: Move the cubes into a circle.

a. How many cubes are there? (Assessing for conservation.)

b. Please show me how to count these cubes that are now in a circle.

T: Put one more cube in your circle. How many cubes do you have now?

What did the student do?	What did the student say?

EUREKA MATH

Mid-Module Assessment Task Standards Addressed	Topics A–C

Know number names and the count sequence.

K.CC.1 Count to 100 by ones and by tens.

K.CC.3 Write numbers from 0 to 20. Represent a number of objects with a written numeral 0–20 (with 0 representing a count of no objects).

Count to tell the number of objects.

K.CC.4 Understand the relationship between numbers and quantities; connect counting to cardinality.

b. Understand that the last number name said tells the number of objects counted. The number of objects is the same regardless of their arrangement or the order in which they were counted.

c. Understand that each successive number name refers to a quantity that is one larger.

K.CC.5 Count to answer "how many?" questions about as many as **20** things arranged in a line, a rectangular array, or a circle, or as many as 10 things in a scattered configuration; given a number from 1–20, count out that many objects.

Work with numbers 11–19 to gain foundations for place value.

K.NBT.1 Compose and decompose numbers from 11 to 19 into ten ones and some further ones, e.g., by using objects or drawings, and record each composition or decomposition by a drawing or equation (e.g., 18 = 10 + 8); understand that these numbers are composed of ten ones and one, two, three, four, five, six, seven, eight, or nine ones.

Evaluating Student Learning Outcomes

A Progression Toward Mastery is provided to describe and quantify steps that illuminate the gradually increasing understandings that students develop *on their way to proficiency*. In this chart, this progress is presented from left (Step 1) to right (Step 4). The learning goal for students is to achieve Step 4 mastery. These steps are meant to help teachers and students identify and celebrate what the students CAN do now while pointing the way toward what they need to work on next.

A Progression Toward Mastery				
Assessment Task Item and Standards Assessed	**STEP 1 Little evidence of reasoning without a correct answer.** **(1 point)**	**STEP 2 Evidence of some reasoning without a correct answer.** **(2 point)**	**STEP 3 Evidence of some reasoning with a correct answer or evidence of solid reasoning with an incorrect answer.** **(3 point)**	**STEP 4 Evidence of solid reasoning with a correct answer.** **(4 point)**
Topic A **K.NBT.1** **K.CC.1**	Student shows little evidence of counting ability or understanding. Almost non-responsive.	Student shows evidence of beginning to understand counting beyond 10 but counts the quantity incorrectly (lacks organization, inconsistent 1:1 correspondence, etc.).	Student correctly counts 10 straws into a pile, and then 6 straws, but is unable to count to 16.	Student correctly: ▪ Counts 10 straws into a pile, and then 6 straws. ▪ Counts from 1 to 16. ▪ Counts the Say Ten way starting with the group of 10 (ten, ten 1, ten 2, ten 3, ten 4, ten 5, ten 6).
Topic B **K.NBT.1** **K.CC.3**	Student shows little evidence of understanding how to represent a teen number or use Hide Zero cards. Student writes the number 16 incorrectly.	Student shows a beginning understanding of representing teen numbers and using Hide Zero cards but is unable to answer correctly. Student writes the number 16 incorrectly.	Student correctly counts 13 cubes and accurately uses the Hide Zero cards, but produces an incorrect quantity to represent the 1 in 13. OR Student identifies a group of 10 as representing the 1 in 13 but cannot use the Hide Zero cards accurately. Student writes the numeral 16 correctly.	Student correctly: ▪ Counts 13 cubes and selects both the 10 and 3 Hide Zero cards to accurately make 13. ▪ Identifies a group of 10 as being representative of the 1 in the numeral 13. ▪ Writes the numeral 16.

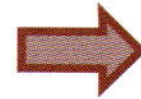

A Progression Toward Mastery				
Assessment Task Item and Standards Assessed	**STEP 1 Little evidence of reasoning without a correct answer. (1 point)**	**STEP 2 Evidence of some reasoning without a correct answer. (2 point)**	**STEP 3 Evidence of some reasoning with a correct answer or evidence of solid reasoning with an incorrect answer. (3 point)**	**STEP 4 Evidence of solid reasoning with a correct answer. (4 point)**
Topic C **K.CC.4b** **K.CC.4c** **K.CC.5** **K.NBT.1**	Student shows little evidence of understanding how to make or count objects in arrays and circles.	Student shows evidence of beginning to understand counting arrays and circles but is unable to do so accurately and consistently.	Student arranges and counts each array and circle correctly but cannot add one more and identify the new quantity. Student recounts to know that it is 12. OR Student adds one more and identifies the new quantity but struggles with one or more of the counting array tasks.	Student correctly: ▪ Counts 12 cubes. ▪ Arranges and counts each array and knows the total is 12 without recounting. ▪ Arranges and counts in a circle and knows the total is 12 without recounting. ▪ Adds 1 more to the quantity and determines the new quantity with or without recounting.

Class Record Sheet of Rubric Scores: Module 5				
Student Names:	**Topic A:** Count 10 Ones and Some Ones	**Topic B:** Compose Numbers 11–20 from 10 Ones and Some Ones; Represent and Write Teen Numbers	**Topic C:** Decompose Numbers 11–20, and Count to Answer "How Many?" Questions in Varied Configurations	**Next Steps:**

Topic D

Extend the Say Ten and Regular Count Sequence to 100

K.CC.1, K.CC.2, K.CC.3, K.CC.4c, K.CC.5, K.NBT.1, 1.NBT.1

Focus Standards:		K.CC.1	Count to 100 by ones and by tens.
		K.CC.2	Count forward beginning from a given number within the known sequence (instead of having to begin at 1).
Instructional Days:		5	
Coherence	**-Links from:**	GPK–M5	Addition and Subtraction Stories and Counting to 20
	-Links to:	G1–M2	Introduction to Place Value Through Addition and Subtraction Within 20

Topic D leads students beyond teen numbers up to 100 (**K.CC.1**). They begin by counting up and down to 100 both the regular way (ten, twenty, thirty, ...) and the Say Ten way (ten, 2 tens, 3 tens, ...). In Lessons 16 to 18, their work from 11 to 19 sets the foundation for success as they realize the number sequence of 1–9 is repeated over and over again within each decade as they count to 100. Students begin by counting within and then across decades (e.g., 28, 29, 30, 31, 32) (**K.CC.2**). Students also write some of the numbers ranging from 21 to 100 in Lessons 15 to 17, which goes beyond the Kindergarten standard to the Grade 1 standard **1.NBT.1** . Writing numerals 21 to 100 is included here because of the wider range of activities they make possible; students readily accept this challenge, which is not assessed. The final lesson of this topic is an optional exploration of decomposing numbers to 100 on the Rekenrek.

A Teaching Sequence Toward Mastery of Extending the Say Ten and Regular Count Sequence to 100	
Objective 1:	**Count up and down by tens to 100 with Say Ten and regular counting.** **(Lesson 15)**
Objective 2:	**Count within tens by ones.** **(Lesson 16)**
Objective 3:	**Count across tens when counting by ones through 40.** **(Lesson 17)**
Objective 4:	**Count across tens by ones to 100 with and without objects.** **(Lesson 18)**
Objective 5:	**Explore numbers on the Rekenrek. (Optional)** **(Lesson 19)**

Lesson 15

Objective: Count up and down by tens to 100 with Say Ten and regular counting.

Suggested Lesson Structure

Fluency Practice	(11 minutes)
Application Problem	(7 minutes)
Concept Development	(24 minutes)
Student Debrief	(8 minutes)
Total Time	**(50 minutes)**

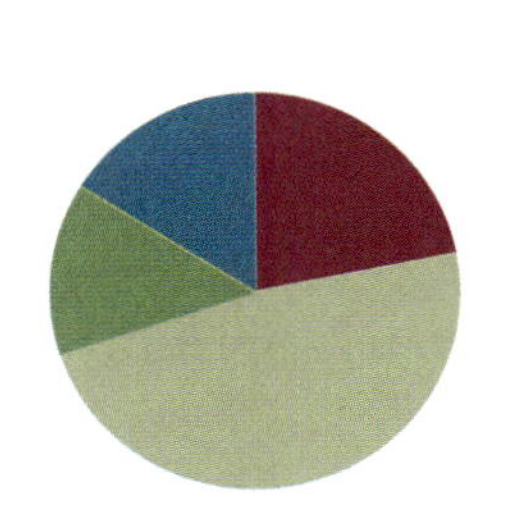

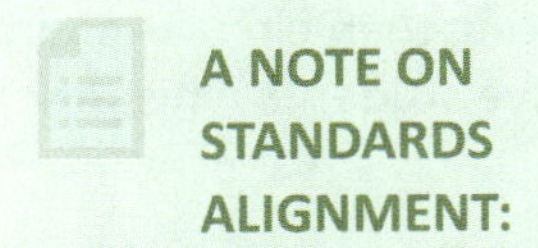

A NOTE ON STANDARDS ALIGNMENT:

In this lesson, students write multiples of 10 through 100, which bridges Kindergarten content of writing numbers to 20 (**K.CC.3**) to Grade 1 content of writing numbers to 120 (**1. NBT.1**).

Fluency Practice (11 minutes)

- Write Teen Numbers with Circular Configurations **K.CC.3** (3 minutes)
- Teen Circular-Counting **K.CC.5** (5 minutes)
- Hide Zero for Teen Numbers **K.NBT.1** (3 minutes)

Write Teen Numbers with Circular Configurations (3 minutes)

Materials: (T) Pre-drawn circular configurations (S) Personal white board

Note: Now that counting teen numbers in circular configurations has been introduced, the goal is to develop accuracy. Encourage students to select a starting point they can remember, so they know when to stop.

T: (Project 13 stars in a circular configuration.) On your personal white board, write the number of stars that you see.
S: (Students write 13.)
T: Say the number the Say Ten way.
S: Ten 3.
T: Say the number the regular way.
S: 13.

Repeat the process for 3 or 4 other teen numbers.

Teen Circular-Counting (5 minutes)

Materials: (S) Teen circular-counting (Fluency Template)

Note: This activity is a step up in complexity from the previous one, because counting out a set is more difficult than counting an existing set. Whisper counting and marking the starting point facilitates accuracy in counting teen numbers in a circular configuration.

After distributing teen circular-counting, have students say each number the regular way and the Say Ten way. Then, have students whisper count as they draw more shapes to match the number indicated.

Hide Zero for Teen Numbers (3 minutes)

Materials: (T) Large Hide Zero cards (Lesson 6 Template 1)

Note: This activity reinforces the grade level standard requiring students to understand that teen numbers are composed of ten ones and some additional ones.

T: (Place the 7 card on the 10 card to show 17.) Say the number.
S: 17.
T: Say the number the Say Ten way.
S: Ten 7.

Break apart the cards into 10 and 7.

Repeat this process for additional teen numbers.

NOTES ON MULTIPLE MEANS OF ENGAGEMENT:

Scaffold the Application Problem for English language learners by giving them sentence starters to help them express how they tackled the challenge. For example, "I put _____ dots of chocolate on the donut."

Application Problem (7 minutes)

Materials: (S) Donuts (Template 1), 14 cubes

Mr. Perry is decorating donuts. He puts 14 little dots of chocolate in rows. Show him an idea about how to put the 14 dots in a circle on his donut. Use the cubes first, and then draw the chocolate dots on his donut. Show the total number of dots of chocolate with a number bond and the Hide Zero cards.

Note: This problem serves as an opportunity for students to apply their recent work with organizing and counting objects in linear and circular configurations. Using Hide Zero cards supports the understanding of 14 as ten ones and 4 ones.

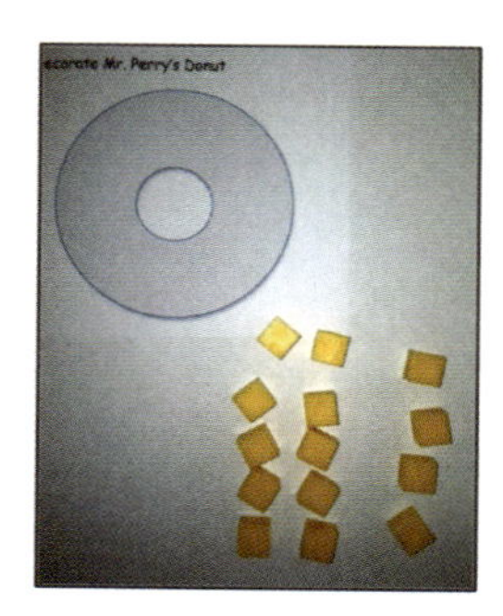
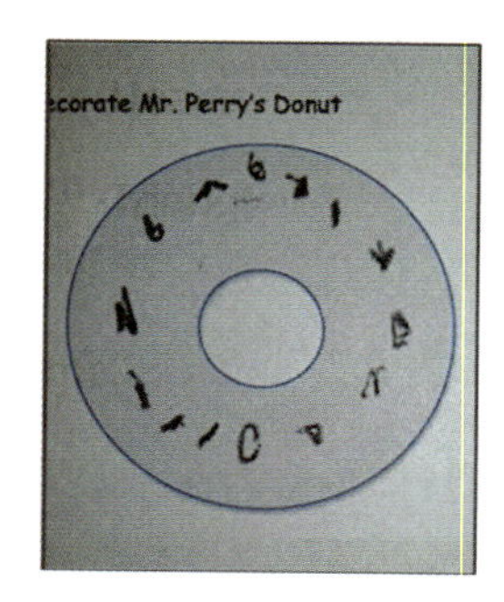

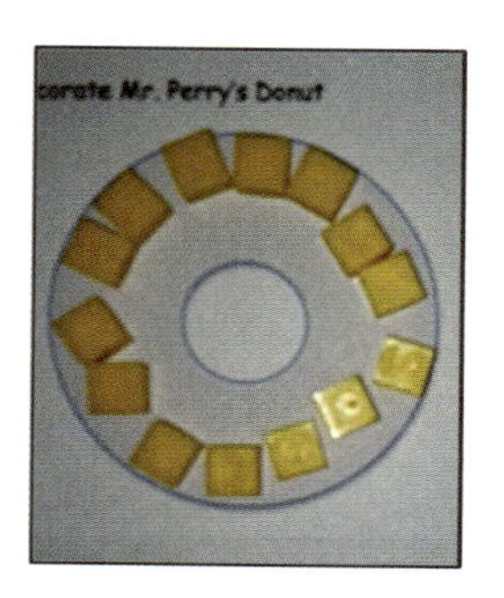

Concept Development (24 minutes)

Materials: (T) 100-bead Rekenrek (S) Set of 10 small 10-frame cards (Template 2)

T: (Invite students to the carpet, and display the Rekenrek.) Count the beads as I move them. (Slide each bead from right to left.)

S: 1, 2, 3, 4, 5, 6, 7, 8, 9, 10.

T: How many beads are in this row?

S: 10.

T: (Point to the beads in the second row.) How many beads are in this row?

S: 10.

T: How can you tell there are ten beads?

S: I see 5 red beads and 5 white beads, and 5 and 5 is 10. → It looks just like the first row.

T: So, each row has how many beads?

S: 10.

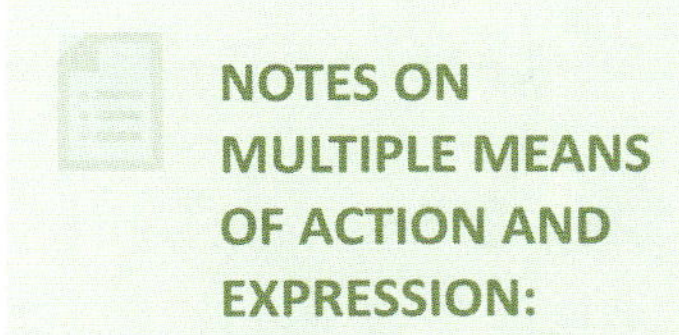

NOTES ON MULTIPLE MEANS OF ACTION AND EXPRESSION:

Scaffold the lesson for students who are working below grade level by having them work in a small group with the Rekenrek. Lead them in counting the Say Ten way while they move the row of beads.

T: Let's count all the beads. Should we count by ones or by tens? Which way is faster?

S: By tens!

T: Let's count by tens. (Slide each row from right to left as students count.)

S: 10, 20, 30, 40, 50, 60, 70, 80, 90, 100.

T: Now, let's count back. (From the bottom, sliding each row from left to right.)

S: 100, 90, 80, 70, 60, 50, 40, 30, 20, 10.

Have students return to their seats, and pass out ten 10-frame cards to each child.

T: Lay your 10-frame cards out at the top of your table.

T: Let's count them the Say Ten way.

S: Ten, 2 tens, 3 tens, 4 tens, 5 tens, 6 tens, 7 tens, 8 tens, 9 tens, 10 tens.

T: And now count them the regular way.

S: 10, 20, 30, 40, 50, 60, 70, 80, 90, 100.

T: I will say a number the Say Ten way. Pull down that many cards in front of you.

T: 3 tens.

S: (Show 3 cards.)

T: Count up by tens, and tell me how many.

S: 10, 20, 30.

T: Use your finger and write 30 on your table.

NOTES ON MULTIPLE MEANS OF ENGAGEMENT:

Challenge students who are working above grade level by placing the ten card and two ones on the table. Have them count by tens starting with twelve (12, 22, 32, 42, 52, and so on).

T: Now, slide each card back to the top of your table, and count down by ten as you do so.

S: 30, 20, 10.

T: Here's a new number. 8 tens.

S: (Show eight cards.)

T: Count up by tens, and tell me how many.

S: 10, 20, 30, 40, 50, 60, 70, 80.

T: Use your finger and write 80 on the table.

T: Slide each card back, and count down by ten as you go.

S: 80, 70, 60, 50, 40, 30, 20, 10.

Repeat with the other tens.

Name Charleigh Date 4-4-14

Count up by tens and write the numbers.

●●●●● ●●●●●	10
(2 ten-frames)	20
(3 ten-frames)	30
(4 ten-frames)	40
(5 ten-frames)	50
(6 ten-frames)	60
(7 ten-frames)	70
(8 ten-frames)	80
(9 ten-frames)	90
(10 ten-frames)	100

Problem Set (6 minutes)

Students should do their personal best to complete the Problem Set within the allotted time.

Note: This Problem Set asks students to write numbers greater than 20, which is a Grade 1 standard (**1.NBT.1**). If students are not ready for this step, consider having them use numeral cards or simply tell the amount pictured.

After completing the Problem Set, have students fold after 50 to see and analyze the same "stairs" from Lesson 11 as one more ten is placed on each row as pictured to the right below. While students work, encourage them to count both in the regular way and the Say Ten way.

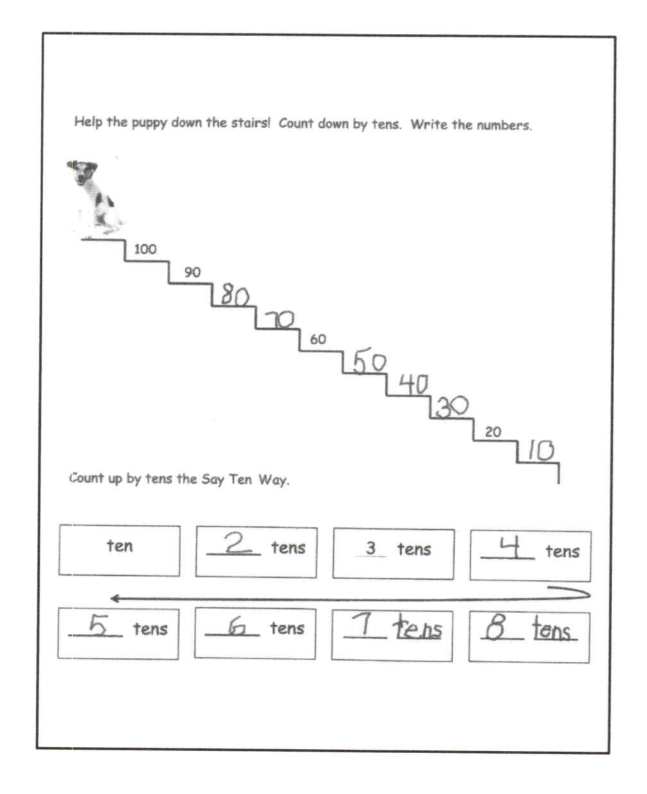

Help the puppy down the stairs! Count down by tens. Write the numbers.

100 90 80 70 60 50 40 30 20 10

Count up by tens the Say Ten Way.

ten	2 tens	3 tens	4 tens
5 tens	6 tens	7 tens	8 tens

Student Debrief (8 minutes)

Lesson Objective: Count up and down by tens to 100 with Say Ten and regular counting.

The Student Debrief is intended to invite reflection and active processing of the total lesson experience.

Invite students to review their solutions for the Problem Set. They should check work by comparing answers with a partner before going over answers as a class, taking turns reading the numbers forward and back. Look for misconceptions or misunderstandings that can be addressed in the Student Debrief. Guide students in a conversation to debrief the Problem Set and process the lesson.

Any combination of the questions below may be used to lead the discussion.

- How would the picture of the stairs be different if you were counting by ones?
- What kinds of things could we count by tens?
- Why is it helpful to count by tens?
- Practice more counting on the Rekenrek.

Exit Ticket (3 minutes)

After the Student Debrief, instruct students to complete the Exit Ticket. A review of their work will help with assessing students' understanding of the concepts that were presented in today's lesson and planning more effectively for future lessons. The questions may be read aloud to the students.

Name ______________________ Date ____________

Count up by tens, and write the numbers.

	10
	20
	50

Help the puppy down the stairs! Count down by tens. Write the numbers.

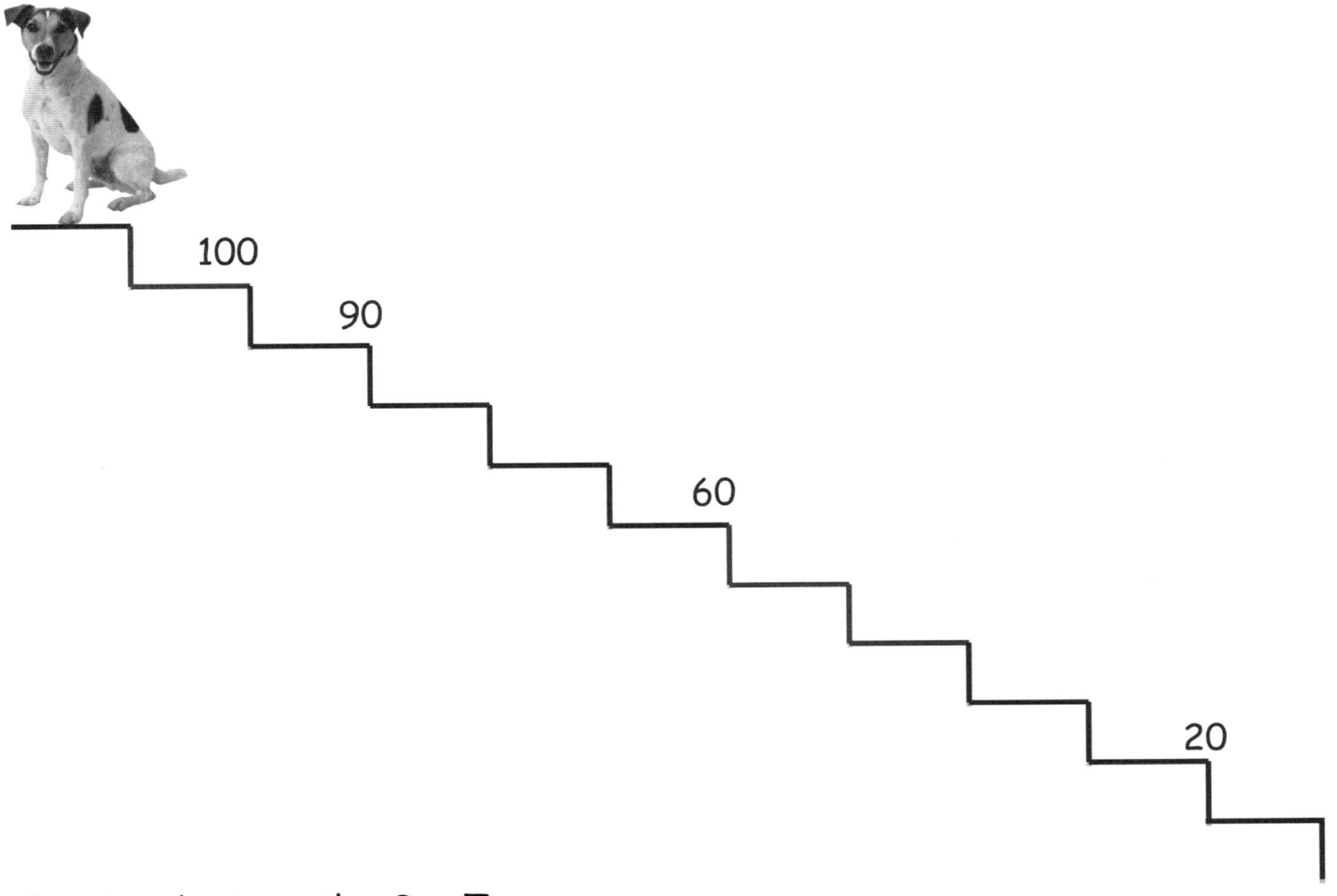

Count up by tens the Say Ten way.

ten	____ tens	3 tens	____ tens
____ tens	____ tens	____ ____	____ ____

Name ______________________________ Date ______________

Count up and down by 10. Write the numbers.

1 ten-frame	10
2 ten-frames	
3 ten-frames	
4 ten-frames	
5 ten-frames	
4 ten-frames	40
3 ten-frames	
2 ten-frames	
1 ten-frame	

Count down and up by 10 the Say Ten way.

↓		
	100	10 tens
	90	tens
	80	tens
	70	7 tens
	60	tens

↑		
	50	tens
	40	4 tens
	30	tens
	20	tens
	10	1 ten

EUREKA MATH

Name ______________________ Date ____________

Count down by 10, and write the number on top of each stair.

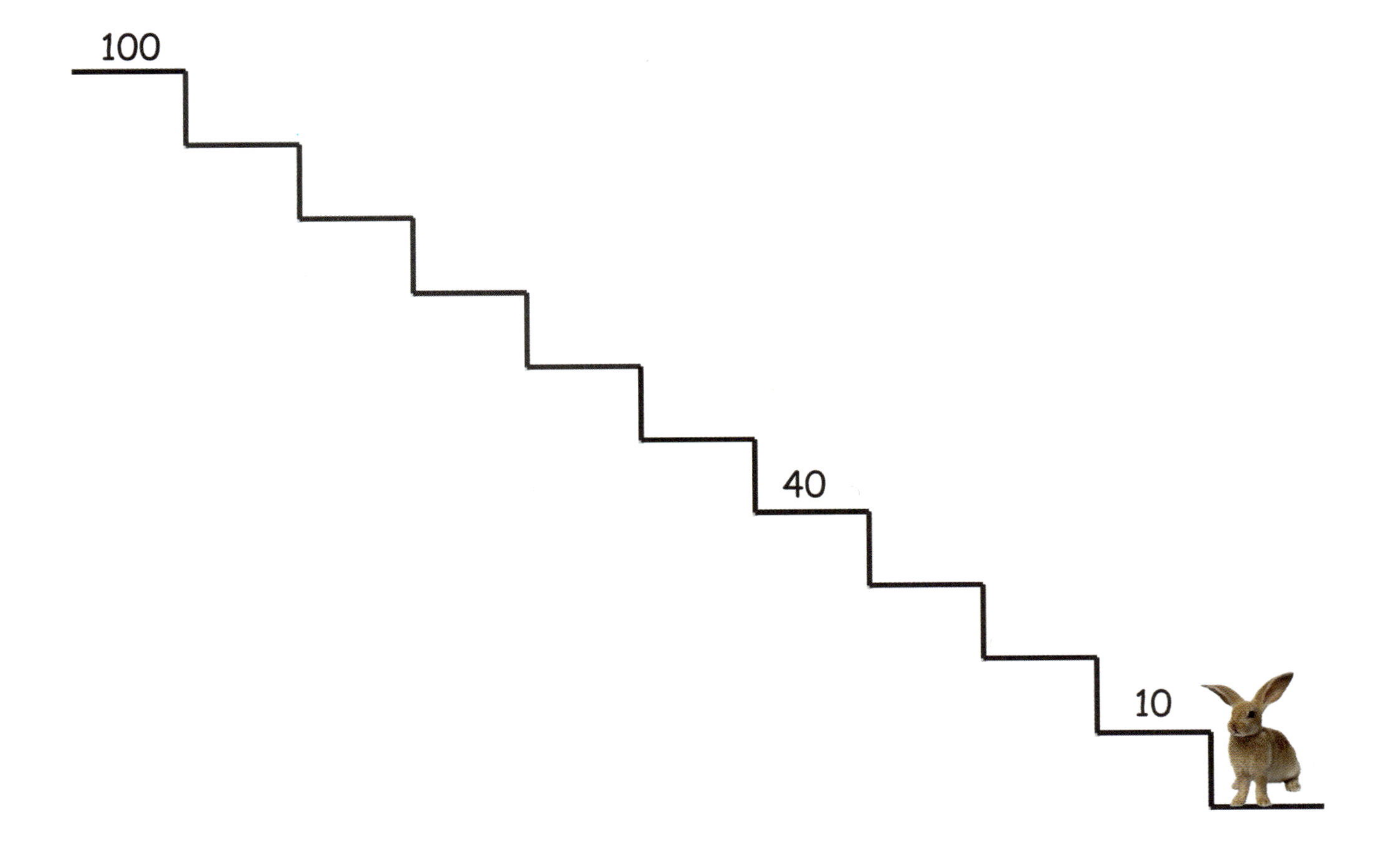

Count down the Say Ten way. Write the missing numbers.

	100	
		9 tens
	80	______ tens
	70	______ tens
		6 tens
		______ tens
	40	4 tens
		______ tens
		______ tens
		______ ten

Name ______________________________ Date ______________

Whisper count and draw in more shapes to match the number.

14

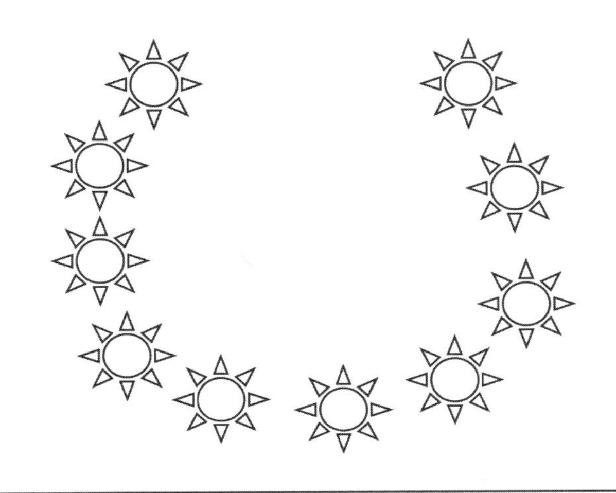

15

17

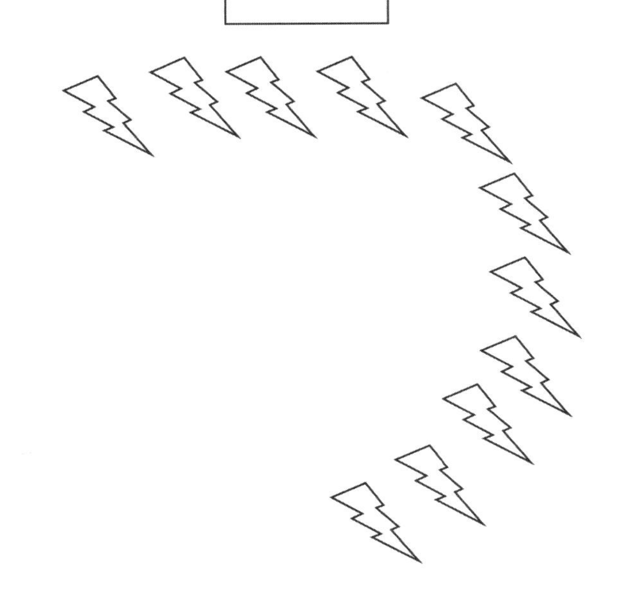

teen circular-counting

Whisper count and draw in more shapes to match the number.

16

19

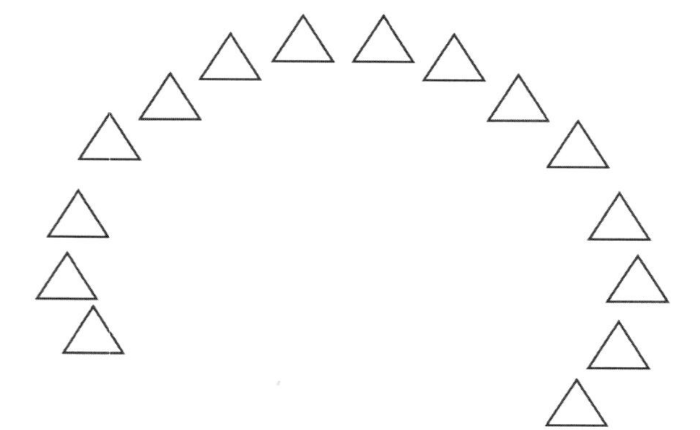

13

20

teen circular-counting

Decorate Mr. Perry's Donut

Decorate Mr. Perry's Donut

Decorate Mr. Perry's Donut

Decorate Mr. Perry's Donut

____

donuts

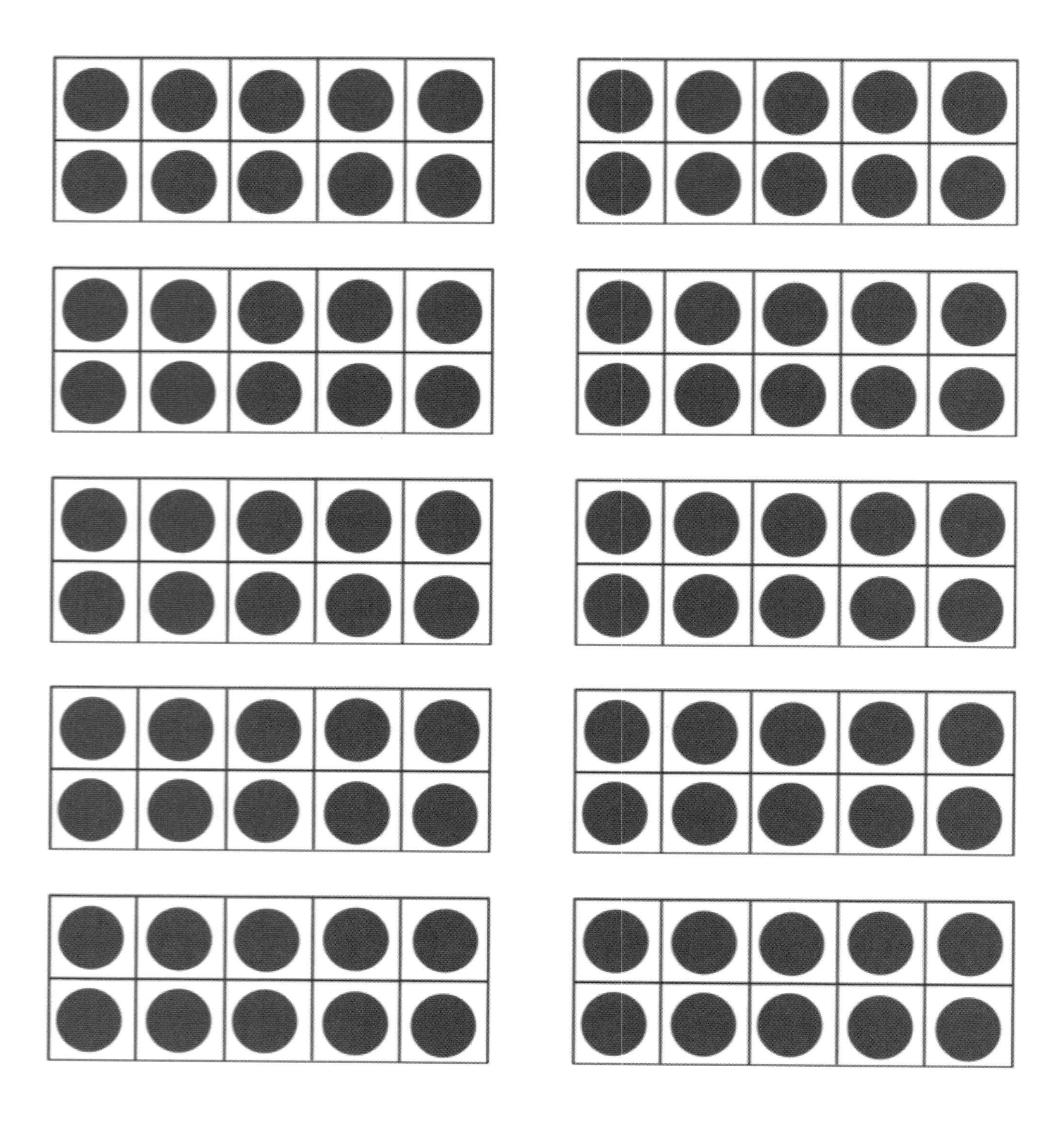

small 10-frame cards

Lesson 16

Objective: Count within tens by ones.

Suggested Lesson Structure

Fluency Practice	(12 minutes)
Application Problem	(5 minutes)
Concept Development	(25 minutes)
Student Debrief	(8 minutes)
Total Time	**(50 minutes)**

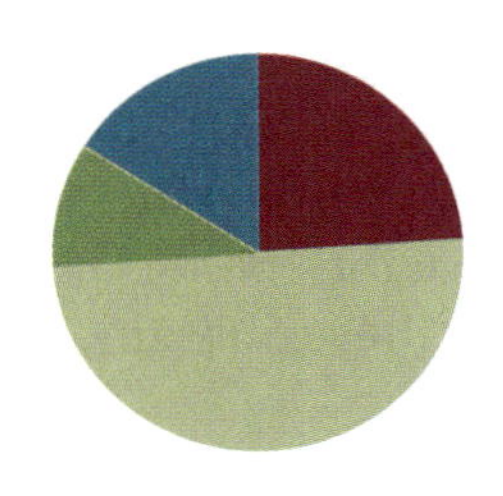

A NOTE ON STANDARDS ALIGNMENT:

In this lesson, students write numbers through 100, which bridges Kindergarten content of writing numbers to 20 (**K.CC.3**) to Grade 1 content of writing numbers to 120 (**1.NBT.1**).

Fluency Practice (12 minutes)

- Hide Zero for Teen Numbers **K.NBT.1** (7 minutes)
- Count by Tens the Say Ten WayN**K.CC.1** (2 minutes)
- Count with Ten-Frame Cards **K.CC.1** (3 minutes)

Hide Zero for Teen Numbers (7 minutes)

Materials: (S) Hide Zero cards: 1 Hide Zero 10 card (Lesson 6 Template 2) and 5-group cards 1–9 (Lesson 1 Fluency Template 2), interesting counters

Note: This activity provides practice with counting out 11–20 objects. Circulate around the classroom as students work, and observe how they organize their objects as they count. For students who are struggling to count accurately, consider suggesting they count out a pile of ten first, before counting out the additional ones. Some students might benefit from arranging their objects in a 5-group formation to match the cards.

Give each pair of students a set of Hide Zero cards, and have them place the number 10 in the middle. One partner gets 4 of the cards numbered 1–9, and the other partner gets the remaining 5 cards. The player with 5 cards puts one of his cards down on the ten. The other partner counts out that many interesting counters (shells, rocks, pennies). They then reverse roles.

Count by Tens the Say Ten Way (2 minutes)

Materials: (T) 100-bead Rekenrek

Note: This activity allows students to see the rows of ten increase and decrease as they count the Say Ten way.

T: (Show 10 on the Rekenrek.) Say the number you see.

S: Ten.

T: (Show 2 tens on the Rekenrek.) Say the number the Say Ten way.
S: 2 tens.

Work toward 100 and back to zero, occasionally changing direction.

Count with Ten-Frame Cards (3 minutes)

Materials: (S) Small 10-frame cards (Lesson 15 Template 2)

Note: This activity provides a visual representation that each ten is composed of ten ones. Students make the connection between pictorial and abstract numbers as they count the Say Ten way.

T: Place a 10-frame card in front of you.
S: (Students place a 10-frame card in front of them.)
T: Say the number.
S: Ten.
T: Place another 10-frame card in front of you.
S: (Students place a second 10-frame card in front of them.)
T: Say the number the Say Ten way.
S: 2 tens.

Continue with this possible sequence: 3 tens, 4 tens, 5 tens, 6 tens, 7 tens, 8 tens, 9 tens, and 10 tens.

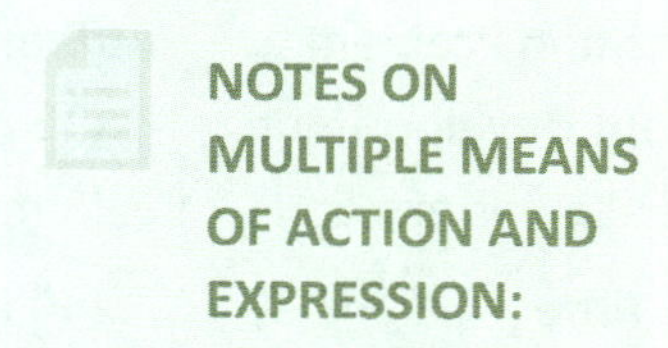

NOTES ON MULTIPLE MEANS OF ACTION AND EXPRESSION:

Let students who are working above grade level work independently or at a center in a small group. Give them many 10-frame cards since they may be able to go far beyond the rest of the class.

Application Problem (5 minutes)

Materials: (S) 2-hand cards (Template)

The students in Pre-Kindergarten are making handprints. 7 students are putting their handprints on a poster board. How many fingers will show on the poster? Use the 2-hand cards to help find out.

Note: This Application Problem is designed to help students make the natural connection between tens and their fingers.

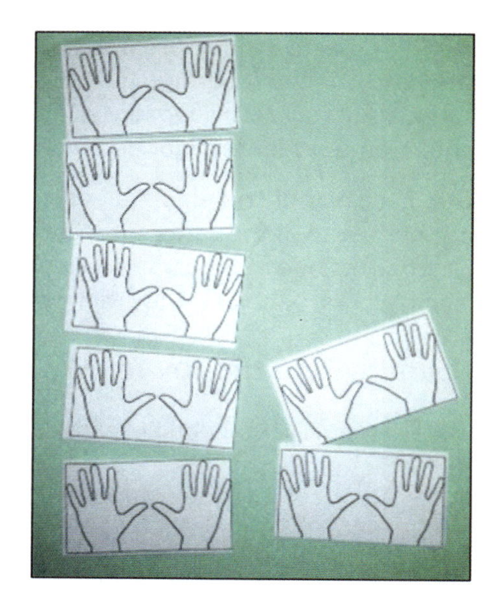

Concept Development (25 minutes)

Materials: (T) 10 pieces of tagboard (S) Small 10-frame cards (Lesson 15 Template 2), 9 counters

Demonstrate the following before having students do it with a partner:

Students count up from 0 to 9 as they place counters on their table in vertical 5-groups. When done, have them raise their hands to receive a 10-frame. They remove the nine counters the moment they are given the 10-frame. They then count from 10 to 19 while placing counters on the table as before. Then, hand them a new 10-frame as they remove the 9 counters, and have them count from 20 to 29 while placing the counters down. Do not mention trading or regrouping. For now, just tell the students that when they have counted to 29—or 39 or 49 or 59, etc.—to clear off all the ones, and they are given a new card of 10 ones. Show students how what they know about counting to 9 will help them count much larger numbers! The Say Ten way really shows that correlation.

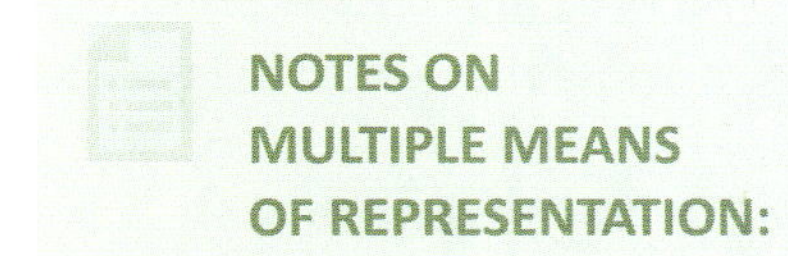

Support English language learners by alternating between Say Ten counting and regular counting. When the students are using their 10-frames and counters, have them whisper count. Puppets can help diffuse performance anxiety. One partner places the counters while the other partner controls the puppet, which counts.

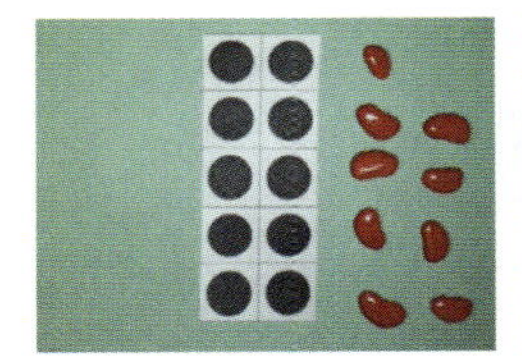
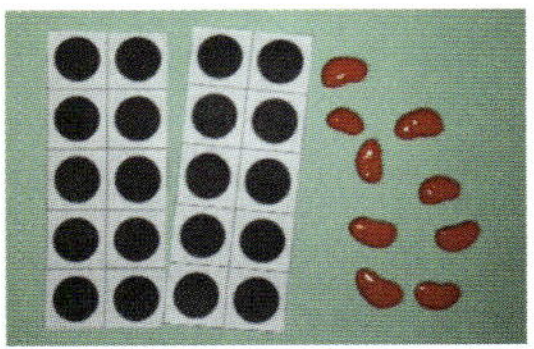

Group Activity:

T: (Create a path by laying the pieces of tagboard across the floor like stepping stones. Have fun creating a story with students about what is at the end of the path.) There's a magic pot at the end of this path, and if you can reach it, you can wish for anything you want! But to get there, you have to count in order from 30 to 39, or 40 to 49, or 50 to ...?

S: 59.

T: From 60 to ...?

S: 69.

T: Who would like to try to reach the magic pot? We'll help you count so you can get there.

T: (Choose a student, and then write 30 on the board.) Let's help Miles count, starting at 30.

S: (As student steps on each "stone.") 30, 31, 32, 33, 34, 35, 36, 37, 38, 39.

T: He made it! What did you wish for? (Allow a quick response.)

T: Who would like to go next?

T: (Choose another student, and then write 50 on the board.) Let's help Victoria get to the magic pot!

S: 50, 51, 52, 53, 54, 55, 56, 57, 58, 59.

T: Victoria made it to the pot! What did you wish for?

Give 2 to 3 students a chance to walk the path to the magic pot, changing the start number each time to a larger number. Students count chorally and get excited by counting to larger numbers.

Afterward, remove 5 stepping stones. Start counting to the magic pot from 35 to 39, 45 to 49, and 75 to 79. Next, put 2 stepping stones back, and start counting to the pot from 23, 53, 83, and 93. Again, only count up to the number with nine in the ones place. Students will be blurting out and wanting to say the multiple of ten, but if they do, it means they cannot get to the magic pot! This creates suspense and enhances students' desire to know those numbers, which are covered in Lesson 18.

Problem Set (5 minutes)

Now that students have worked with the numbers orally and with concrete materials, on the Problem Set they model mathematics with the abstract number.

Students should do their personal best to complete the Problem Set within the allotted time.

Note: This Problem Set asks students to write numbers greater than 20, which is a Grade 1 standard (**1.NBT.1**). If students are not ready for this step, consider having them use numeral cards or simply say the amount pictured.

Student Debrief (8 minutes)

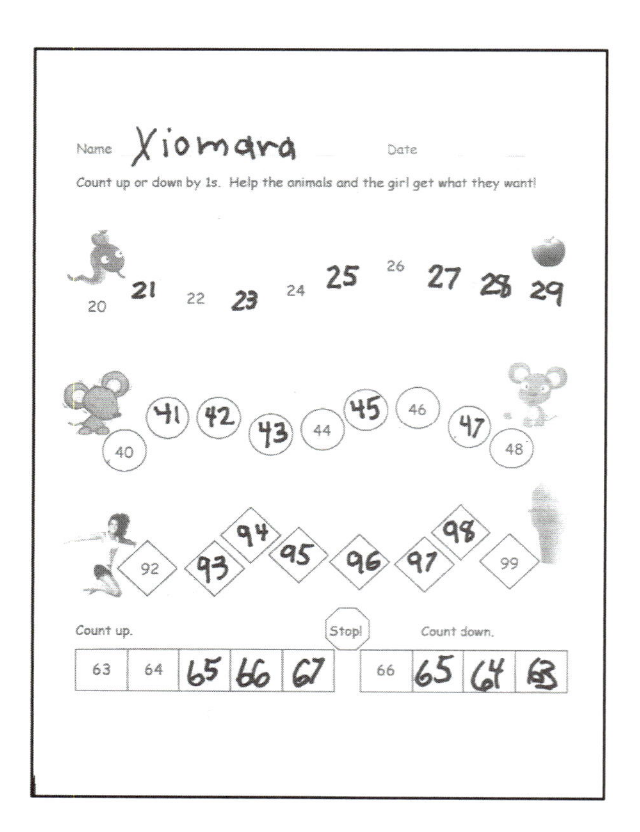

Lesson Objective: Count within tens by ones.

The Student Debrief is intended to invite reflection and active processing of the total lesson experience.

Invite students to review their solutions for the Problem Set. They should check work by comparing answers with a partner before going over answers as a class, taking turns reading the numbers forward and back. Look for misconceptions or misunderstandings that can be addressed in the Student Debrief. Guide students in a conversation to debrief the Problem Set and process the lesson.

Any combination of the questions below may be used to lead the discussion.

- Look at the numbers in the first row on your Problem Set. What is the same about the numbers? What is different?
- Use the Rekenrek to practice more counting within a sequence. Possibly count from 63 to 69, 72 to 79, and 84 to 89.

Exit Ticket (3 minutes)

After the Student Debrief, instruct students to complete the Exit Ticket. A review of their work will help with assessing students' understanding of the concepts that were presented in today's lesson and planning more effectively for future lessons. The questions may be read aloud to the students.

Name ________________________________ Date ______________

Count up or down by 1s. Help the animals and the girl get what they want!

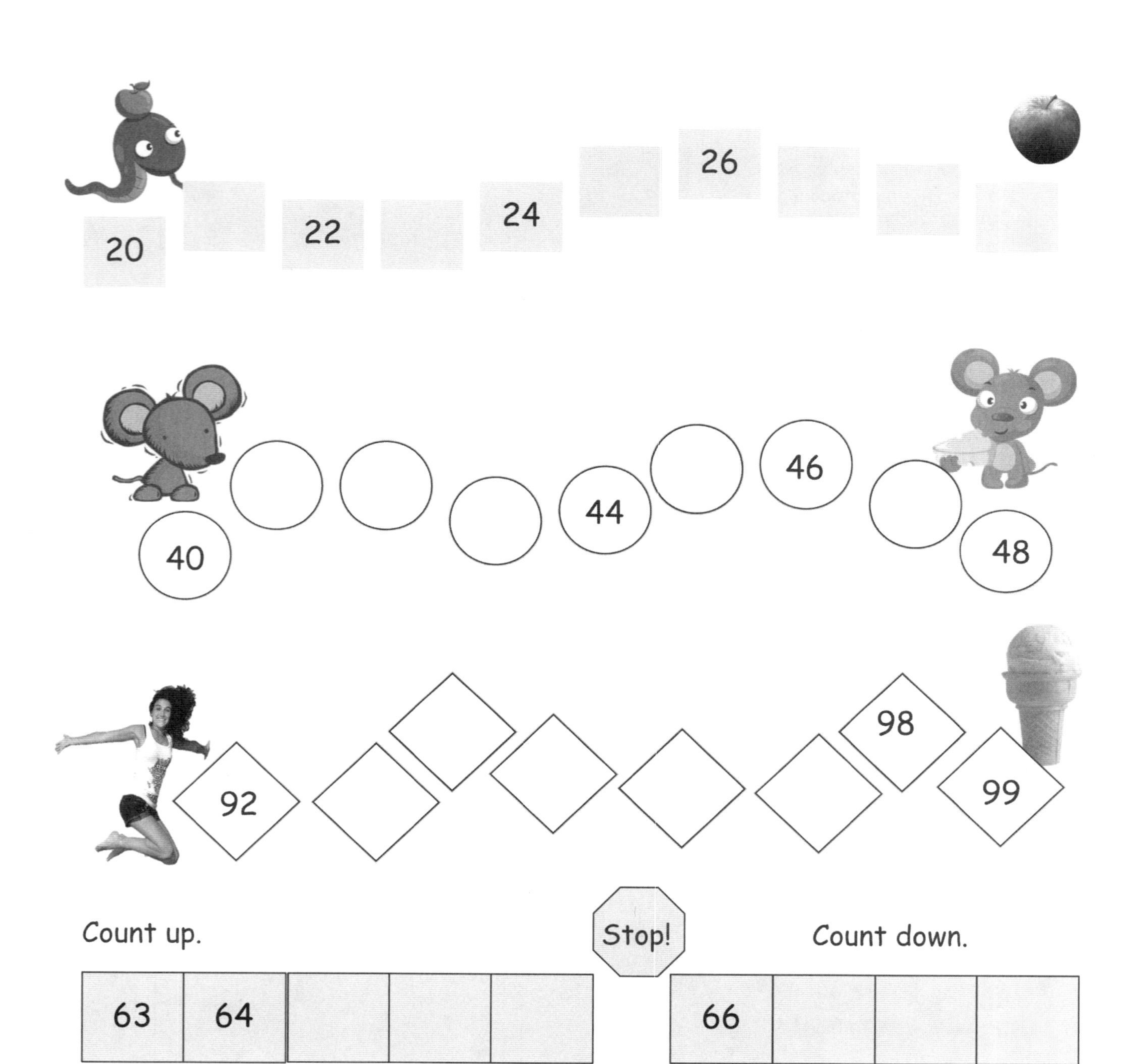

Name ______________________________ Date ______________

Help the cow get to the barn by counting by 1s.

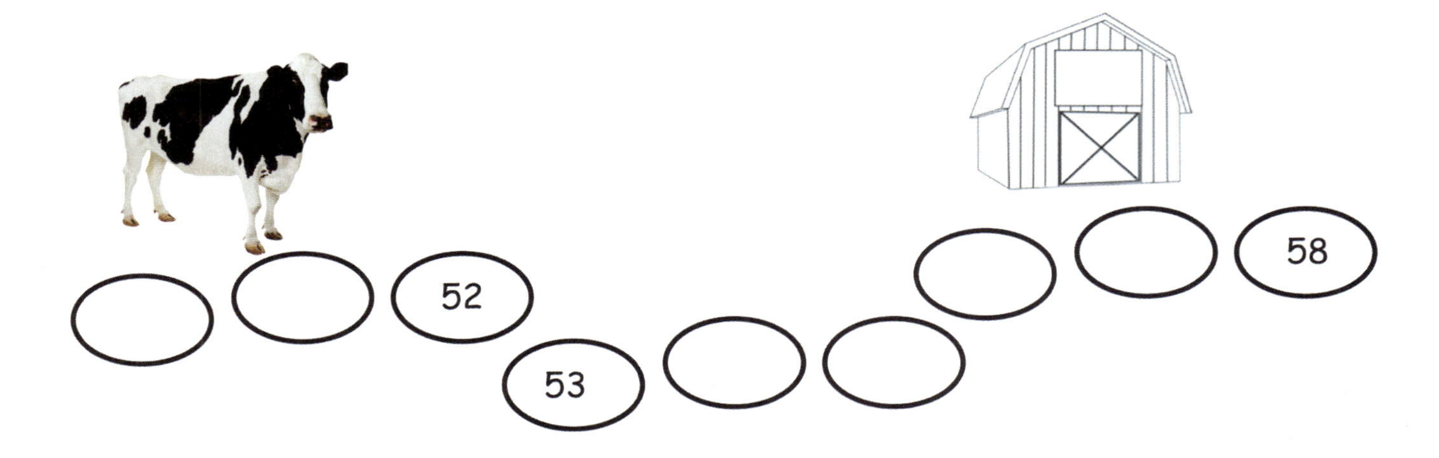

Help the boy get to his present. Count up by 1s. When you get to the top, count down by 1s.

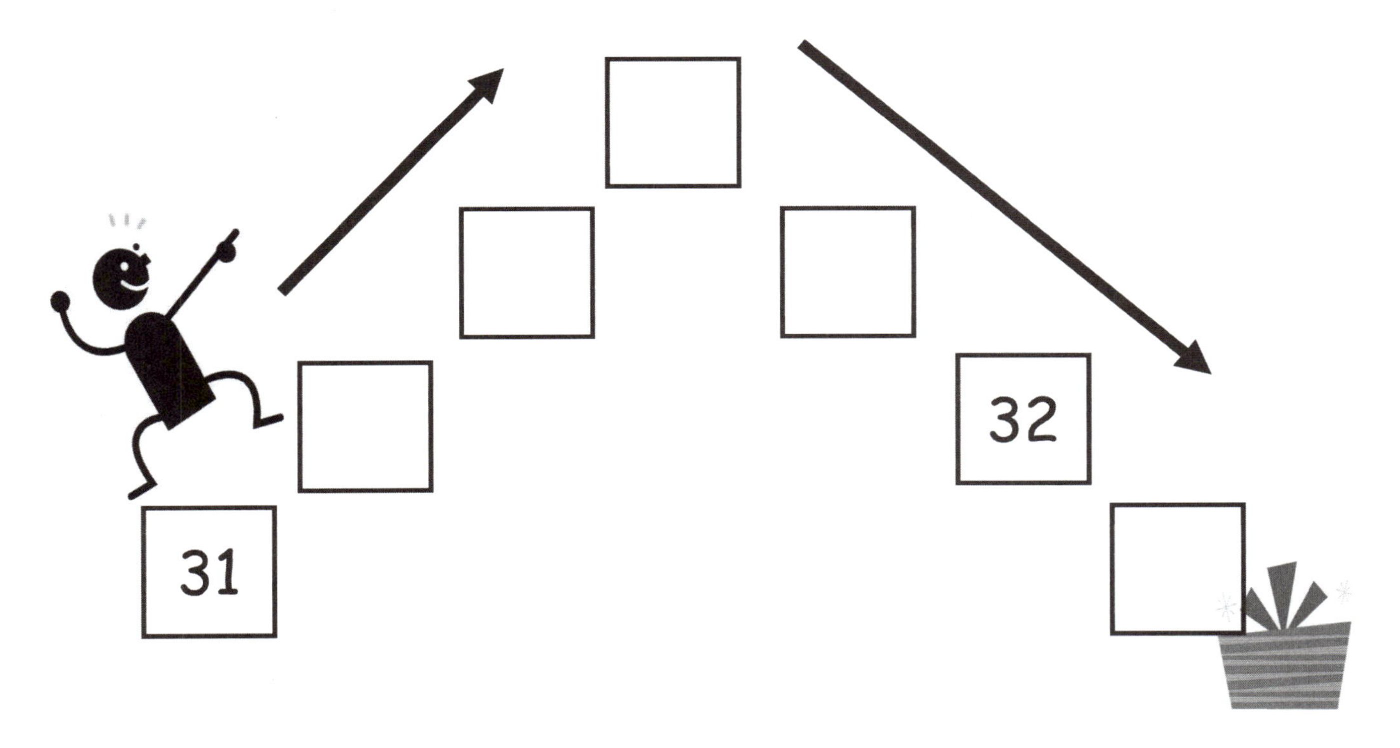

Name ______________________________ Date ______________

Help the rabbit get his carrot. Count by 1s.

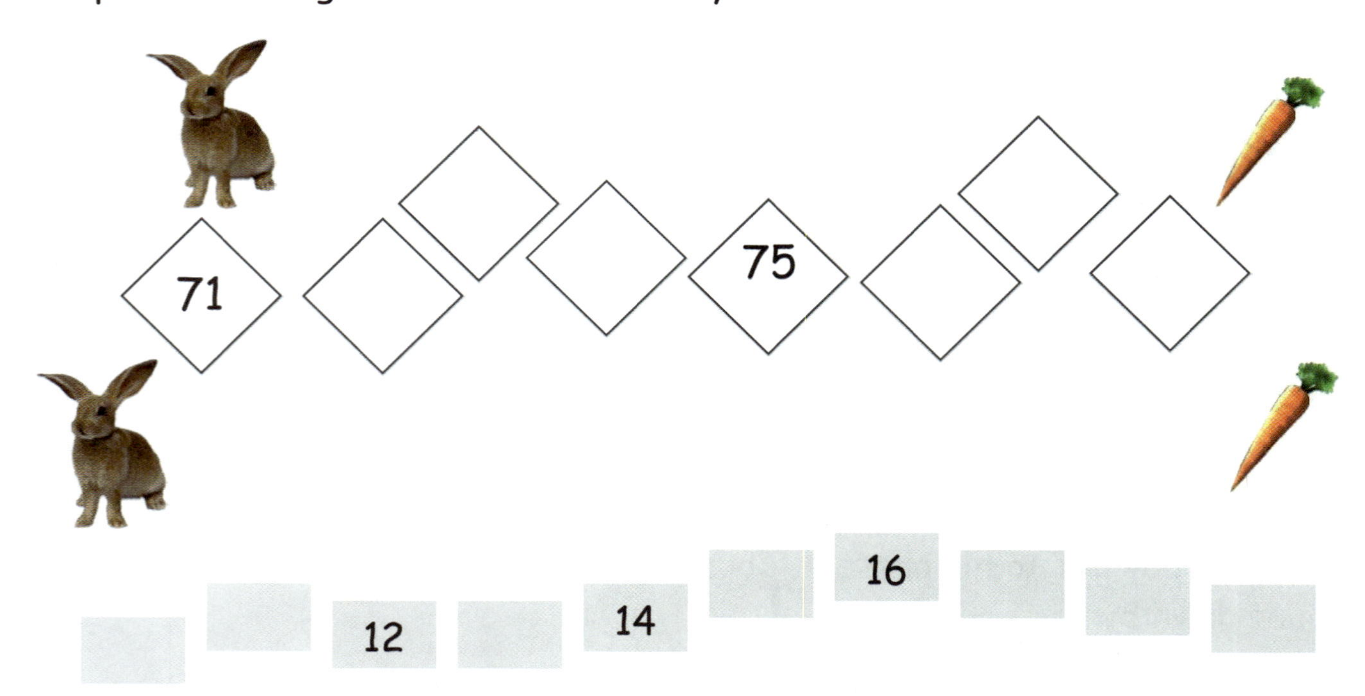

Count up by 1s, then down by 1s.

↑			↓
		89	
	84		

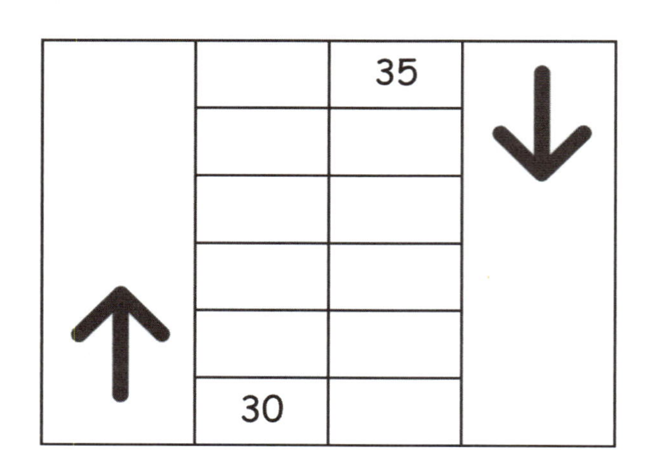

Help the boy mail his letter. Count up by 1s. When you get to the top, count down by 1s.

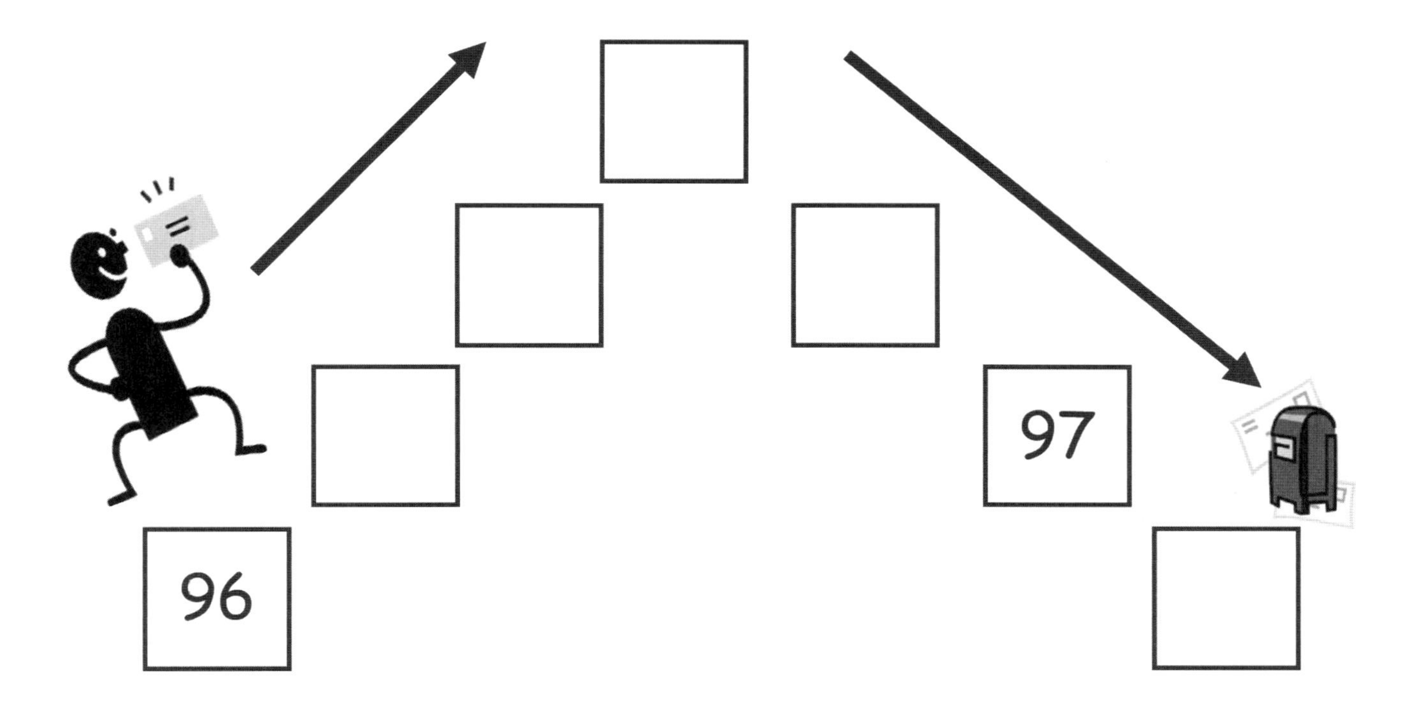

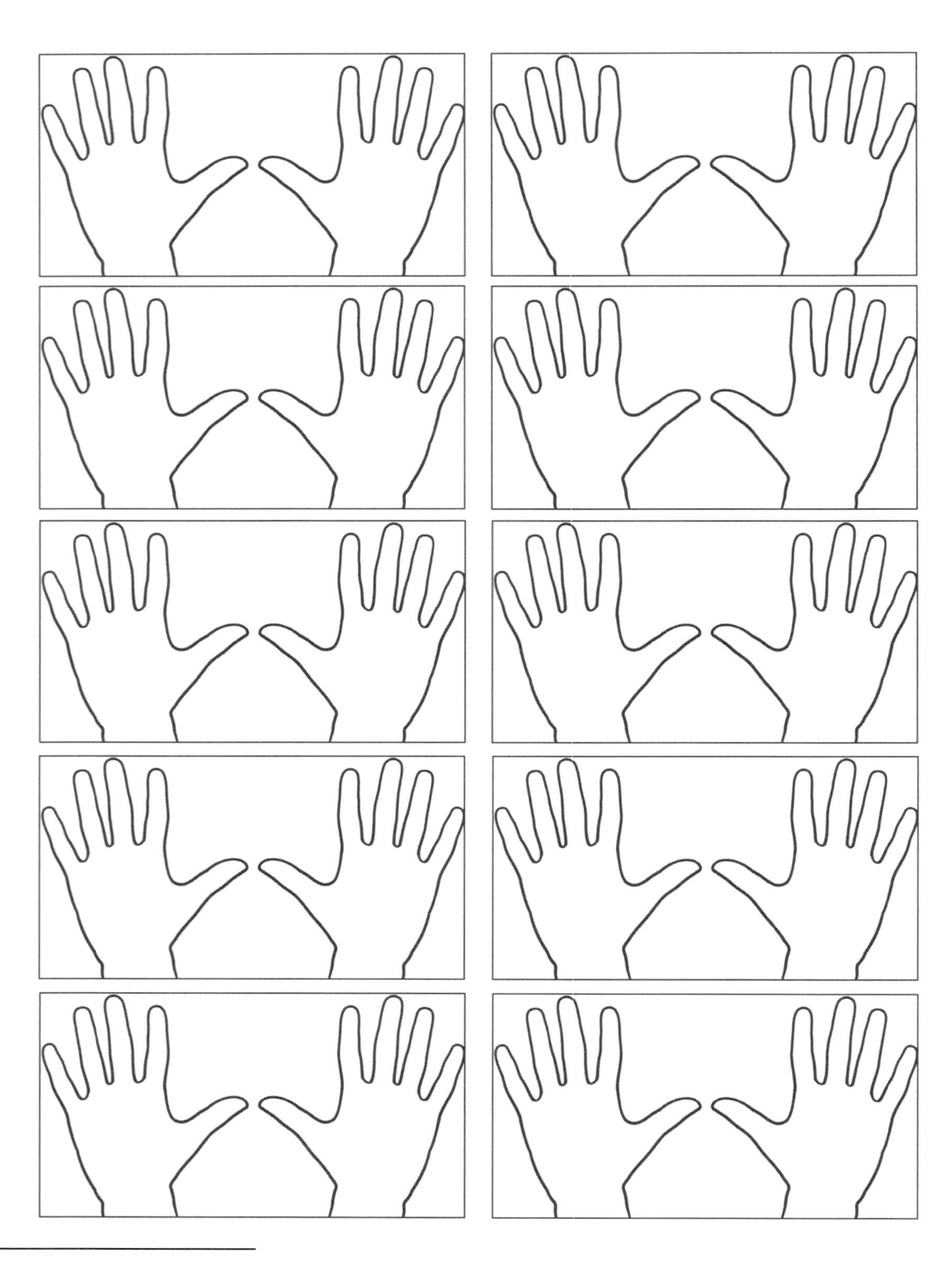

2-hand cards

Lesson 17

Objective: Count across tens when counting by ones through 40.

Suggested Lesson Structure

- Application Problem (7 minutes)
- Fluency Practice (10 minutes)
- Concept Development (25 minutes)
- Student Debrief (8 minutes)

Total Time (50 minutes)

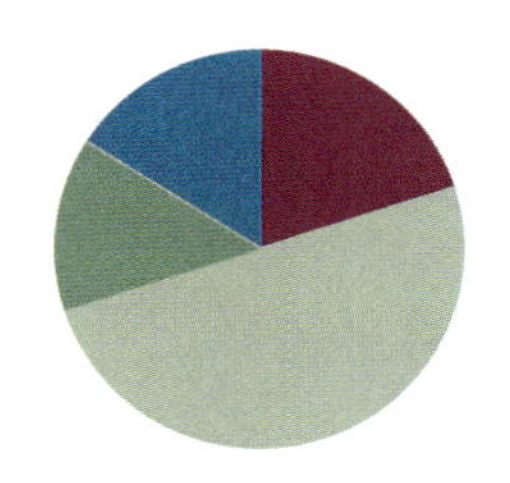

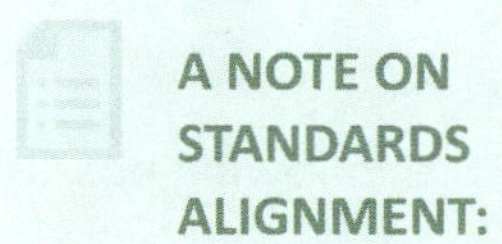

A NOTE ON STANDARDS ALIGNMENT:

In this lesson, students write numbers through 100, which bridges Kindergarten content of writing numbers to 20 (**K.CC.3**) to Grade 1 content of writing numbers to 120 (**1.NBT.1**).

Application Problem (7 minutes)

Sammy's mom has 10 apples in a bag. Some are red and some are green. What might be the number of each color apple in her bag? There is more than one possible answer. See how many different answers can be found. Show the answers with number bonds. Label the parts as R and G.

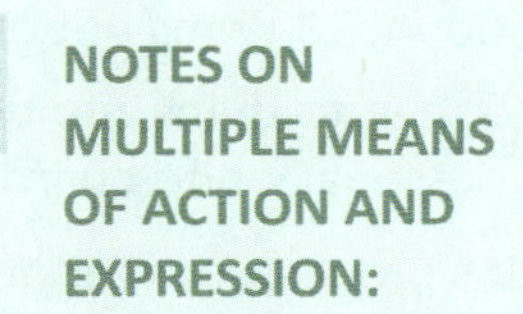

NOTES ON MULTIPLE MEANS OF ACTION AND EXPRESSION:

Challenge students who are working above grade level to model all nine possible solutions for the Application Problem and to explain both orally and in writing how all nine possibilities are a response to the same problem.

Note: In this lesson, the Application Problem precedes the Fluency Practice because the fluency activities lead directly into the counting of the lesson.

Fluency Practice (10 minutes)

- 5-Group Flashes: Partners to 5 **K.OA.5** (4 minutes)
- Count Out Teen Numbers **K.CC.1** (4 minutes)
- Count Within Tens **K.CC.1** (2 minutes)

5-Group Flashes: Partners to 5 (4 minutes)

Materials: (T) Large 5-group cards (Lesson 1 Fluency Template 1)

Note: Reviewing compositions of 5 leads to proficiency with the Core Fluency for the grade, **K.OA.5**, add and subtract within 5.

T: (Show 4 dots.) How many dots do you see?
S: 4.
T: How many more to make 5?
S: 1.
T: Say the addition sentence.
S: 4 + 1 = 5.

Continue with the following possible sequence: 1, 3, 2, 5, 0, 4, 2.

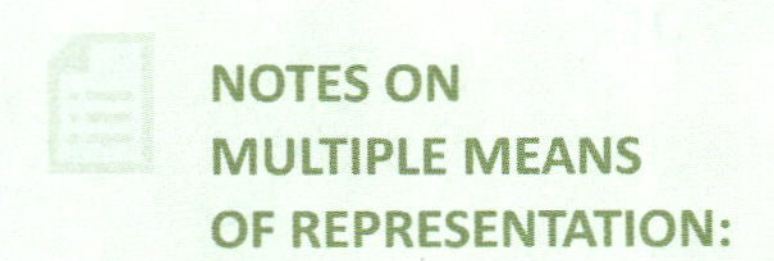

NOTES ON MULTIPLE MEANS OF REPRESENTATION:

5-Group Flashes foster English language learners' number sense and ability to speak about math in English. Review number words by counting the dots, if necessary. Tailor the sequence according to students' needs, repeating flashes when necessary.

Count Out Teen Numbers (4 minutes)

Materials: (S) personal white board, 1 bag of about 20 objects (per pair)

Note: This activity provides students with concrete practice decomposing teen numbers into ten ones and some additional ones.

MP.2

T: Count 13 items out of your bag.
T: Separate them into two parts—one part with 10 and another part. Write the number on your personal white board.

Repeat this process for four or five other amounts.

Count Within Tens (2 minutes)

T: Let's count starting at 20.

Note: This activity gives students practice counting by ones within the decades to prepare them to count across the decades in today's Concept Development.

Guide students, counting from 20 to 29, occasionally changing directions. Repeat for 50–59 and 80–89.

Concept Development (25 minutes)

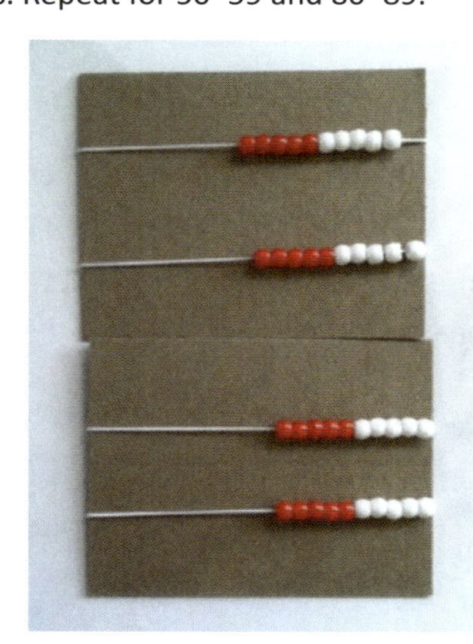

Materials: (S) Personal Rekenrek (from Lesson 10)

T: Put your Rekenrek together with your partner's.
T: Move all your beads to the right-hand side.
T: Count your beads by ones. Partner A, move the first row. Both of you whisper each number as you move your beads from right to left.
S: (Moving beads with partner.) 1, 2, 3, 4, 5, 6, 7, 8, 9, 10.
T: Say the number.
S: 10.

T: Partner B moves the beads of the second row one at a time. What is the first number we will say? Say it the Say Ten way.
S: Ten 1.
T: How do we say the number the regular way?
S: 11.
T: Count the second row starting with eleven. Move your beads one at a time, and whisper the numbers.
S: (Moving the beads.) 11, 12, 13, 14, 15, 16, 17, 18, 19, 20.
T: What is the number the Say Ten way?
S: 2 tens.
T: Now, it's Partner A's turn. Move one bead on the next row. What is the number the Say Ten way?
S: 2 tens 1.
T: Say it the regular way.
S: 21.
T: Keep counting the regular way.
S: (Moving the beads.) 22, 23, 24, 25, 26, 27, 28, 29, 30.
T: What is the number the Say Ten way?
S: 3 tens.

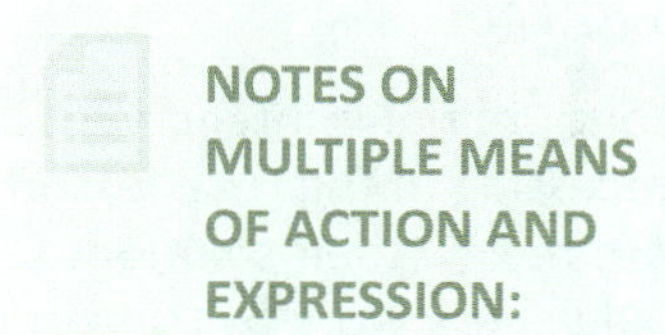

Counting with the Rekenrek is great for students working below grade level who will benefit from practicing one-to-one correspondence, the support of a peer, and the lesson's frequent checks for understanding. To avoid miscounting, encourage deliberate counting through song or rhythm.

Continue to 40 in this manner. Then, ask students to count to 40 on their own with their partner. To add excitement to this exercise, students can speak the last bead of each row loudly.

Problem Set (7 minutes)

Students should do their personal best to complete the Problem Set within the allotted time.

Note: In this Problem Set, students write numbers to 100, which bridges to the Grade 1 standard, **1.NBT.1**. The Kindergarten standard requires students to write numbers only to 20.

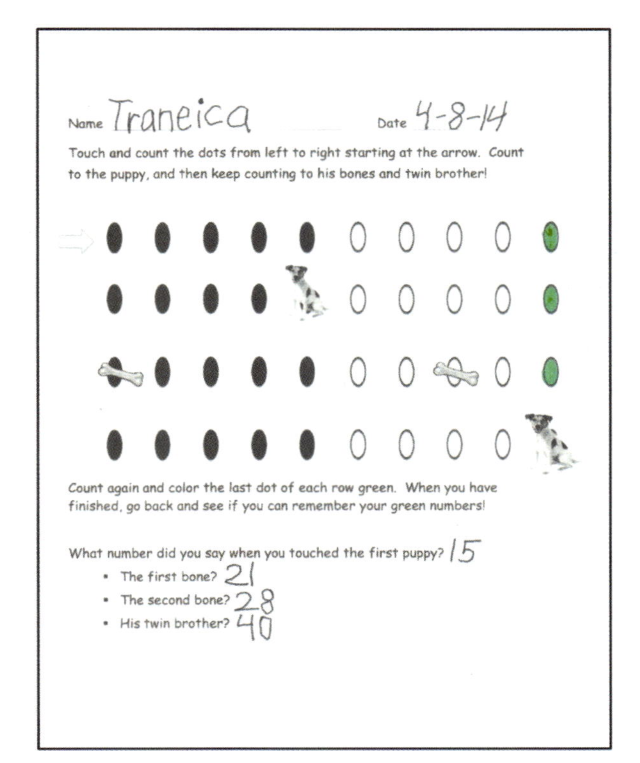
Name Traneica Date 4-8-14

Touch and count the dots from left to right starting at the arrow. Count to the puppy, and then keep counting to his bones and twin brother!

Count again and color the last dot of each row green. When you have finished, go back and see if you can remember your green numbers!

What number did you say when you touched the first puppy? 15
- The first bone? 21
- The second bone? 28
- His twin brother? 40

Student Debrief (8 minutes)

Lesson Objective: Count across tens when counting by ones through 40.

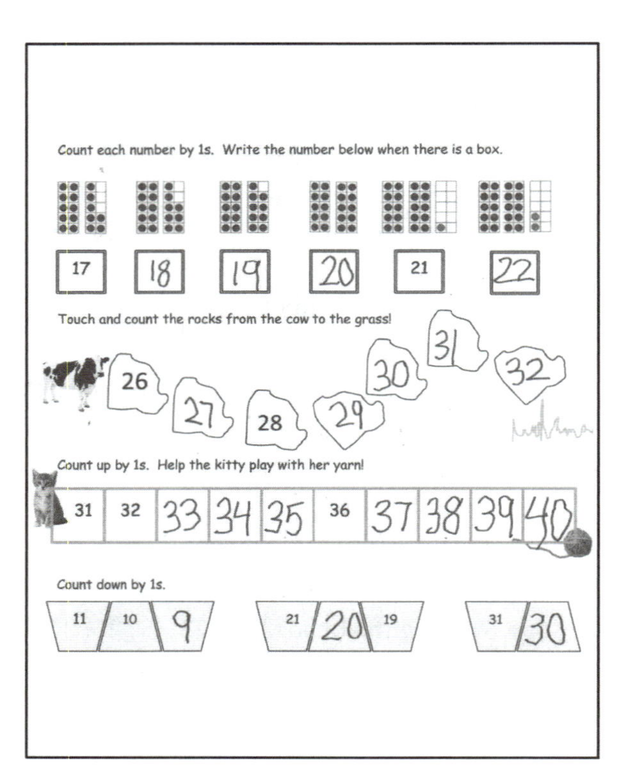

Count each number by 1s. Write the number below when there is a box.
17
18
19
20
21
22
Touch and count the rocks from the cow to the grass!
26
27
28
29
30
31
32
Count up by 1s. Help the kitty play with her yarn!
31 32 33 34 35 36 37 38 39 40
Count down by 1s.
11 10 9
21 20 19
31 30

The Student Debrief is intended to invite reflection and active processing of the total lesson experience. Invite students to review their solutions for the Problem Set. They should check work by comparing answers with a partner before going over answers as a class, taking turns reading the numbers forward and back. Look for misconceptions or misunderstandings that can be addressed in the Student Debrief. Guide students in a conversation to debrief the Problem Set and process the lesson.

Any combination of the suggestions below may be used to lead the discussion.

- Touch and count each series of numbers, pointing out that students read from left to right as they do when reading.
- Read each series of numbers in a different voice, like an elf, like a giant, like a witch, as a crescendo, etc. Adding drama makes the learning memorable and fun!
- Count across ten from various starting points using the Rekenrek.

Exit Ticket (3 minutes)

After the Student Debrief, instruct students to complete the Exit Ticket. A review of their work will help with assessing students' understanding of the concepts that were presented in today's lesson and planning more effectively for future lessons. The questions may be read aloud to the students.

Name ______________________________ Date _____________

Touch and count the dots from left to right starting at the arrow. Count to the puppy, and then keep counting to his bones and twin brother!

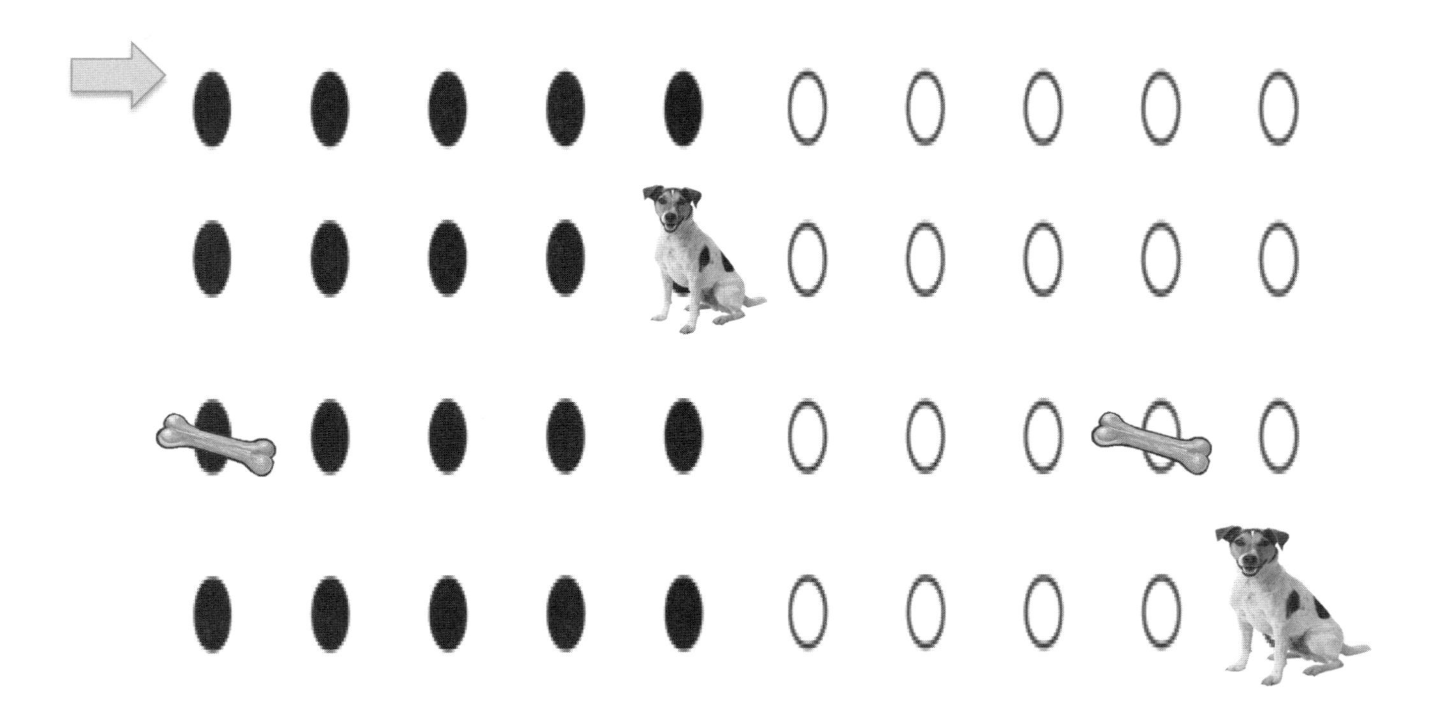

Count again and color the last dot of each row green. When you have finished, go back and see if you can remember your green numbers!

What number did you say when you touched the first puppy?

- The first bone?
- The second bone?
- His twin brother?

Count each number by 1s. Write the number below when there is a box.

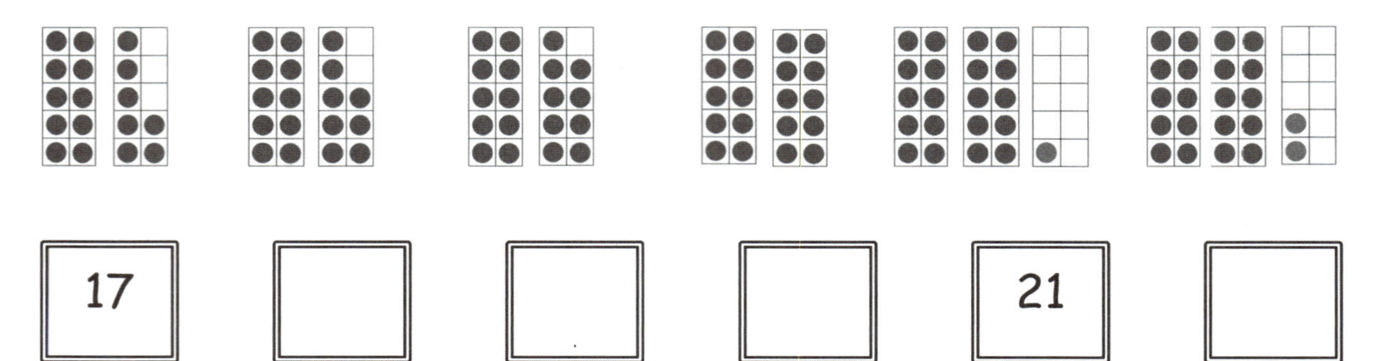

17				21	

Touch and count the rocks from the cow to the grass!

Count up by 1s. Help the kitty play with her yarn!

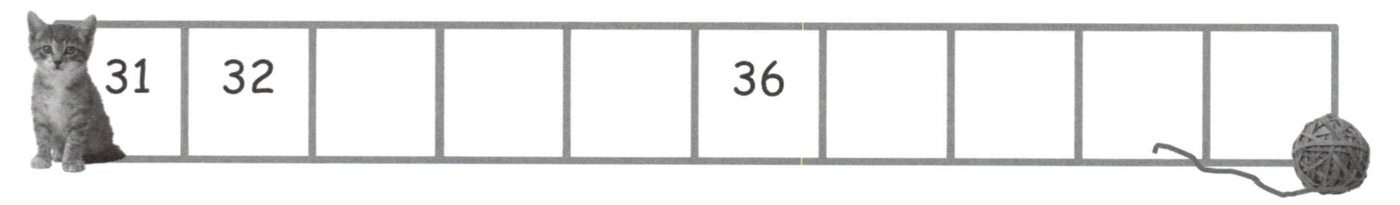

Count down by 1s.

Name ______________________________ Date ______________

Touch and count carefully. Cross out the mistake, and write the correct number.

Example:

3

1, 2, ~~9~~, 4, 5

20	21	22	23	24	25	29

30	31	32	33	43	35	36

25	26	27	28	29	29	31

34	35	36	37	38	39	44

Name ______________________________ Date ______________

Draw more to show the number.

Example:

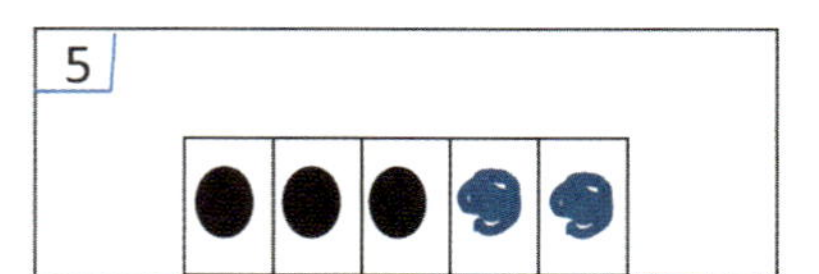

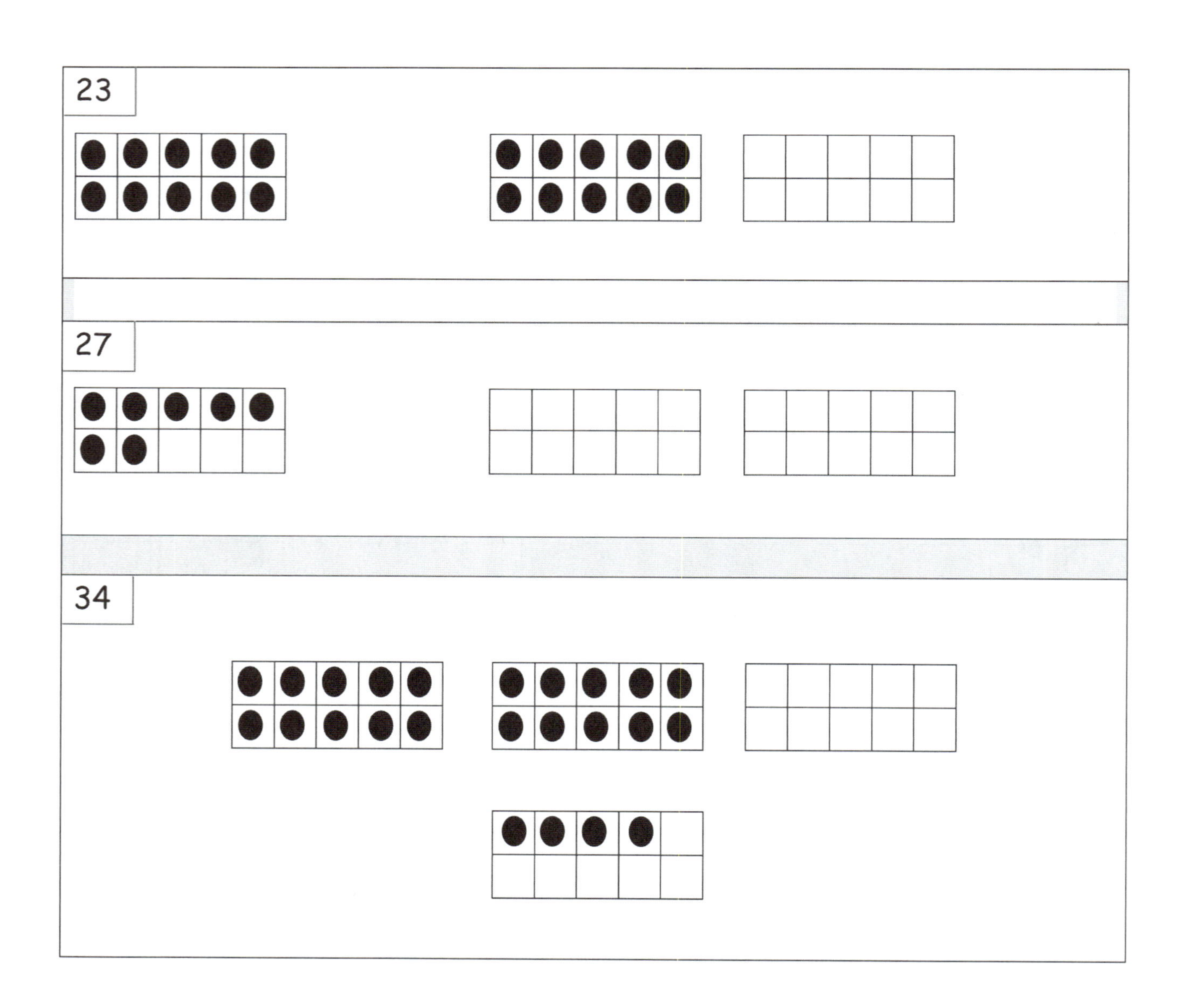

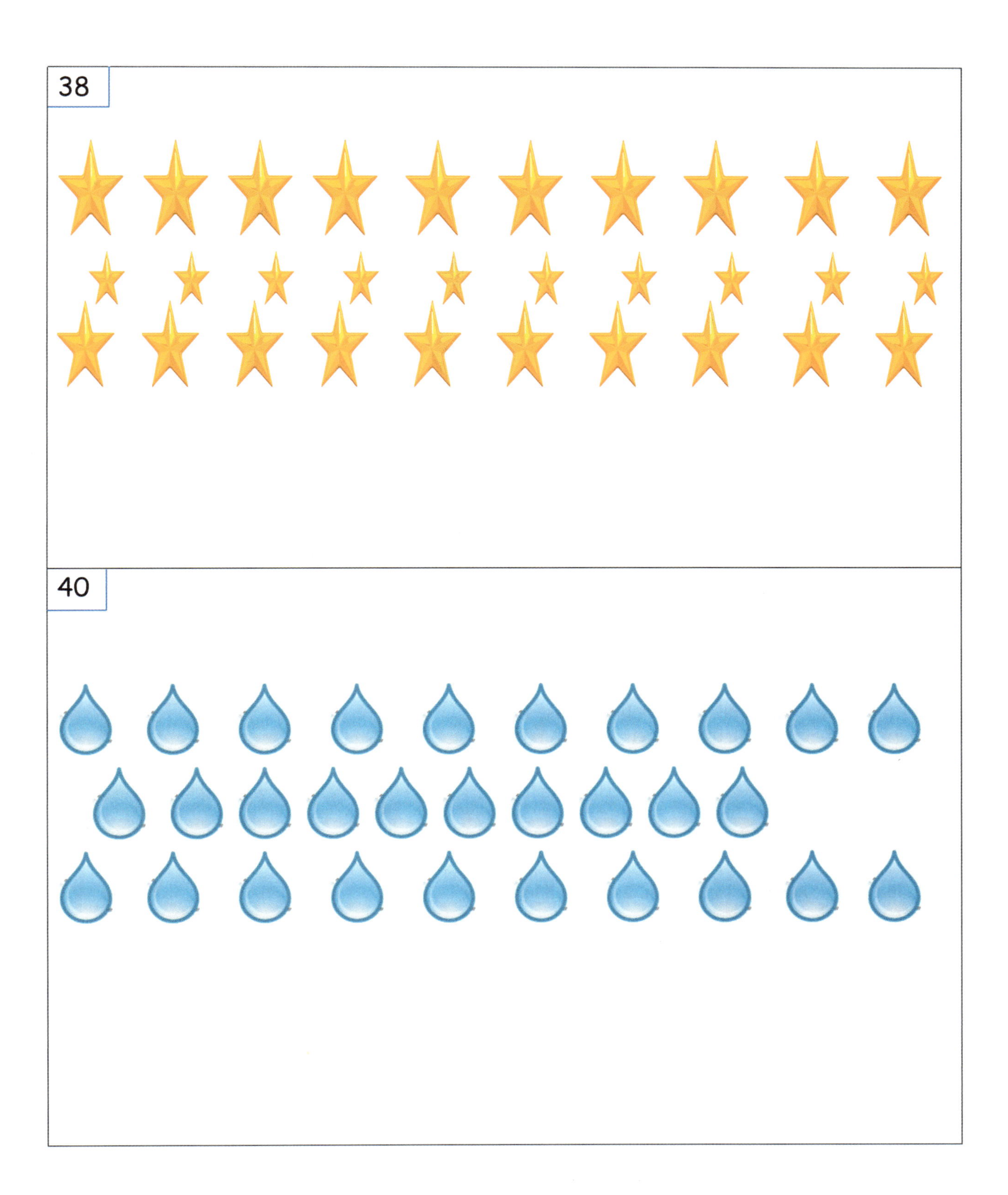
38
40

Lesson 18

Objective: Count across tens by ones to 100 with and without objects.

Suggested Lesson Structure

- Application Problem (7 minutes)
- Fluency Practice (11 minutes)
- Concept Development (24 minutes)
- Student Debrief (8 minutes)

Total Time (50 minutes)

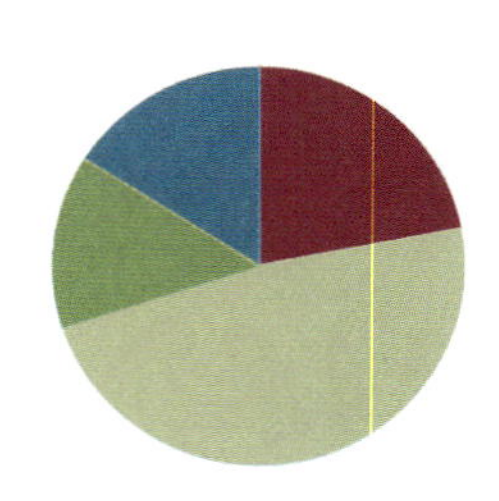

A NOTE ON STANDARDS ALIGNMENT:

In this lesson, students write numbers through 100, which bridges Kindergarten content of writing numbers to 20 (**K.CC.3**) to Grade 1 content of writing numbers to 120 (**1. NBT.1**).

Application Problem (7 minutes)

Susan is putting 9 flowers in 2 vases. Draw a picture to show a way she might do that. Make a number bond and a number sentence to match the idea. (Extension: See if there is another way to put the flowers in the vases.)

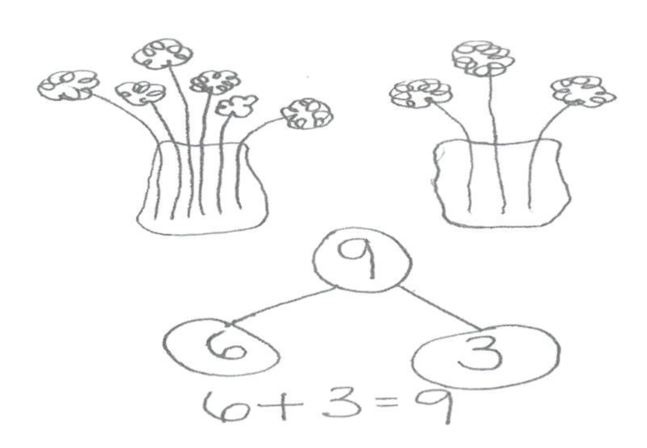

When students have finished, have them compare their work with another student. Are their ways of showing the flowers the same? Why or why not? How is the flower problem similar to the apple problem from yesterday?

Note: In this lesson, the Application Problem precedes the Fluency Practice because the fluency activities lead directly into the counting of the lesson.

Fluency Practice (11 minutes)

- 5-Group Flashes: Partners to 10 **K.CC.2** (3 minutes)
- Teen Number Bonds **K.CC.1** (4 minutes)
- Count on the Rekenrek **K.CC.4** (4 minutes)

5-Group Flashes: Partners to 10 (3 minutes)

Materials: (T) Large 5-group cards (Lesson 1 Fluency Template 1)

Note: The 5-group formation facilitates speed and accuracy in recognizing partners of 10.

T: (Show 9 dots.) How many dots do you see?

S: 9.

T: How many more does 9 need to be 10?

S: 1.

Continue with the following possible sequence: 1, 5, 8, 2, 3, 7, 6, 1, 4, 3, 5, 2, 9.

Teen Number Bonds (4 minutes)

Materials: (S) number bond (Lesson 7 Template)

Note: This activity reinforces part–whole relationships within teen numbers.

MP.4

T: (Project the number bond with parts of 10 objects and 6 objects.) Say the larger part.

S: 10.

T: Say the smaller part.

S: 6.

T: Count the whole, or total, with me.

S: 1, 2, 3, 4, 5, 6, 7, 8, 9, 10, 11, 12, 13, 14, 15, 16.

Continue with the following possible sequence: 10 and 7, 10 and 3, 10 and 1, 10 and 8, 10 and 4.

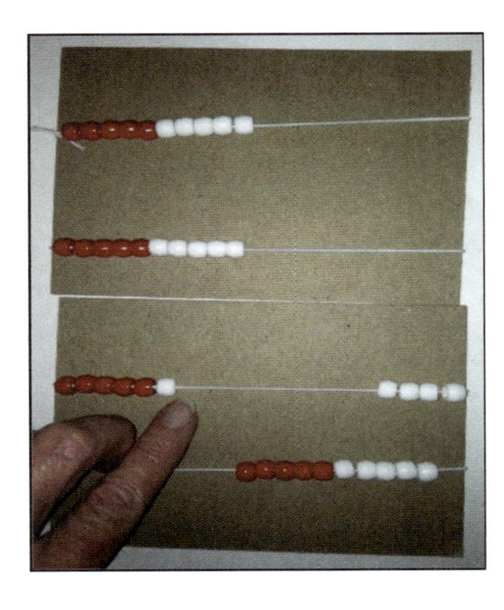

Count on the Rekenrek (4 minutes)

Materials: (S) Personal Rekenrek (from Lesson 10)

Note: Manipulating their own Rekenreks allows students to work at a comfortable pace. Saying "buzz" at the end of each row delightfully draws attention to the grouping of ten on the Rekenrek.

T: Put your Rekenrek together with your partner's. Whisper count with your partner up to 40 on your Rekenrek. Take turns moving the beads with each new row. Buzz before you say the first number of each row.

Concept Development (24 minutes)

Materials: (T) 100-bead Rekenrek (S) 9 small 10-frame cards (Lesson 15 Template 2), 2 empty 10-frame cards (Template), 20 counters, blank paper to use as a hiding paper for the Problem Set

T: (Count by tens to 40 by sliding four rows on the Rekenrek.) Count with me.

S: 10, 20, 30, 40.

T: Now, count by ones. (Slide one bead at a time as students count.)

S: 41, 42, 43, 44, 45, 46, 47, 48, 49, 50.

T: What is the number the Say Ten way?

S: 5 tens.

T: (Slide one more bead.) Tell me the number the Say Ten way.

S: 5 tens 1.

T: Tell me the number the regular way.

S: 51.

T: (Slide the bead back so that only 50 beads are showing.) How many beads are there now?

S: 50.

T: (Slide one bead back so that 49 are showing.) How many beads are there, the Say Ten way?

S: 4 tens 9.

T: How many, the regular way?

S: 49.

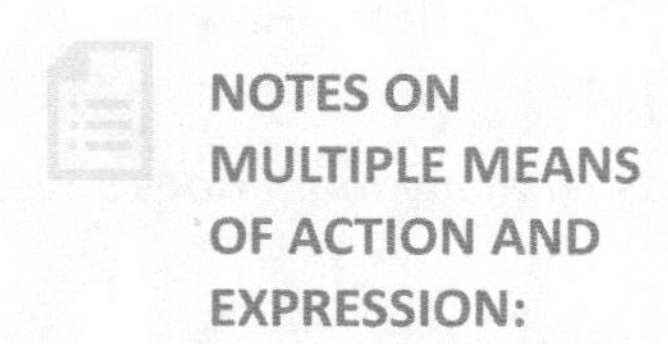

NOTES ON MULTIPLE MEANS OF ACTION AND EXPRESSION:

Use English language learners' culture to teach them the names of the numbers. For instance, couple *twenty* with *veinte* and *thirty* with *treinta* for native Spanish speakers. Building on students' culture and language while teaching helps native English speakers as well by expanding their horizon and exposing them to other cultures and languages.

Repeat this process from different starting points within 100, going back and forth across the ten.

T: Now, let's show and count numbers a different way. Lay out 10-frame cards as we count the Say Ten way.

S: (Slowly counting as students lay out the cards.) Ten, 2 tens, 3 tens, 4 tens, 5 tens.

T: Now, let's count the regular way by tens. Touch each card as we count it.

S: 10, 20, 30, 40, 50.

T: Place the two empty 10-frames down after 50.

T: Count on from 50, placing one counter at a time as we say each number. Let's start the Say Ten way.

S: (Placing a counter each time they count.) 5 tens 1, 5 tens 2, 5 tens 3, ... , 6 tens.

T: Now, let's count that the regular way, starting at 51. Touch each counter as you count.

S: 51, 52, 53, ... , 60.

T: Place one more counter on the next 10-frame. Say the number the Say Ten way.

S: 6 tens 1.

T: What is the number the regular way?

S: 61.

T: What is one more than 60?

S: 61.

T: Take one counter off. What is the number the Say Ten way?

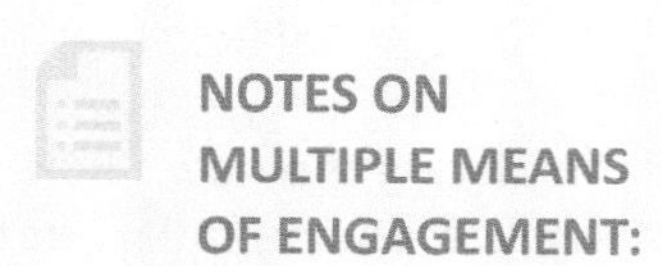

NOTES ON MULTIPLE MEANS OF ENGAGEMENT:

Challenge students working above grade level by providing them with opportunities to extend the lesson. For instance, after counting by ones, have students skip-count from 28 by twos, by threes, and by fives, using the Rekenrek on their own. For very advanced students, ask them to write their answers before the teacher moves the beads to encourage their counting in their heads rather than relying on the visual support!

S: 6 tens.

T: What is the number the regular way?

S: 60.

T: Take away one more counter. What is the number the Say Ten way?

S: 5 tens 9.

T: Say the number the regular way.

S: 59.

Repeat this process starting from different numbers within 100, focusing on crossing over to the next ten and then back (e.g., 69, 70, 71, 70, 69).

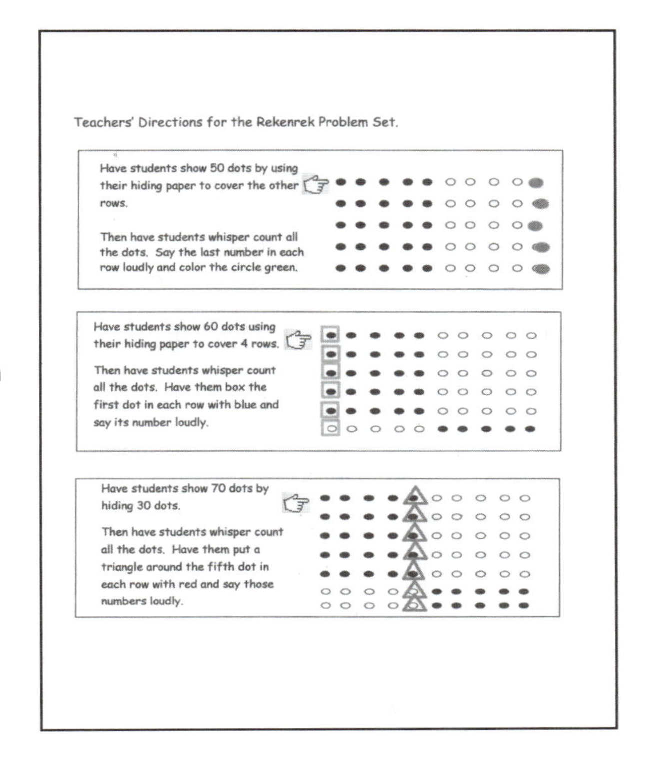

Teachers' Directions for the Rekenrek Problem Set.

Have students show 50 dots by using their hiding paper to cover the other rows.

Then have students whisper count all the dots. Say the last number in each row loudly and color the circle green.

Have students show 60 dots using their hiding paper to cover 4 rows.

Then have students whisper count all the dots. Have them box the first dot in each row with blue and say its number loudly.

Have students show 70 dots by hiding 30 dots.

Then have students whisper count all the dots. Have them put a triangle around the fifth dot in each row with red and say those numbers loudly.

Problem Set (7 minutes)

Students should do their personal best to complete the Problem Set within the allotted time.

Note: Do not show students the directions paper included in the materials for the lesson and pictured above to the right. It would give away the answers. The Rekenrek template is used by the students for the Problem Set and Homework.

Have students continue the patterns to the larger numbers, identifying the number for each triangle, box, and green circle.

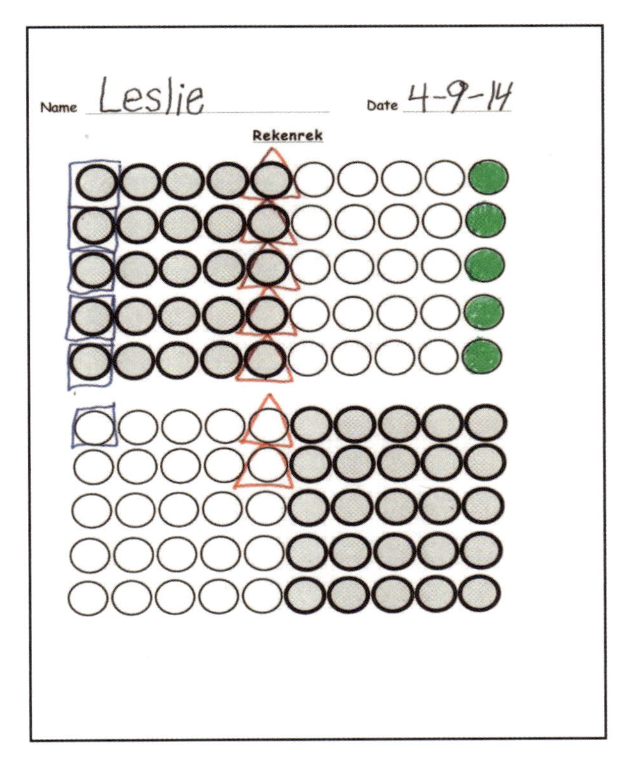

Student Debrief (8 minutes)

Lesson Objective: Count across tens by ones to 100 with and without objects.

The Student Debrief is intended to invite reflection and active processing of the total lesson experience.

Invite students to review their solutions for the Problem Set. They should check work by comparing answers with a partner before going over answers as a class, taking turns reading the numbers forward and back. Look for misconceptions or misunderstandings that can be addressed in the Student Debrief. Guide students in a conversation to debrief the Problem Set and process the lesson.

Any combination of the questions below may be used to lead the discussion.

- What is one more than 19? What is one more than 29?
- Count from 79 to 90. From 61 to 71.
- Who can come up and show one more than 30 on the Rekenrek? One more than 80?
- What did you get better at (learn, understand, do better) today?

Exit Ticket (3 minutes)

After the Student Debrief, instruct students to complete the Exit Ticket. A review of their work will help with assessing students' understanding of the concepts that were presented in today's lesson and planning more effectively for future lessons. The questions may be read aloud to the students.

Teachers' Directions for the Rekenrek Problem Set

Have students show 50 dots by using their hiding paper to cover the other rows.

Then, have students whisper count all the dots. Say the last number in each row loudly, and color the circle green.

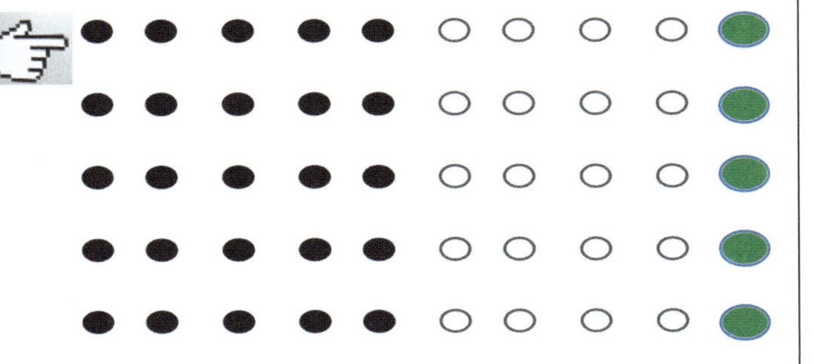

Have students show 60 dots using their hiding paper to cover 4 rows.

Then, have students whisper count all the dots. Have them box the first dot in each row with blue and say its number loudly.

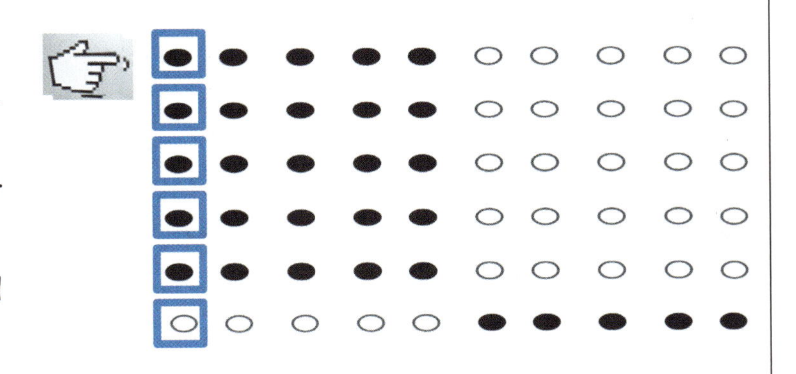

Have students show 70 dots by hiding 30 dots.

Then, have students whisper count all the dots. Have them put a triangle around the fifth dot in each row with red and say those numbers loudly.

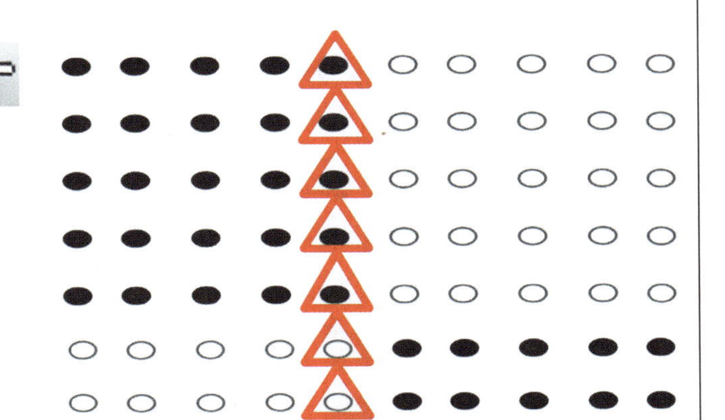

Name ______________________________ Date ______________

Rekenrek

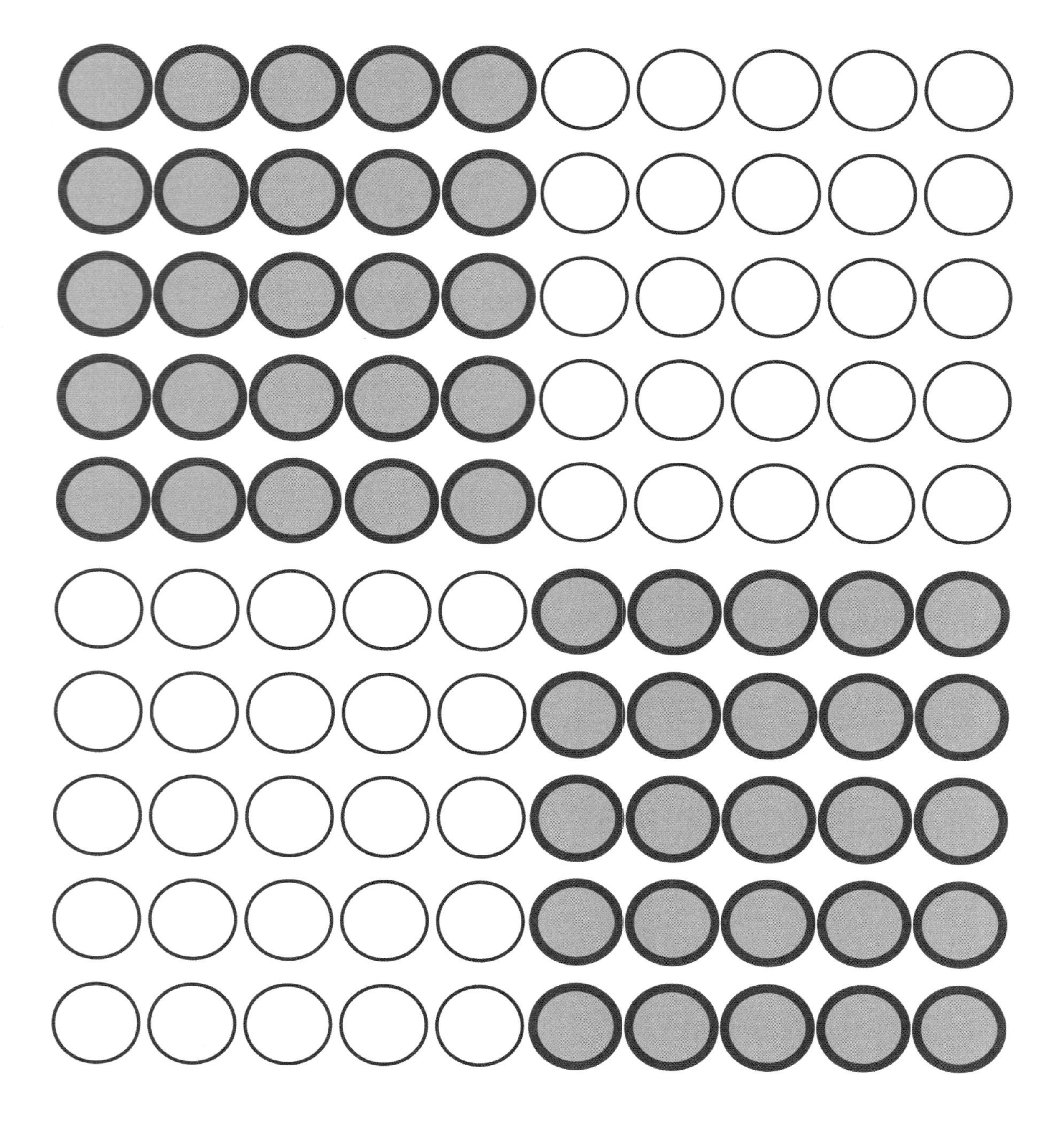

Name ______________________________ Date ______________

Touch and whisper count the circles by 1s to 100. Say the last number in each row loudly, and color the circle purple. Do your best. Your teacher may call time before you are finished.

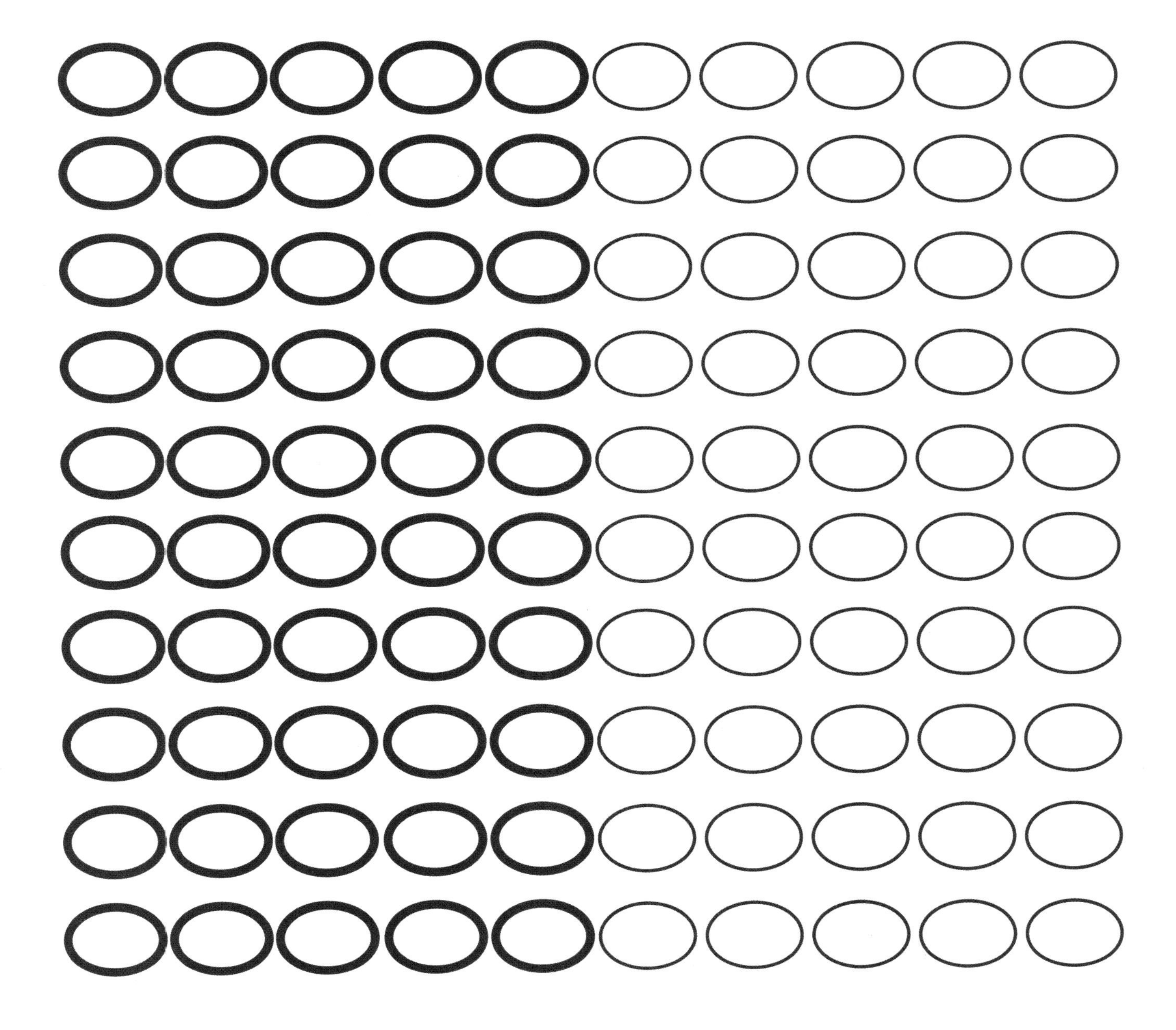

Directions for Rekenrek Homework

Use your Rekenrek (attached), hiding paper (an extra paper to hide some of the dots), and crayons to complete each step listed below. Read and complete the problems with the help of an adult.

Hide to show just 40 on your Rekenrek dot paper. Touch and count the circles until you say 28. Color 28 green.

- Touch and count each circle from 28 to 34.
- Color 34 (the 34th circle) with a red crayon.

Hide to show just 60 on your Rekenrek dot paper. Touch and count the circles until you say 45. Color 45 yellow.

- Touch and count each circle from 45 to 52.
- Color 52 with a blue crayon.

Hide to show just 90 on your Rekenrek dot paper. Touch and count the circles until you say 83. Color 83 purple.

- Touch and count down from 83 to 77.
- Color 77 with a red crayon.

Show 100.

- Touch and count, starting at 1.
- Say the last number in each row loudly. Color the circle black.

Name ______________________________ Date ______________

Rekenrek

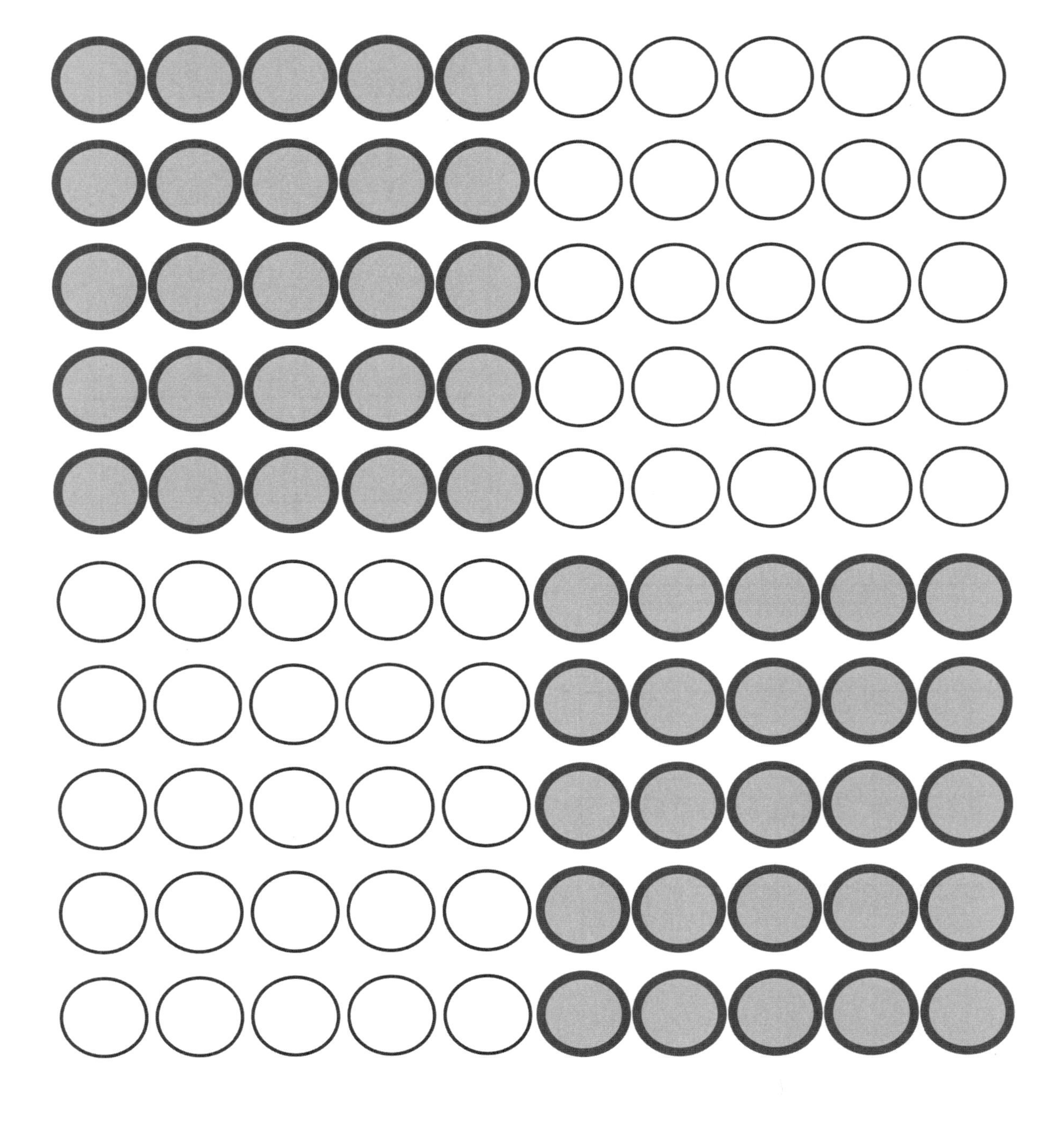

empty 10-frame cards

Lesson 19

Objective: Explore numbers on the Rekenrek. (Optional)

Suggested Lesson Structure

■ Application Problem	(7 minutes)
■ Fluency Practice	(10 minutes)
■ Concept Development	(25 minutes)
■ Student Debrief	(8 minutes)
Total Time	**(50 minutes)**

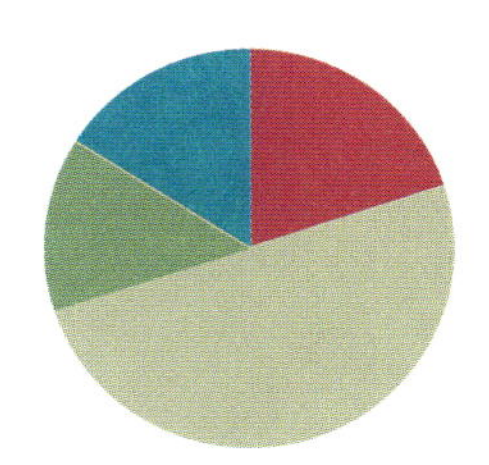

A NOTE ON STANDARDS ALIGNMENT:

In this lesson, students explore decomposing numbers to 100. To begin, they simply decompose numbers to 10 and see the relevance of that to teen numbers. Next, they sit with a partner and decompose numbers to 40 as tens and ones (**1.NBT.2**). They then represent numbers on two Rekenreks with a friend and realize that there is a teen number hiding inside this larger number by pulling apart their two Rekenreks! The exploration is meant to be playful, generating excitement about decomposing numbers.

Application Problem (7 minutes)

The light is out, and it's dark. Peter knows that he left 7 blue and green beads for his crafts on his desk. But he can't see how many are blue or how many are green in the dark! Draw a picture to show what the colors of his beads might be when he turns on the light.

When students have finished, have them compare their work with another student. Is their way of showing the beads the same? Why or why not? How is this problem like the problems in previous lessons with the flowers and the apples?

Note: In this lesson, the Application Problem precedes the Fluency Practice because the fluency activities lead directly into the counting of the lesson.

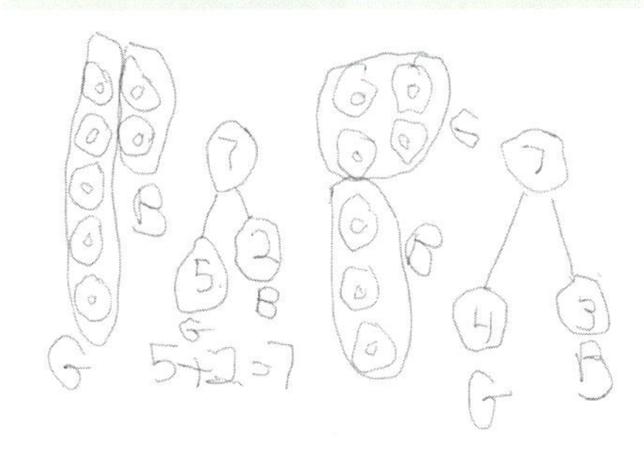

Fluency Practice (10 minutes)

- Number Bonds of 7 **K.OA.3** (3 minutes)
- Count to 100 by Ones **K.CC.1** (3 minutes)
- Hide Zero for Numbers to 100 **K.CC.1** (4 minutes)

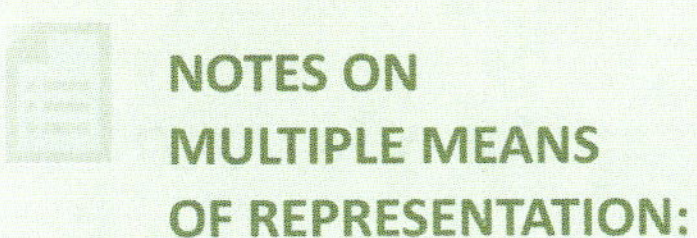

NOTES ON MULTIPLE MEANS OF REPRESENTATION:

Teach English language learners to ask questions such as, "How is your work different from mine?" in order to extend partner shares. Model asking different types of questions and have students practice until they feel confident to try with a partner.

Number Bonds of 7 (3 minutes)

Materials: (S) Personal Rekenrek (from Lesson 10)

Note: This fluency activity gives students an opportunity to develop increased familiarity with decompositions of seven and practice seeing part–whole relationships.

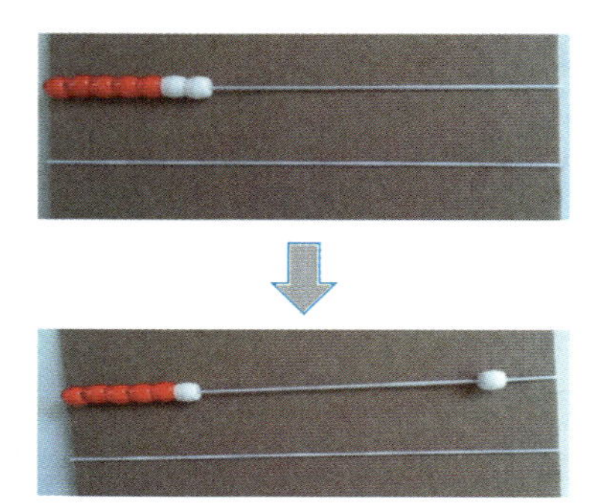

T: Show ten beads only. (Students push a row of ten behind.)

T: Hide 3 white beads behind your board.

T: The total number of beads you see is ...?

S: 7.

T: Push over 1 bead to the right to make 2 parts. Tell your partner the number bond. Part ______, part ______, total 7.

S: Part 6, part 1, total 7.

Continue one bead at a time stating the related bond. Keep the Rekenreks at 7 for the Concept Development component of this lesson.

Count to 100 by Ones (3 minutes)

Materials: (S) Rekenrek dot paper (Fluency Template 1)

Note: This activity targets the grade-level standard of counting to 100 by ones.

Students count to 100 (or as high as they can in 3 minutes) by touching the beads on the Rekenrek dot paper. Have them say "buzz" after the last number of each row.

Hide Zero for Numbers to 100 (4 minutes)

Materials: (T) Hide Zero cards: 1 Hide Zero 10 card (Lesson 6 Template 2) and 5-group cards 1–9 (Lesson 1 Fluency Template 2), Hide Zero cards 20–100 (Fluency Template 2)

Note: This activity connects identifying numbers the Say Ten way and students' growing understanding of place value, a Grade 1 standard they explore in today's Concept Development.

T: (Hold the 30 card and 7 card so they show 37.) Say the number.

S: 37.

T: Say the number the Say Ten way.

S: 3 tens 7.

T: (Break apart the cards into 30 and 7.)

Repeat the process for four or five other numbers between 20 and 100.

Concept Development (25 minutes)

Materials: (S) Personal Rekenrek (from Lesson 10)

Exploration 1

T: Show me 7 again on your Rekenrek.

T: Take the bottom ten beads of your Rekenrek out of hiding. Push them over to the left under your 7.

T: How many beads are on the left?

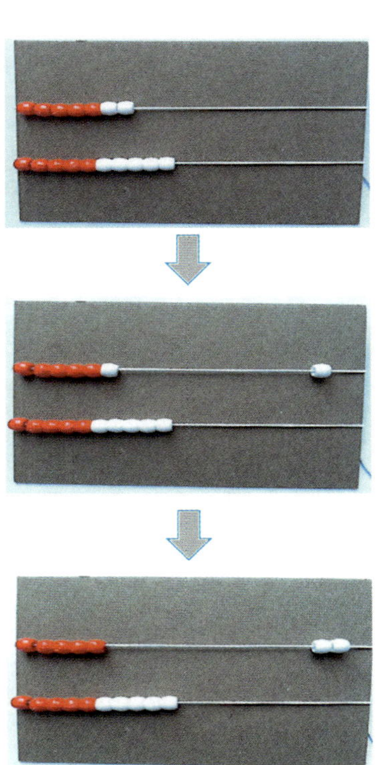

S: Seventeen.

T: Today, let's work the Say Ten way.

S: Ten 7.

T: Move 1 bead from your 7 over to the right like we did in our fluency activity.

T: Total 16. The two parts are ...?

S: 10 and 6.

T: Move another bead. Total 15. The parts are ...?

S: 10 and 5.

T: Move another!

S: Total 14. The parts are 10 and 4.

T: Keep going! (Give students a moment to work through the teen numbers.)

Exploration 2

T: Now, sit with a partner. Partner B, take all your beads out of hiding, and put your Rekenrek under your partner's. Partner A, show ten 7 again.

T: Using both Rekenreks, how many beads do you have on the left now? Tell me the Say Ten way.

S: 3 tens 7.

T: Move 1 bead from the 7 to the right. How many beads are on the left?

S: 3 tens 6.

T: Move a bead.

S: 3 tens 5.

T: Move a bead.

S: 3 tens 4.

Have students work with base numbers other than 7 within the twenties and thirties. Then, three students can sit together and work with numbers within the forties and sixties. The decomposition of the larger numbers is **1.NBT.2**, Understand Place Value. This playful work lets students get a sense of these important understandings that the decomposition of the numbers 1–9 and the teens give. Avoid part–whole language in Exploration 2 and in the Problem Set. Simply let students' natural knowledge see the connection between the *base number*, the teens, and the larger numbers.

Problem Set (7 minutes)

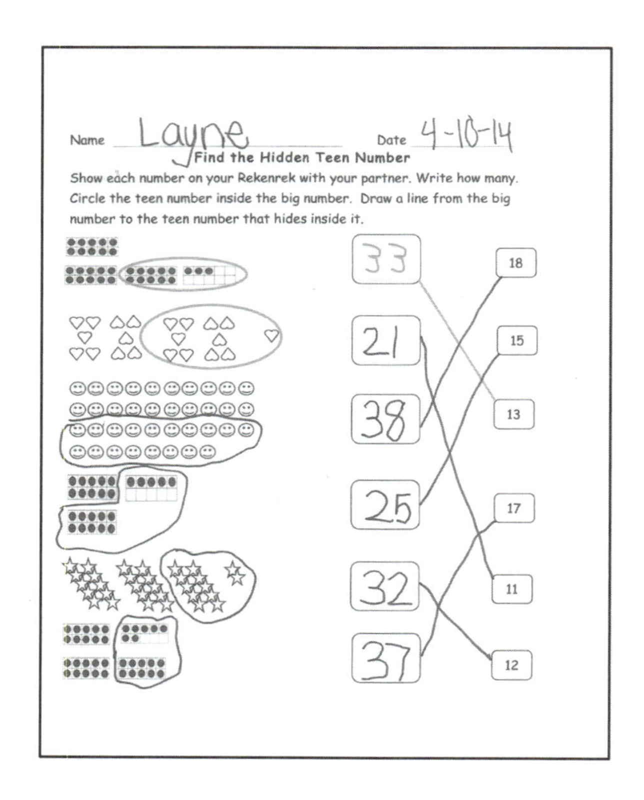

Before doing the Problem Set, guide students to see that they can also isolate the teen numbers when working with their partners.

T: Where is our teen number? It is in Partner A's Rekenrek! While the top row shows 7, the top Rekenrek shows 17. The teen numbers are hiding inside larger numbers just like 7 was inside 17. Pretend you are breaking the number, pulling hard at the Rekenreks to break that number apart.

This is of course beyond the grade-level standard (**1.NBT.2**), but it illustrates the idea that numbers can be broken into parts—the Rekenreks make it so easy to show that! Keep it playful.

Students should do their personal best to complete the Problem Set within the allotted time.

Student Debrief (8 minutes)

Lesson Objective: Explore numbers on the Rekenrek. (Optional)

The Student Debrief is intended to invite reflection and active processing of the total lesson experience.

Invite students to review their solutions for the Problem Set. They should check work by comparing answers with a partner before going over answers as a class, taking turns reading the numbers forward and back. Look for misconceptions or misunderstandings that can be addressed in the Student Debrief. Guide students in a conversation to debrief the Problem Set and process the lesson.

Guide students to see that their work with the first row of numbers on their Rekenreks, 1 to 10, helps them work with bigger numbers, just like when they count from 1 to 9 it helps them to count all the way to 100.

Any combination of the questions below may be used to lead the discussion.

- What did your teen number bonds help you see about the larger numbers?
- When you make a teen number in parts, what do you notice? Which is always larger, the parts or the total (or whole)?
- What happens if the top row on your Rekenrek is a part? What is the other part?
- What else could be a part of a larger number?
- When you circled teen numbers on the Problem Set, you were finding a part. What part did you find in the first problem?
- How does finding parts help you to understand large numbers better?

Exit Ticket (3 minutes)

After the Student Debrief, instruct students to complete the Exit Ticket. A review of their work will help with assessing students' understanding of the concepts that were presented in today's lesson and planning more effectively for future lessons. The questions may be read aloud to the students.

Name ______________________________ Date ______________

Find the Hidden Teen Number

Show each number on your Rekenrek with your partner. Write how many. Circle the teen number inside the big number. Draw a line from the big number to the teen number that hides inside it.

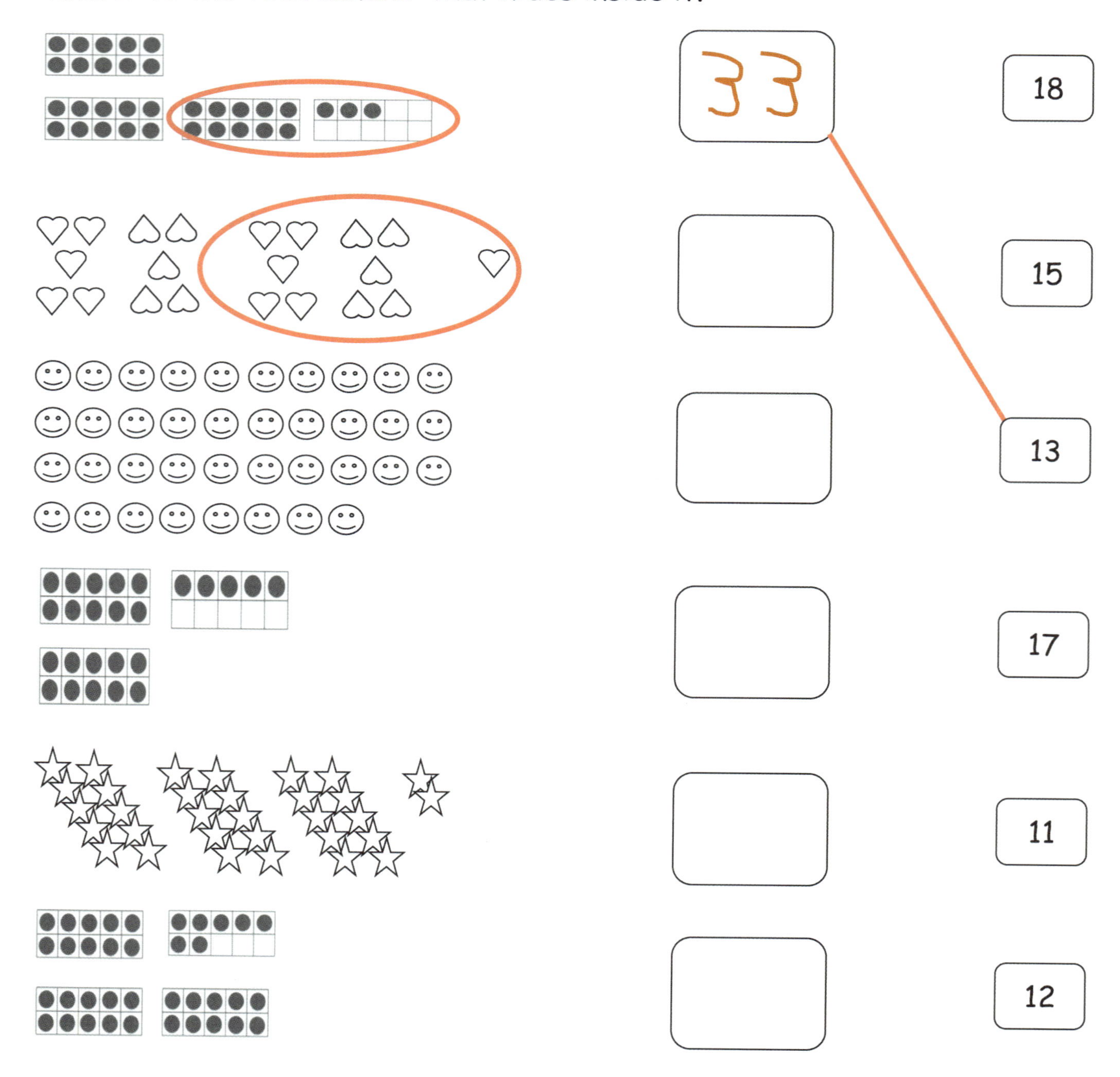

Name ______________________________ Date ____________

Show the number on your Rekenrek with your partner. In the box, write the number that tells how many objects there are. Circle the teen number you see. Write the teen number in the other box.

Name ______________________________ Date ______________

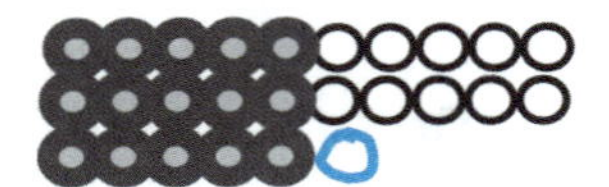

Write the number you see. Now, draw one more, then write the new number.

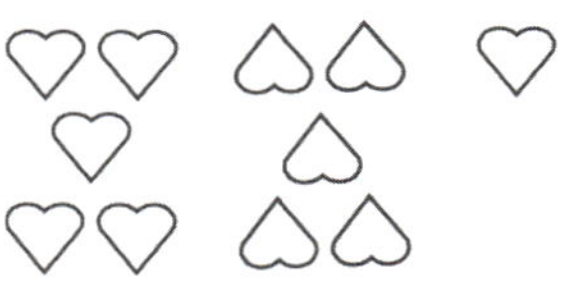
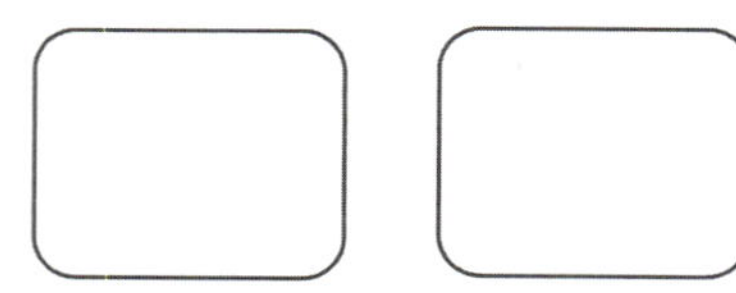

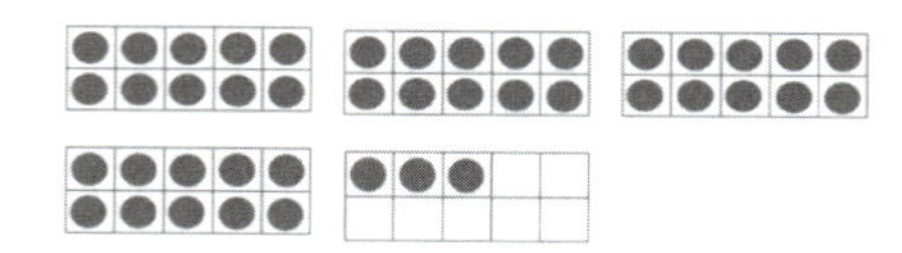

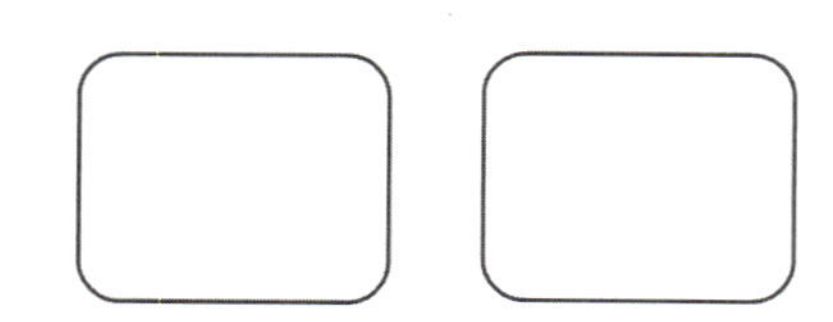

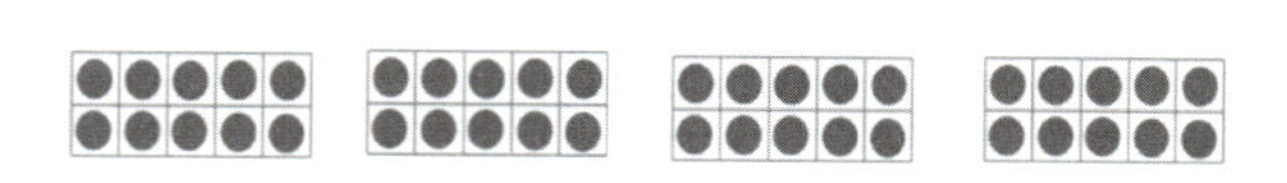
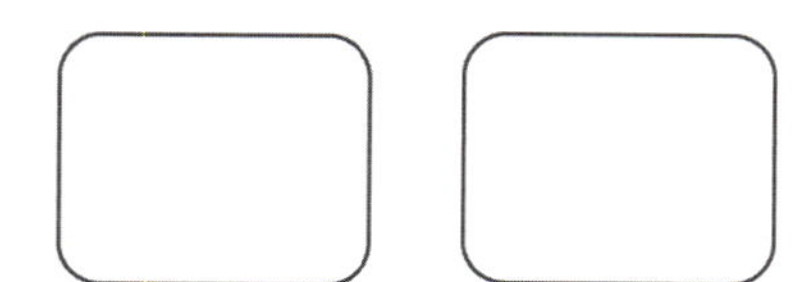

Rekenrek-

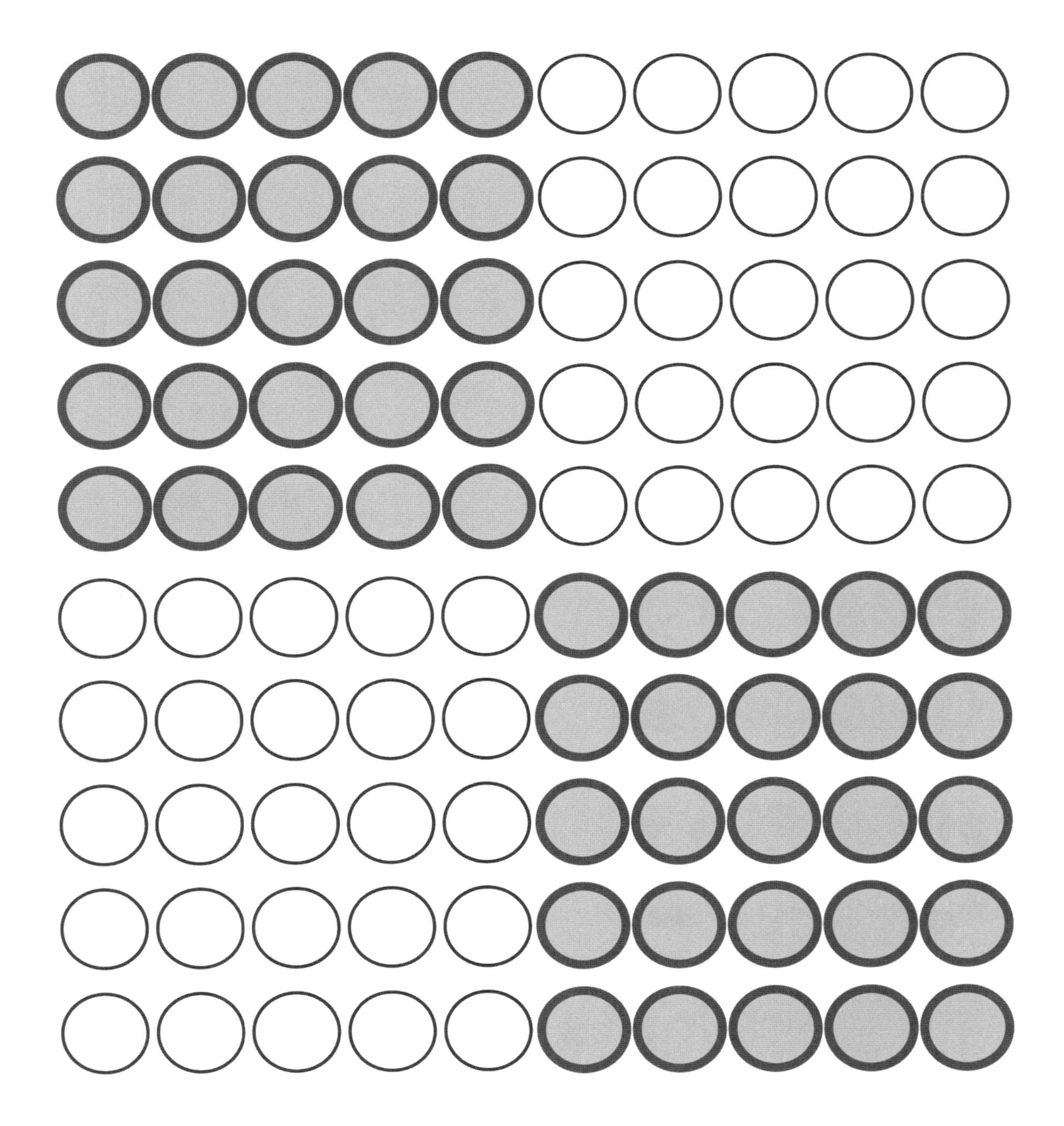

Rekenrek dot paper

2 0	3 0
4 0	5 0
6 0	7 0
8 0	9 0

Hide Zero cards 20–100

1	0	0

Hide Zero cards 20–100

Topic E

Represent and Apply Compositions and Decompositions of Teen Numbers

K.CC.5, K.NBT.1, K.CC.1, K.CC.2, K.CC.3, K.CC.4c, K.CC.6, 1.OA.8, 1.NBT.3

Focus Standards:		K.CC.5	Count to answer "how many?" questions about as many as 20 things arranged in a line, a rectangular array, or a circle, or as many as 10 things in a scattered configuration; given a number from 1–20, count out that many objects.
		K.NBT.1	Compose and decompose numbers from 11 to 19 into ten ones and some further ones, e.g., by using objects or drawings, and record each composition or decomposition by a drawing or equation (e.g., 18 = 10 + 8); understand that these numbers are composed of ten ones and one, two, three, four, five, six, seven, eight, or nine ones.
Instructional Days:		5	
Coherence	**-Links from:**	GPK–M5	Addition and Subtraction Stories and Counting to 20
	-Links to:	G1–M2	Introduction to Place Value Through Addition and Subtraction Within 20

Topic E's Lesson 20 begins as students represent teen number decompositions and compositions by writing addition sentences. In Lesson 21, students make bonds with materials and hide one of the parts for their partners, who must figure out what the hidden part is. The number bond with a hidden part is represented by the teacher as an addition equation with a missing addend—the hidden part (aligns to **1.OA.8**). In Lesson 22, students compare teen numbers by counting and comparing the extra ones. For example, students decompose 12 into 10 and 2, and 16 into 10 and 6. They compare 2 ones and 6 ones to see that 16 is more than 12 using the structure of the 10 ones (MP.7). This is an application of the Kindergarten comparison standards (**K.CC.6**, **K.CC.7)**, which move into the Grade 1 comparison standard (**1.NBT.3**).

In Lesson 23, students reason about situations to determine if they are decomposing a teen number as 10 ones and some ones, or composing 10 ones and some ones to *find* a teen number. They analyze the number sentences that best represent each situation (**K.NBT.1**). Throughout the lesson, students draw the number of objects presented in the situation (**K.CC.5**).

The module closes with an exploration in which students count teen quantities and represent them in various ways as the teacher gives the prompt, "Open your mystery bag. Show the number of objects in your bag in different ways using the materials you choose." This exercise also serves as a culminating assessment, allowing the student to demonstrate skill and understanding in applying all the learning gained throughout the module.

A Teaching Sequence Toward Mastery of Representing and Applying Compositions and Decompositions of Teen Numbers

Objective 1: **Represent teen number compositions and decompositions as addition sentences.**
(Lesson 20)

Objective 2: **Represent teen number decompositions as 10 ones and some ones, and find a hidden part.**
(Lesson 21)

Objective 3: **Decompose teen numbers as 10 ones and some ones; compare *some ones* to compare the teen numbers.**
(Lesson 22)

Objective 4: **Reason about and represent situations, decomposing teen numbers into 10 ones and some ones and composing 10 ones and some ones into a teen number.**
(Lesson 23)

Objective 5: **Culminating Task—Represent teen number decompositions in various ways.**
(Lesson 24)

Lesson 20

Objective: Represent teen number compositions and decompositions as addition sentences.

Suggested Lesson Structure

Fluency Practice	(12 minutes)
Application Problem	(7 minutes)
Concept Development	(24 minutes)
Student Debrief	(7 minutes)
Total Time	**(50 minutes)**

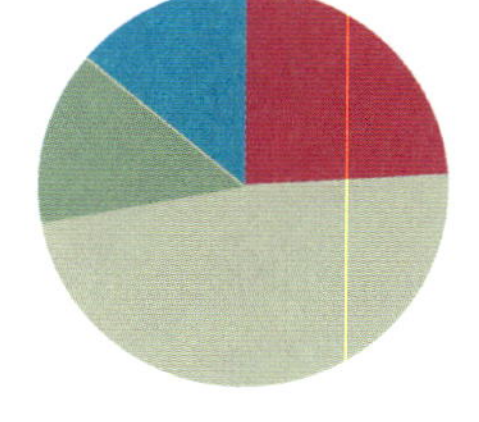

Fluency Practice (12 minutes)

- Dot Cards of Seven **K.CC.5, K.CC.2** (4 minutes)
- Count Crossing Tens **K.CC.1** (4 minutes)
- Group Tens and Ones **K.CC.4** (4 minutes)

Dot Cards of Seven (4 minutes)

Materials: (T) Dot cards of 7 (Lesson 5 Fluency Template 1)

Note: The varied configurations of dots used in this fluency activity allow students to see different ways to decompose 7, strengthening their understanding of part–whole relationships.

T: (Show 7 dots.) How many do you see? (Give students time to count.)
S: 7.
T: How can you see 7 in two parts?
S: (Coming up to the card.) 5 here and 2 here.
T: Say the number sentence.
S: 5 and 2 makes 7.
T: Who sees 7 in two different parts?
S: (Coming up to the card.) I see 3 here and 4 here.
T: Say the number sentence.
S: 3 and 4 makes 7.

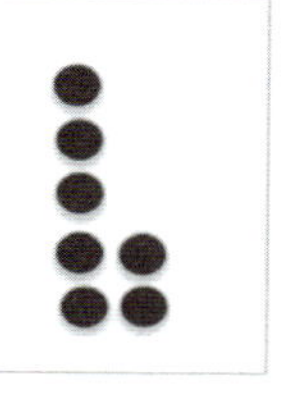

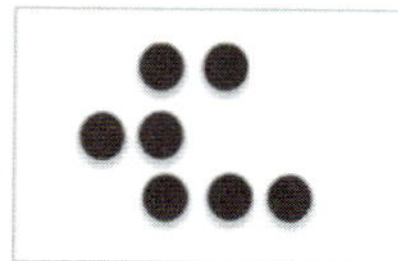

Continue with other dot cards of 7.

Count Crossing Tens (4 minutes)

Materials: (S) Personal Rekenrek (Lesson 10)

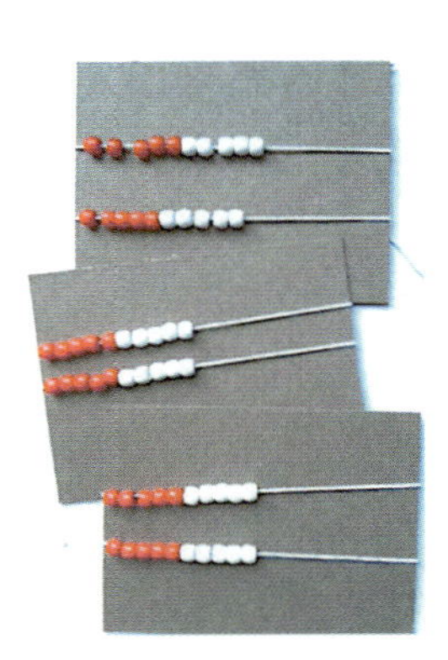

Note: For this activity, it may be preferable to combine six elastics of beads onto one card. However, it may help students develop number sense to use their three individual cards as described below so that students reference where they left off very clearly when counting to 40.

T: Today, we're going to work in groups of 3. Put your personal Rekenreks together, and count your beads. Say "buzz" after you finish a row. Partner A moves the beads of the first Rekenrek, Partner B moves the beads of the second, and Partner C moves the beads of the third.

T: If you finish early, count again. This time, after the color changes, say "buzz."

Group Tens and Ones (4 minutes)

Materials: (T) Prepared images of arrays and circular configurations, large 5-group cards (Lesson 1 Fluency Template 1)

Note: This activity advances the skill of grouping tens and ones by moving on to visual recognition. Counting only by sight pushes students to work efficiently because it is easier to keep track of groups than individual objects.

T: (Project a circular configuration of 12 objects.) Say the number of objects that you see.

S: (Pause while they count.) 12.

T: Say the number the Say Ten way.

S: Ten 2.

Repeat the process for four or five other numbers between 10 and 100, mixing arrays, circular configurations, and 5-group cards.

Although students cannot touch the images, encourage them to track their grouping with hands from afar. They might hold up a finger to mark the starting point in a circular configuration or use an outstretched hand to visually separate a group of ten from remaining stars in an array.

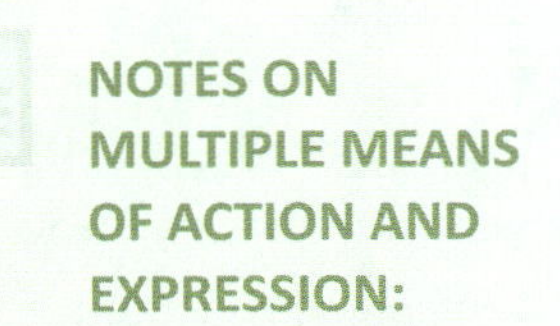

Increase the learning pace for students who are working above grade level by providing extensions to the Application Problem:

- What if each student was given 16 colored pencils and 4 regular pencils? How many pencils are there altogether? *Hint: Use your first drawing to help you solve.*
- How many pencils would two students have altogether? *Hint: Use your first two drawings to help you solve.*

Application Problem (7 minutes)

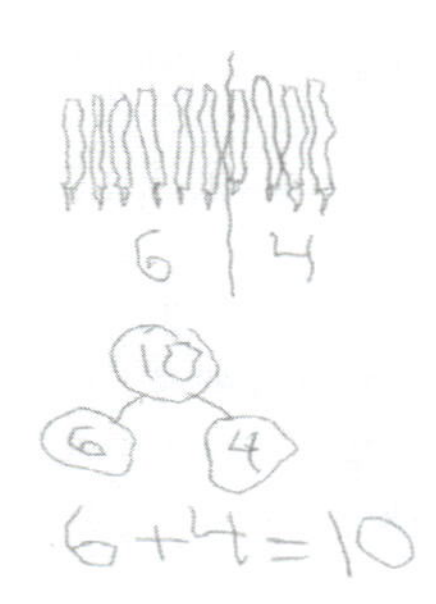

Each student was given 6 colored pencils and 4 regular pencils. How many pencils did each student get? Draw a picture and a number bond, and then write a number sentence.

Concept Development (24 minutes)

Materials: (S) Bag of twenty 2-color beans, number bond (Lesson 7 Template) within a personal white board

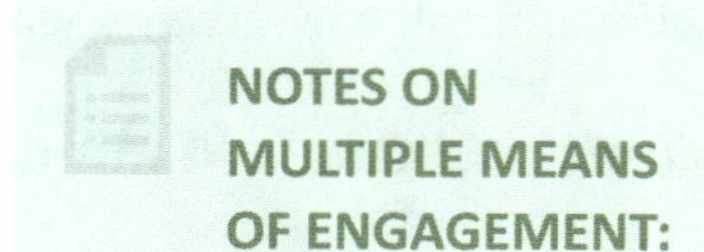

Ask English language learners who are developing language skills to write their answers on their personal white boards, as well as say it aloud. After writing and saying the answer orally, have all students repeat the number sentence aloud again.

T: Put 10 red beans in one part of the number bond. Put 3 white beans in the other part.

T: What is 10 ones and 3 ones?

S: 13 ones.

T: Say the number the Say Ten way.

S: Ten 3.

T: Now, count 13 beans into the place where we show the total or whole amount.

T: So, we have 13 in two parts. What are the parts?

S: 10 and 3.

T: Talk to your partner. When we solved our story problem today, we had two parts. What is another way you already know to show a number in two parts?

S: We can show a number in two parts by making piles of things, like 10 things and 3 things. → We can show the number with a number bond. → We can make a picture. → We can show it with our Hide Zero cards. → We can show it on the Rekenrek. → We can show it with a plus sign.

T: Lots of good ideas. We can show the same idea in so many ways. When we are thinking about 13, what do you think is the clearest way to show the two parts of 10 and 3. Talk to your partner.

S: The number bond. It's so easy to see. → I like to see how big the number is, so counters are my favorite. → I feel big girls and boys do addition, so that's how I want to show it.

T: Each way we show a number in two parts helps us understand our number better. Addition is another way to do that.

T: (Write 10 + 3 = _____ on the board.)

T: What is 10 + 3? Give me a complete number sentence.

S: 10 + 3 = 13.

T: (Write 13 on the board to complete the equation.) Look at your number bond. How many beans do you have in the whole amount?

S: 13.

T: (Write 13 = ______ + ______ on the board.)

T: How many beans are in this part? Let's count.

S: 1, 2, 3, 4, 5, 6, 7, 8, 9, 10.
T: How many beans are in this part?
S: 3.
T: Look at the parts. Complete this number sentence. (Point to 13 = _____ + _____.)
S: 13 = 10 + 3.
T: We started with the whole amount with our beans, so our number sentence also starts with the whole amount.
T: Clear your board. Show 10 red beans and 5 white beans in the two parts.
T: Now, count to find out how many beans you will put to show the total. It needs to match the amount in the parts.
S: (After counting.) 15.
T: Count that many beans into the place where you put your total.
T: (After counting.) What is another way to show the two parts and the total?
S: 10 + 5 = 15.
T: (Write 10 + 5 = 15 on the board.)
T: Do you have the same number of beans in the parts as you have in the place for the total?
S: Yes!
T: When 15 is split into two parts, is it the same as 10 and 5? Then, your number bond is true!
T: Clear your board. This time, use your marker to write 19 where we show the whole. Let's put this number in two parts.
T: Show 10 red beans as one part. (Pause while students place the beans.)
T: Count out the beans you need to put in the other part to get to 19.
S: (After counting.) 9.
T: What is one number sentence that tells about this number bond?
S: 10 + 9 = 19.
T: This time, start with the total, so we really feel that big number splitting into two parts.
S: 19 = 10 + 9.

Continue in this manner with students creating and talking about other teen number bonds and their matching addition sentences

Problem Set (7 minutes)

Students should do their personal best to complete the Problem Set within the allotted time.

Note: Have students complete the bonds and number sentences. Give them access to materials and Hide Zero cards as they do so.

Student Debrief (7 minutes)

Lesson Objective: Represent teen number compositions and decompositions as addition sentences.

Invite students to review their solutions for the Problem Set. They should check work by comparing answers with a partner before going over answers as a class. Look for misconceptions or misunderstandings that can be addressed in the Student Debrief. Guide students in a conversation to debrief the Problem Set and process the lesson.

Any combination of the questions below may be used to lead the discussion.

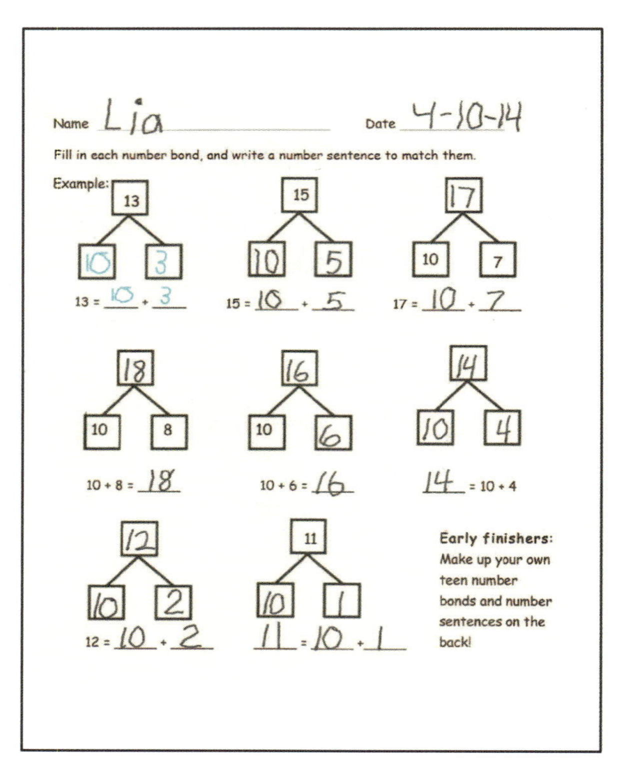
Name Lia Date 4-10-14

Fill in each number bond, and write a number sentence to match them.

Example: 13 → 10, 3 — 13 = 10 + 3

15 → 10, 5 — 15 = 10 + 5

17 → 10, 7 — 17 = 10 + 7

18 → 10, 8 — 10 + 8 = 18

16 → 10, 6 — 10 + 6 = 16

14 → 10, 4 — 14 = 10 + 4

12 → 10, 2 — 12 = 10 + 2

11 → 10, 1 — 11 = 10 + 1

Early finishers: Make up your own teen number bonds and number sentences on the back!

- In a number bond, which number is larger—the whole or a part?
- Explain how the teen numbers are 10 ones and some more ones.
- Look at each number bond as I say the whole. You read the number the Say Ten way; for example, I say 13, and you say ten 3.
- Mental math: I say 16; you say 10 + 6. I say 17; you say ...? I say 19; you say ...?
- Show a row of ten on the Rekenrek, and then slide beads to show the teen numbers. Say the numbers the regular and Say Ten way.
- What are we doing with the parts when we add? Are we putting them together or taking them apart?

Exit Ticket (3 minutes)

After the Student Debrief, instruct students to complete the Exit Ticket. A review of their work will help with assessing students' understanding of the concepts that were presented in today's lesson and planning more effectively for future lessons. The questions may be read aloud to the students.

Name ______________________________ Date ______________

Fill in each number bond, and write a number sentence to match.

Example:

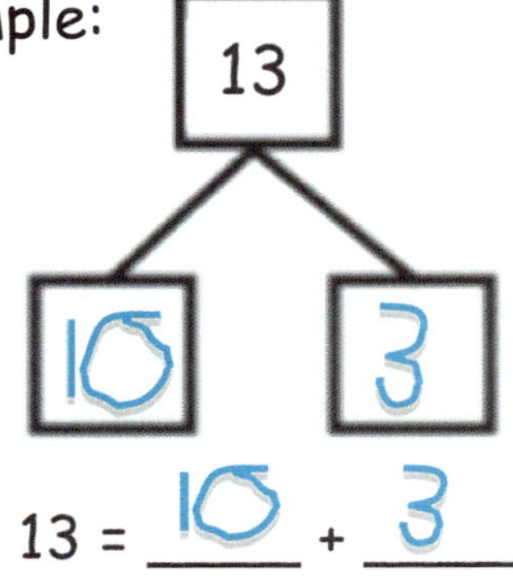

13 = 10 + 3

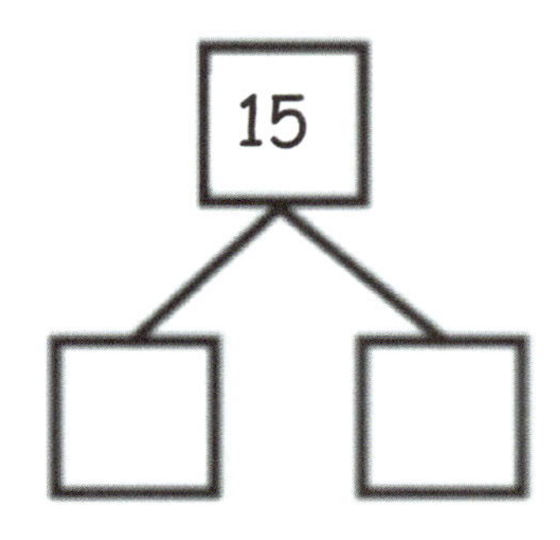

15 = _____ + _____

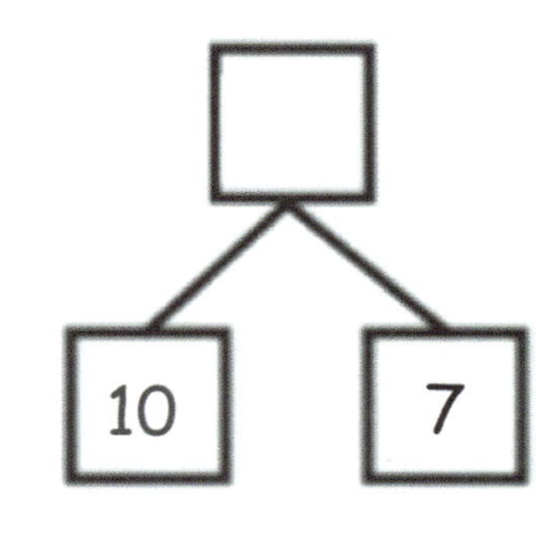

17 = _____ + _____

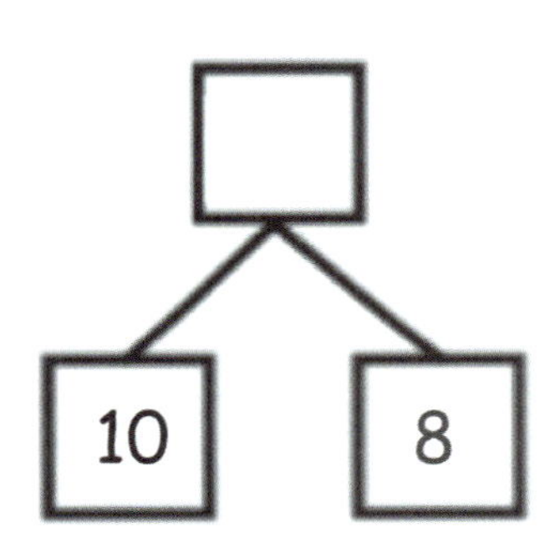

10 + 8 = _____

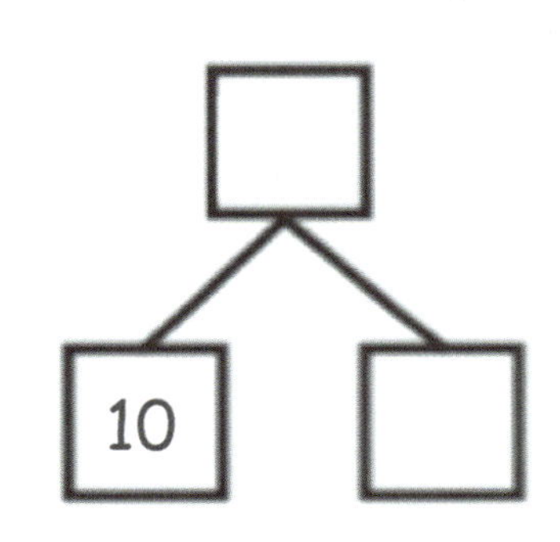

10 + 6 = _____

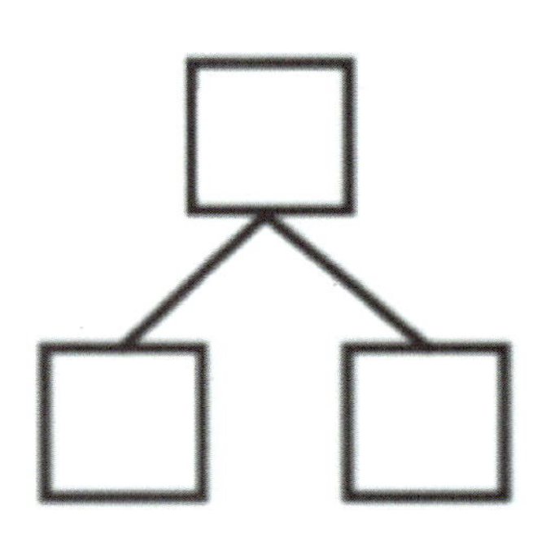

_____ = 10 + 4

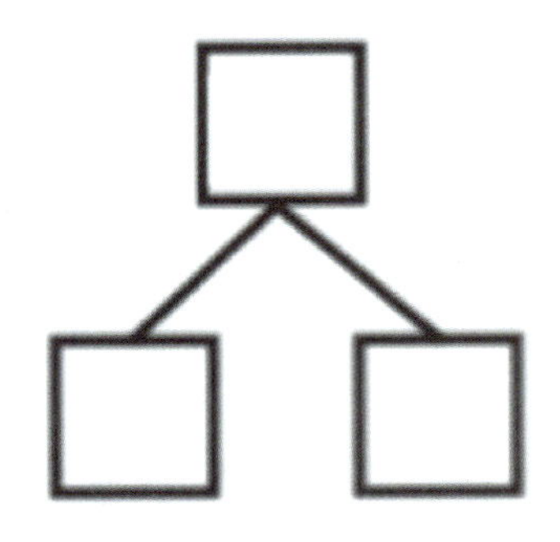

12 = _____ + _____

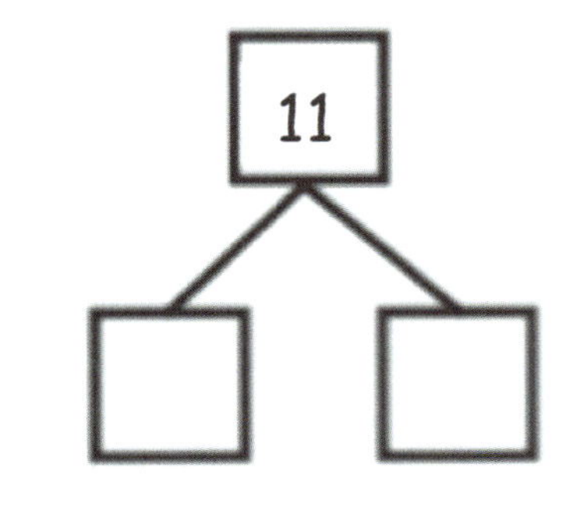

_____ = _____ + _____

Early finishers: Make up your own teen number bonds and number sentences on the back!

Name ______________________________ Date ______________

The first number is the whole. Circle its parts.

5	1	(2)	(3)

12	10	6	2
11	1	10	8
14	4	2	10
18	1	10	8
10	10	1	0
20	10	2	10

Name ______________________________ Date ______________

Draw stars to show the number as a number bond of 10 ones and some ones. Show each example as two addition sentences of 10 ones and some ones.

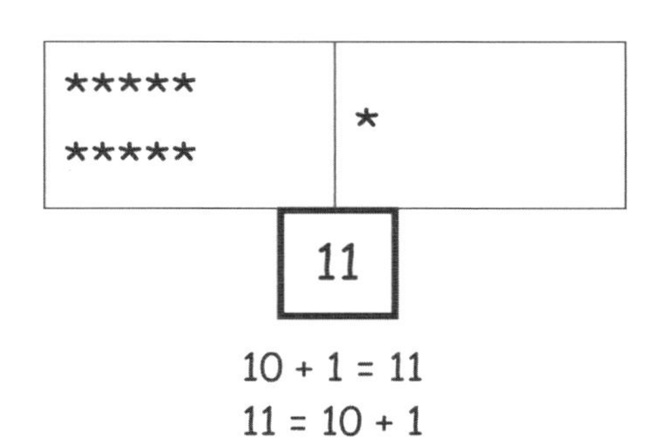

* * * * * * * * * *	

15

	* * * * * * *

17

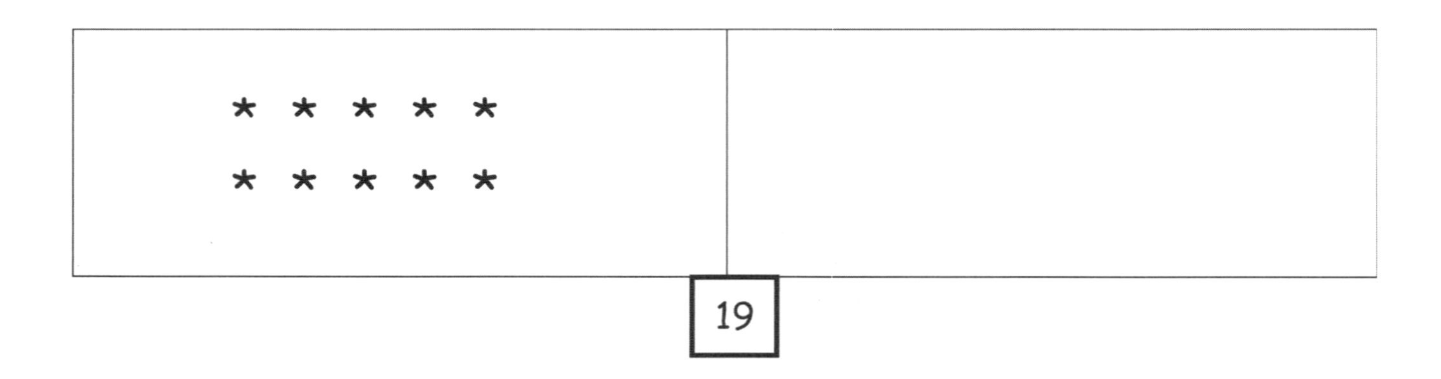
19

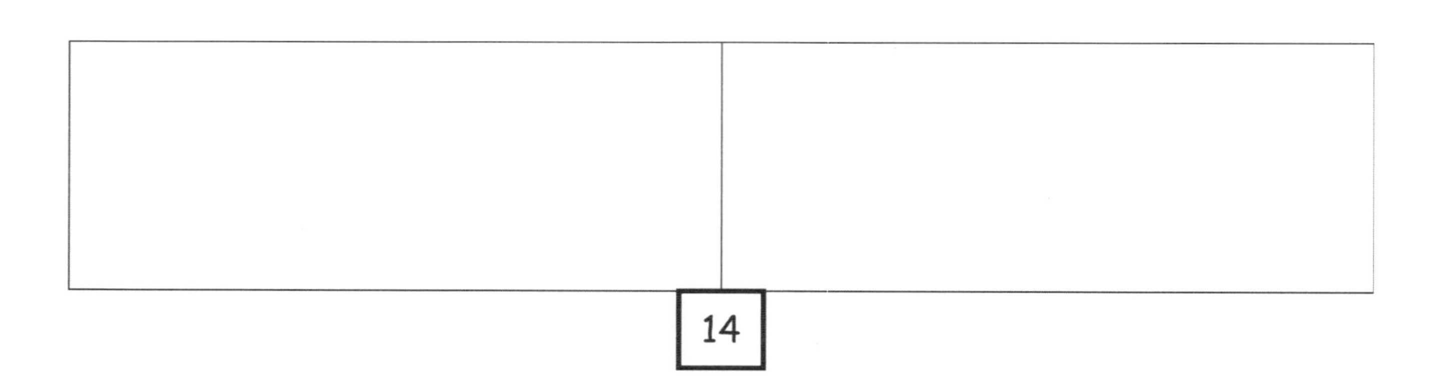
14

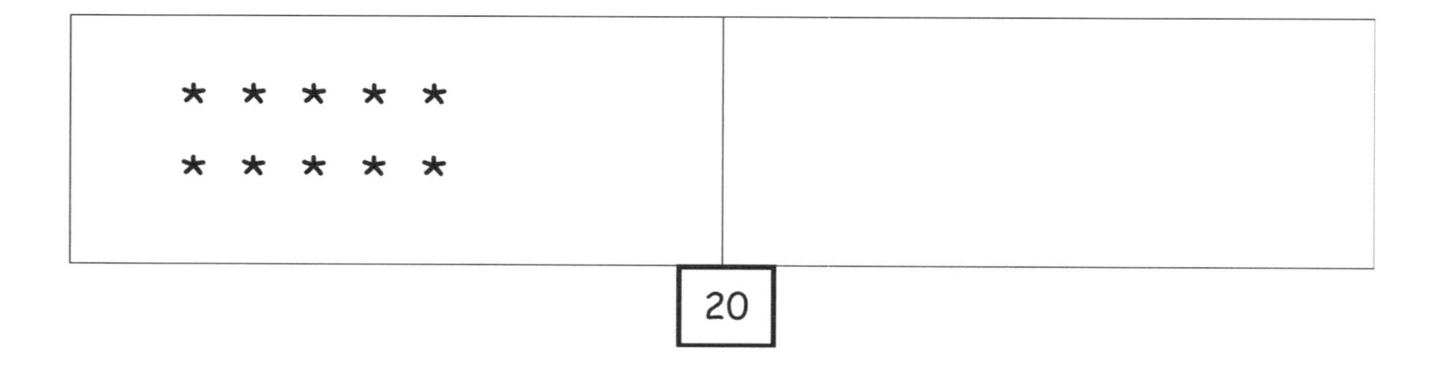
20

Lesson 21

Objective: Represent teen number decompositions as 10 ones and some ones, and find a hidden part.

Suggested Lesson Structure

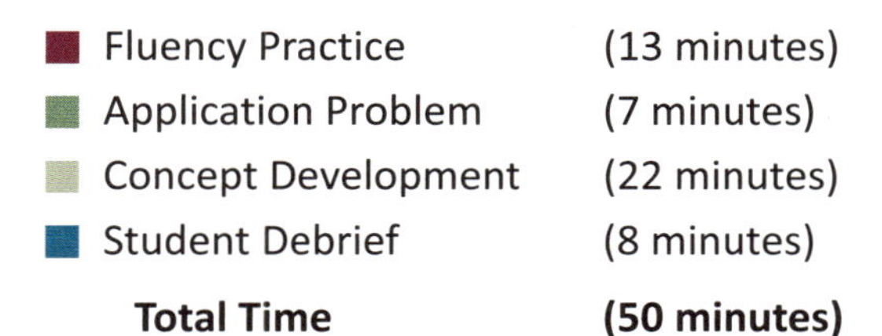

Fluency Practice	(13 minutes)
Application Problem	(7 minutes)
Concept Development	(22 minutes)
Student Debrief	(8 minutes)
Total Time	**(50 minutes)**

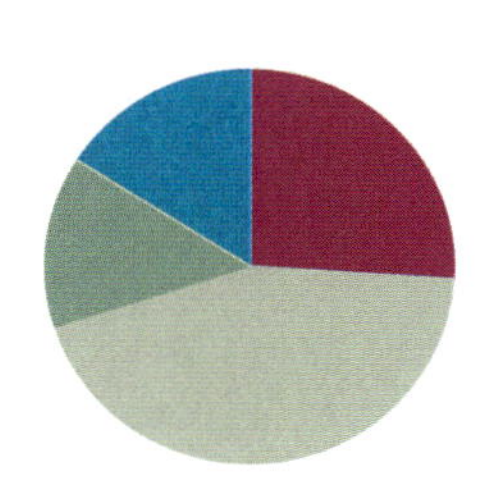

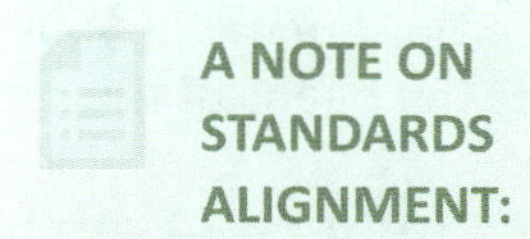

A NOTE ON STANDARDS ALIGNMENT:

In this lesson, students decompose teen numbers into two parts with blocks and hide one of the parts. After guessing what the hidden part is, they then see a number sentence with a *hidden part* such as 12 = 10 + _____. This bridges to Grade 1 content (**1.OA.8**).

Fluency Practice (13 minutes)

- Number Bonds of Seven **K.CC.2** (4 minutes)
- Four Rekenreks **K.CC.1** (5 minutes)
- Count Teen Numbers **K.CC.5** (4 minutes)

Number Bonds of Seven (4 minutes)

Materials: (T) Dot cards of 7 (Lesson 5 Fluency Template 1)

Note: This fluency activity gives students an opportunity to develop increased familiarity with compositions of seven and practice seeing part–whole relationships.

Show a dot card, and indicate 6 and 1 as parts.

T: Say the larger part. (Give students time to count.)
S: 6.
T: Say the smaller part.
S: 1.
T: What is the total number of dots? (Give time to count.)
S: 7.
T: Say the number sentence.
S: 6 and 1 makes 7.
T: (Turn the card around to get 1 and 6.)

Continue with 5 and 2, 7 and 0, 4 and 3.

Four Rekenreks (5 minutes)

Materials: (S) Personal Rekenrek (Lesson 10)

Note: Saying "bop" after each row of 10 provides a pause in counting, both reinforcing the start of a new row of ten and interrupting the count sequence, which helps students when they transition from counting all to count on in Grade 1.

T: Sit in groups of 4. Put your Rekenreks together. Partner A moves the beads of the first row. Partner B moves the beads of the second row, etc. After each number that ends a row, say "bop."

Count Teen Numbers (4 minutes)

Note: Alternating between Say Ten counting and regular counting challenges students to think carefully about each number because they cannot rely on the rote count sequence. By doing so, this reinforces teen numbers as 10 ones and some additional ones. (For example, students must know that 12 comprises 10 ones and 2 ones to recognize that ten 3 would come next if counting forward.)

T: Count from 11 to 20 the Say Ten way.

S: Ten 1, ten 2, ten 3, ten 4, ten 5, ten 6, ten 7, ten 8, ten 9, 2 tens.

T: Count back from 20 to 11 the Say Ten way.

S: 2 tens, ten 9, ten 8, ten 7, ten 6, ten 5, ten 4, ten 3, ten 2, ten 1.

T: Count from 11 to 20 the regular way.

S: 11, 12, 13, 14, 15, 16, 17, 18, 19, 20.

T: Count back from 20 to 11 the regular way.

S: 20, 19, 18, 17, 16, 15, 14, 13, 12, 11.

T: Now, I want you to change the way you count each time. We'll say the first number the Say Ten way. Then, we'll say the next number the regular way. Listen to my example. Ten 1, 12, ten 3, 14, ten 5, 16. Now, it's your turn.

S: Ten 1, 12, ten 3, 14, ten 5, 16, ten 7, 18, ten 9, 20.

T: Count back from 20 to 11, starting with the Say Ten way.

S: 2 tens, 19, ten 8, 17, ten 6, 15, ten 4, 13, ten 2, 11.

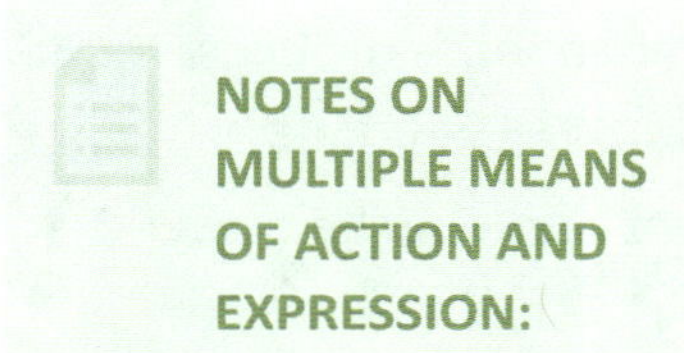

Differentiate the Application Problem for students who are working below grade level by asking them to put the puppies (counters) in a 10-frame.

Ask students who are working above grade level to double the number of puppies in the cage using two 10-frames to show 10 and some more.

Application Problem (7 minutes)

Peter saw 8 puppies at the pet store in a cozy cage. While he was watching them, 2 hid in a little box. How many puppies could Peter see then? Draw a picture, and write a number bond and number sentence to match the story.

Note: This Application Problem is an example of a *take from with result unknown* problem type, which students should be able to solve using objects or manipulatives by the end of Kindergarten.

Concept Development (22 minutes)

Materials: (S) 40 centimeter cubes and number bond (Lesson 7 Template) within a personal white board (per pair)

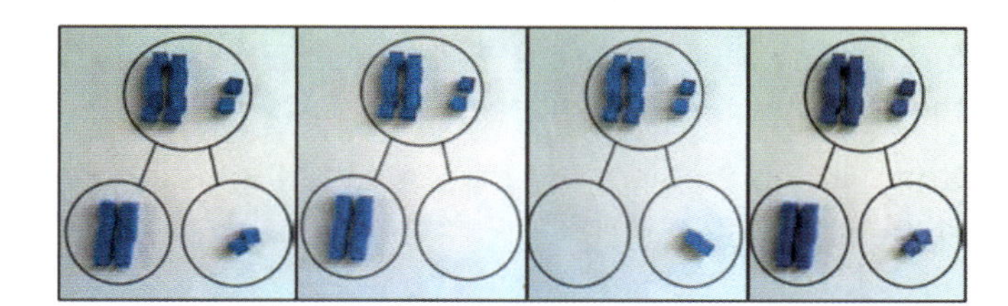

T: Count out 12 cubes, and put them in the place where we show the whole on the number bond.
T: Group 10 ones within that place.
T: What are the parts of 12 you see?
S: 10 and 2.
T: Count out cubes to fill in parts so that the total and the parts are equal.
S: (Students do so.)
T: Fill in this number sentence with me. (On the board, write 12 = ____+ ____.)
S: 12 = 10 + 2.
T: Say the number the Say Ten way.
S: Ten 2.
T: Close your eyes. (Remove the 2 cubes.) Open your eyes. What part is hiding?
S: 2.
T: Fill in this number sentence with me. (Write 12 = 10 + ____ on the board.)
S: 12 = 10 + 2. (Put the cubes back as they say the statement.)
T: Close your eyes. (Remove the 10 cubes.) Open your eyes. What part is hiding?
S: 10 ones!
T: Fill in this number sentence with me. (Write 12 = ____ + 2 on the board.)
S: 12 = 10 + 2.

NOTES ON MULTIPLE MEANS OF REPRESENTATION:

The teen numbers represent a particular challenge for English language learners because the difference between *thirteen* and *thirty* is not easy to hear. Scaffold the lesson for students by providing them with visuals of the teen numbers in both written form and the numeral form. Students also need practice hearing (stress the *teen* of the number by putting a finger near the mouth) and saying *thirteen* and *fourteen* so that they can hear the stress on the *teen* part of the number.

Continue in this manner with other teen numbers. Have students then work in pairs to play Hide and Say the Hidden Part.

- Partner A builds a teen number in the place for the total or whole.
- Partner B models the number as two parts.

- Partner A closes her eyes while Partner B hides one part.
- Partner A writes the complete number sentence (e.g., 14 = 10 + 4). Switch roles.

T: We had a hidden part like in our story problem of the puppies. We didn't know the part that Peter could still see in the cozy cage after the two puppies hid inside the box!

Problem Set (7 minutes)

Students should do their personal best to complete the Problem Set within the allotted time.

Be sure that students have access to materials such as counters, Hide Zero cards, and personal white boards for drawing while using the Problem Set. Encourage them to think about and demonstrate the many ways they can show teen numbers in two parts.

Note: In this Problem Set, students use the centimeter cubes and decompose teen numbers into two parts and then write corresponding equations. 12 = 10 + ______. This bridges to Grade 1 content (**1.OA.8**).

Student Debrief (8 minutes)

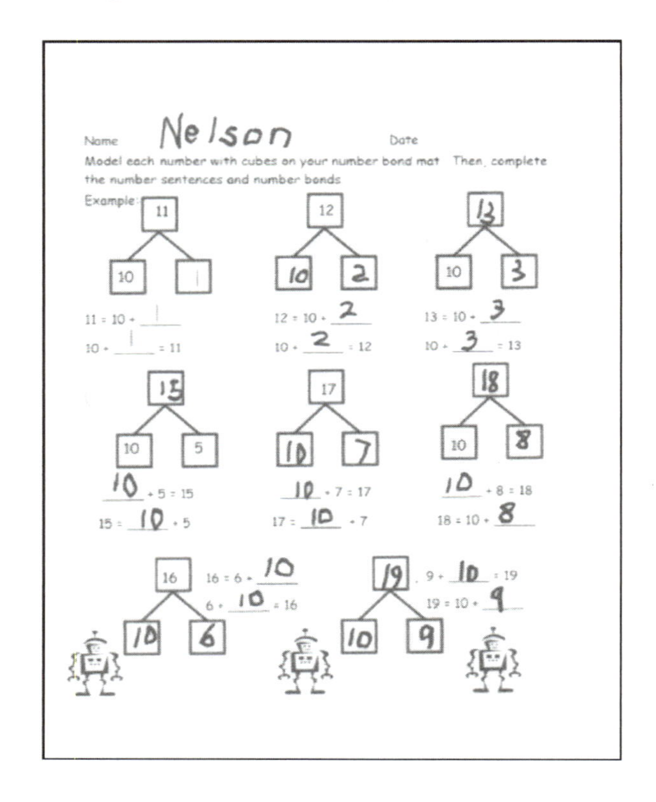

Lesson Objective: Represent teen number decompositions as 10 ones and some ones, and find a hidden part.

The Student Debrief is intended to invite reflection and active processing of the total lesson experience.

Invite students to review their solutions for the Problem Set. They should check work by comparing answers with a partner before going over answers as a class. Look for misconceptions or misunderstandings that can be addressed in the Student Debrief. Guide students in a conversation to debrief the Problem Set and process the lesson.

Any combination of the questions below may be used to lead the discussion.

- What did you get better at today?
- What do you notice from the Problem Set? (An example follows.)

T: Look at the first two number bonds. What is the same and different about these two bonds?

S: Both bonds have 10 ones. → Yeah, but they don't have the same number of extra ones. → One has 2 extra ones, and the other has 3 extra ones. → If you count all the ones together, one is twelve, and one is thirteen. → If we count the Say Ten way, one is ten 2, and one is ten 3. → If you break apart both numbers, there are 10 ones and some ones inside! → The number sentences show that we can write 12 and 13 in number sentences with 10 plus in them.

- What can you explain about the numbers 11, 12, 13, 14, 15, 16, 17, 18, 19? What do they have in common? How are they different?
- What did you learn in this lesson?

Exit Ticket (3 minutes)

After the Student Debrief, instruct students to complete the Exit Ticket. A review of their work will help with assessing students' understanding of the concepts that were presented in today's lesson and planning more effectively for future lessons. The questions may be read aloud to the students.

Name ______________________________ Date ______________

Model each number with cubes on your number bond mat. Then, complete the number sentences and number bonds.

Example:

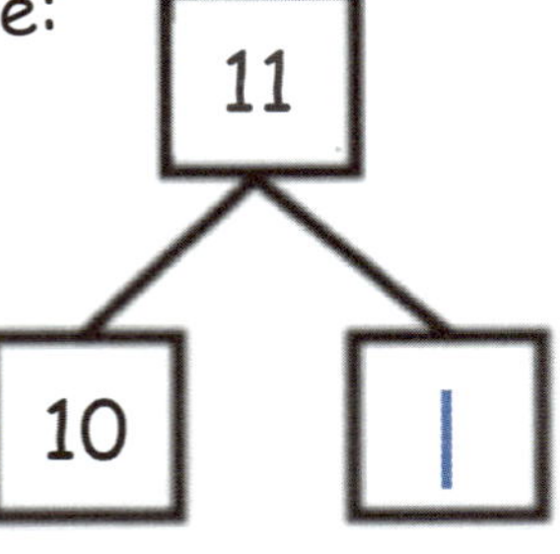

11 = 10 + __|__

10 + __|__ = 11

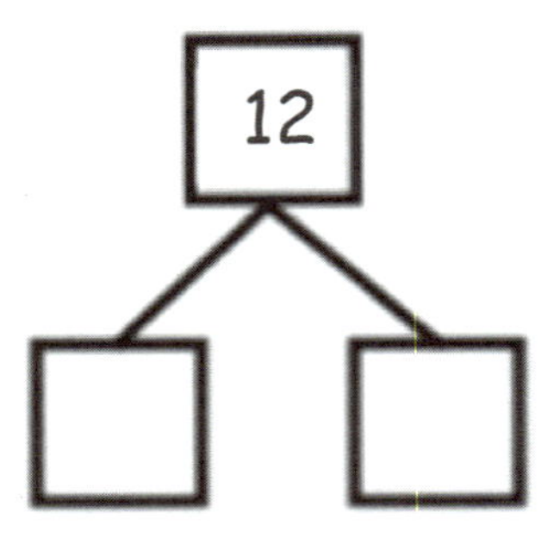

12 = 10 + ______

10 + ______ = 12

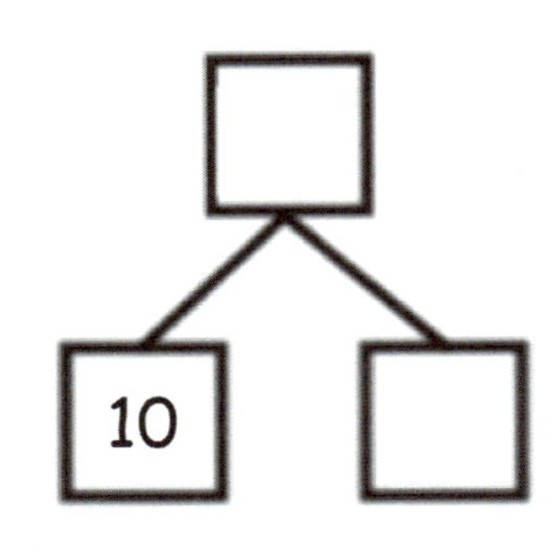

13 = 10 + ______

10 + ______ = 13

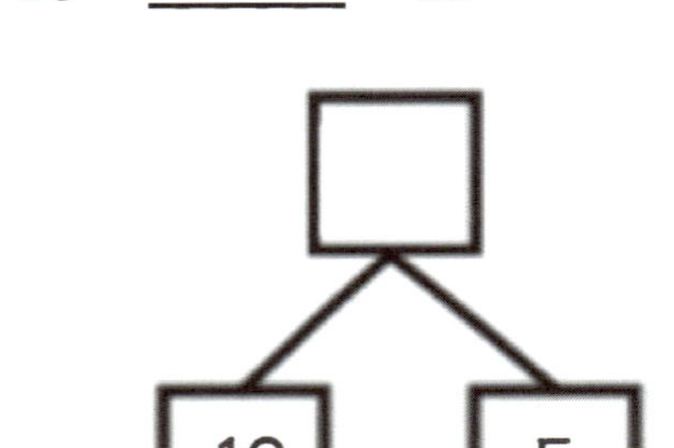

______ + 5 = 15

15 = ______ + 5

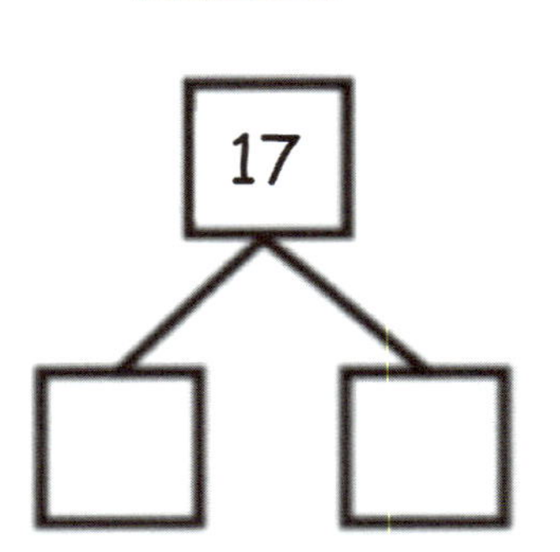

______ + 7 = 17

17 = ______ + 7

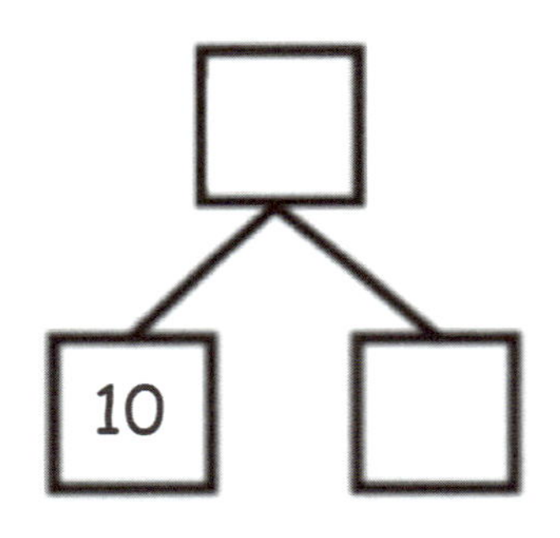

______ + 8 = 18

18 = 10 + ______

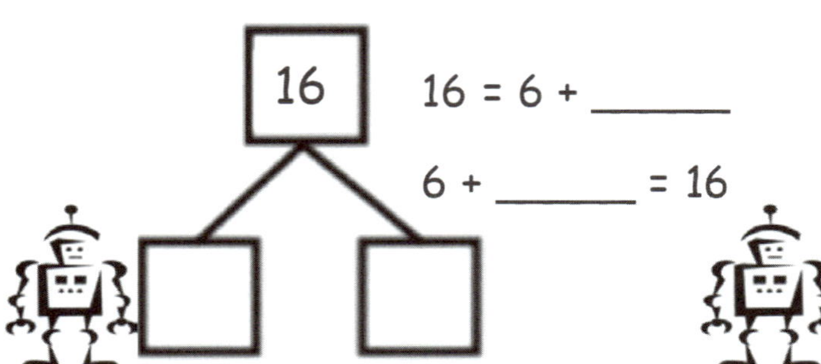

16 = 6 + ______

6 + ______ = 16

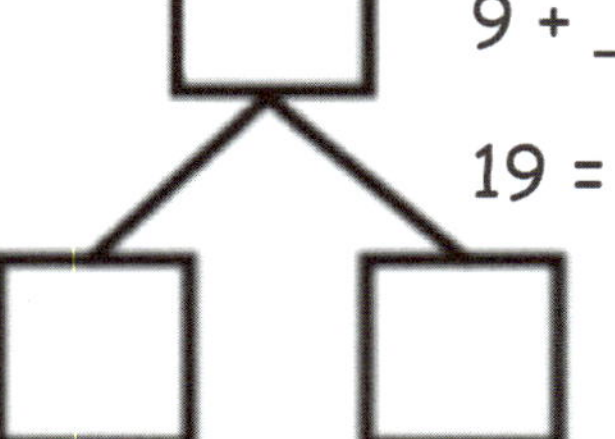

9 + ______ = 19

19 = 10 + ______

Name ______________________________ Date ______________

Complete the number sentences and number bonds. Use your materials to help you.

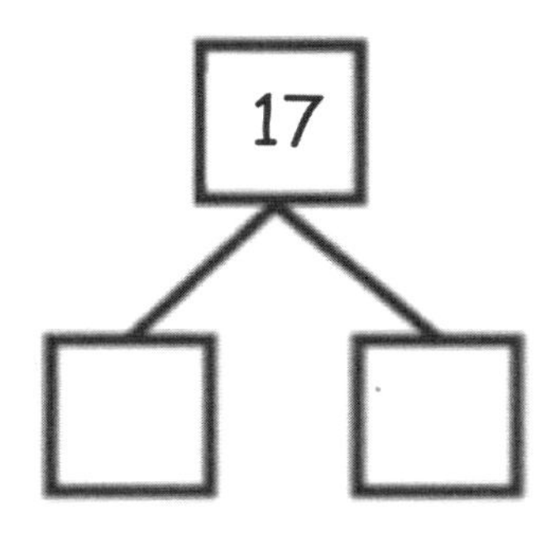

______ + 7 = 17 17 = ______ + 10

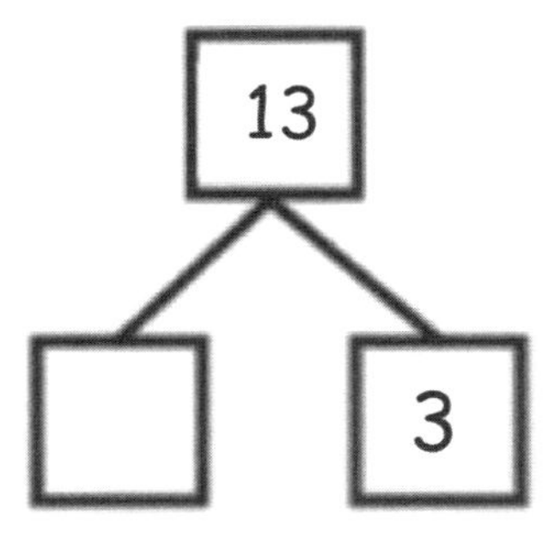

_____ + 3 = ______ 13 = ______ + 10

Name ______________________ Date ____________

Complete the number bonds and number sentences. Draw the cubes of the missing part.

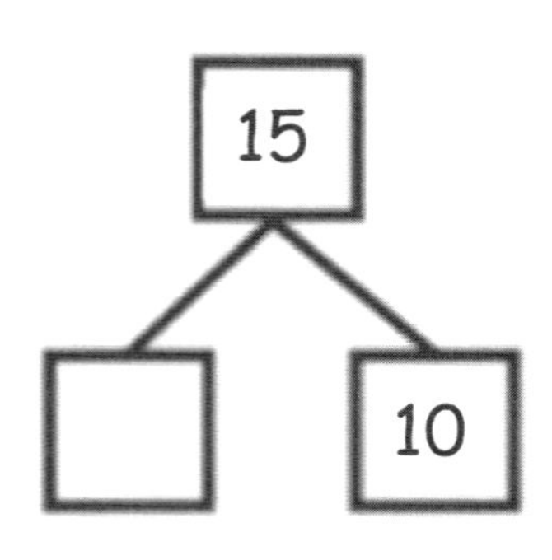

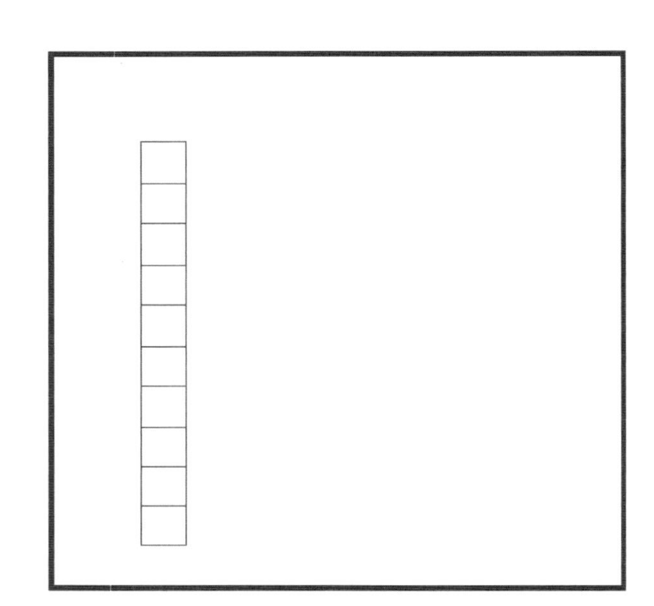

15 = ______ + 10

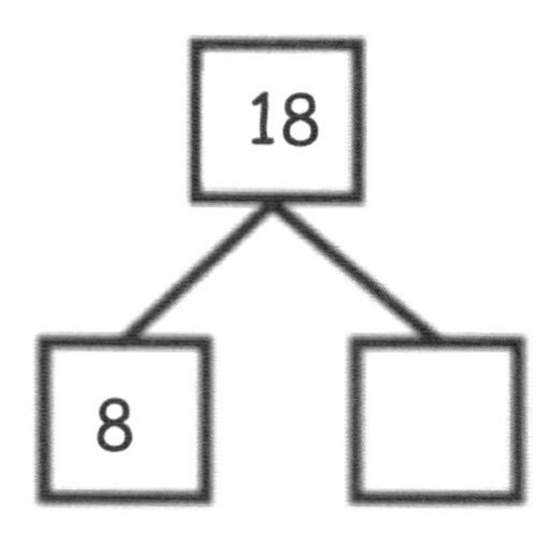

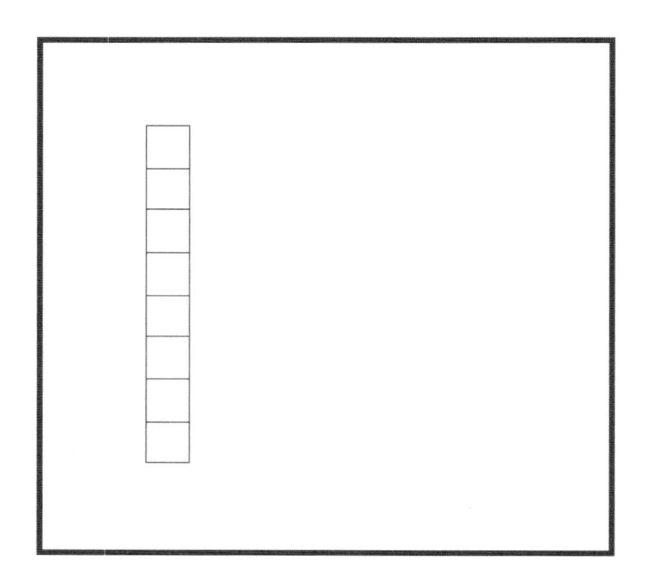

______ + 8 = 18

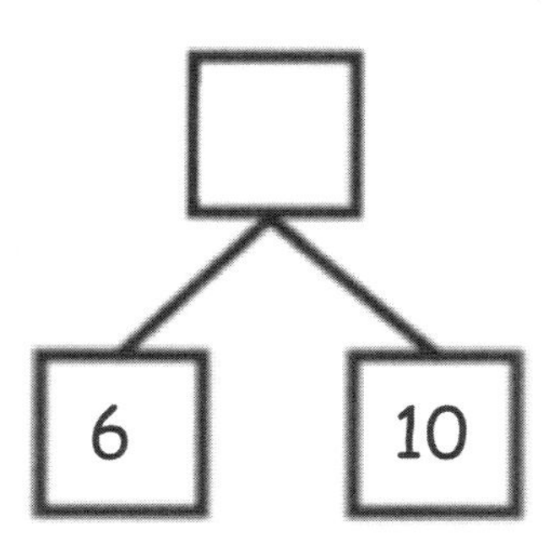

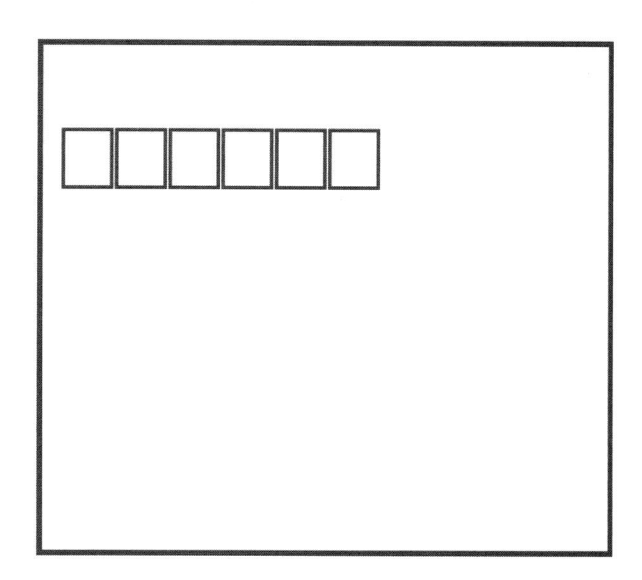

6 + ______ = 16

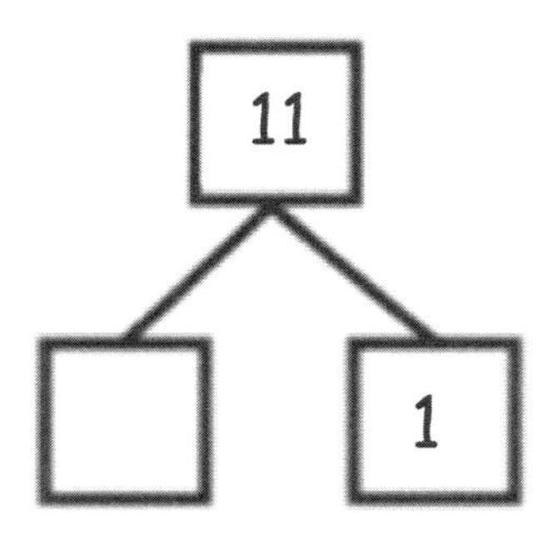

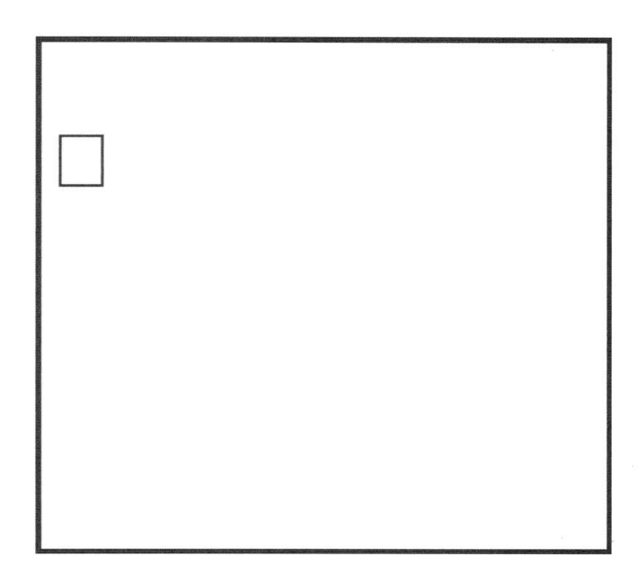

1 + ______ = 11

Lesson 22

Objective: Decompose teen numbers as 10 ones and some ones; compare *some ones* to compare the teen numbers.

Suggested Lesson Structure

Application Problem	(7 minutes)
Fluency Practice	(11 minutes)
Concept Development	(25 minutes)
Student Debrief	(7 minutes)
Total Time	**(50 minutes)**

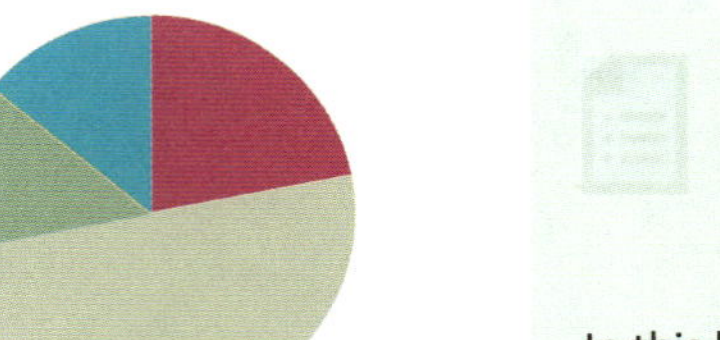

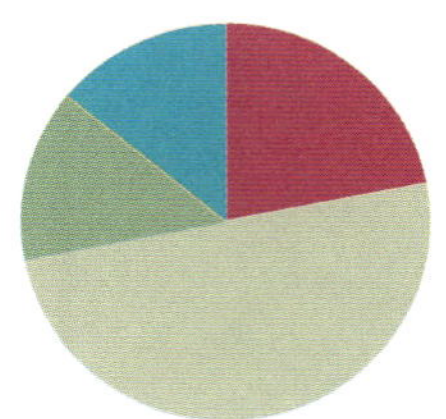

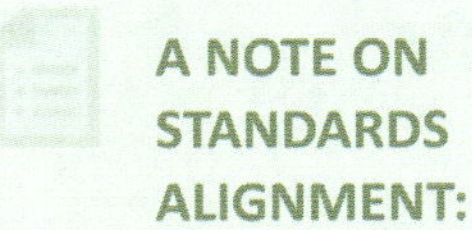

A NOTE ON STANDARDS ALIGNMENT:

In this lesson, students compare numbers 1–9 (**K.CC.6, K.CC.7**) and use their understanding of 10 ones as the structure of the teen numbers (**K.NBT.1** and MP.7) to compare teen numbers. This bridges Kindergarten content to the Grade 1 comparison of numbers (**1.NBT.3**).

Application Problem (7 minutes)

Lisa has 5 pennies in her hand and 2 in her pocket. Matt has 6 pennies in his hand and 2 in his pocket. Who has fewer pennies—Lisa or Matt? How do you know?

Note: This Application Problem reviews comparing numbers within 10, which prepares students to compare teen numbers in today's Concept Development.

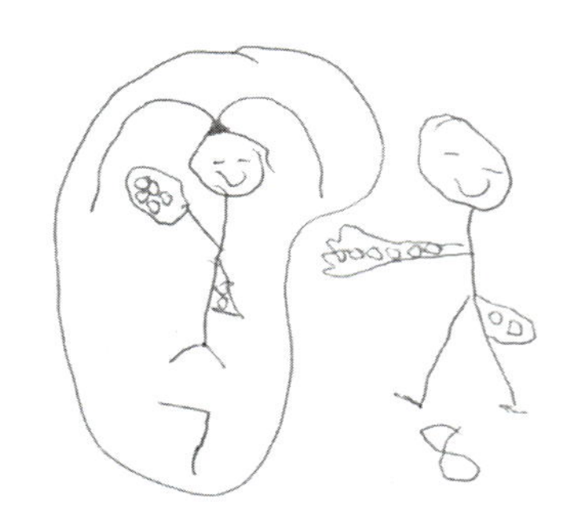

Fluency Practice (11 minutes)

- Dot Cards of Eight **K.CC.5, K.CC.2** (3 minutes)
- Count Teen Numbers **K.NBT.1** (4 minutes)
- Teen Numbers on the Rekenrek **K.NBT.1** (4 minutes)

Dot Cards of Eight (3 minutes)

Materials: (T) Dot cards of 8 (Lesson 6 Fluency Template)

Note: This fluency activity gives students an opportunity to develop increased familiarity with decompositions of eight and practice seeing part–whole relationships.

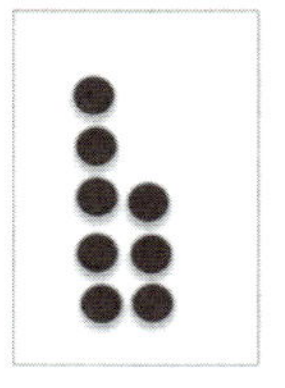

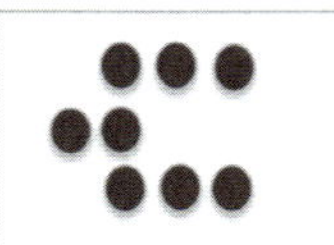

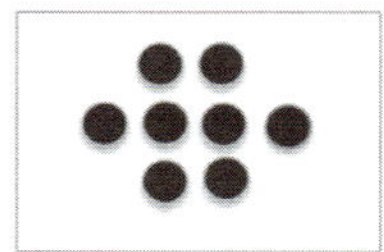

T: (Show a card with 8 dots.) How many dots do you count? Wait for the signal to tell me. Get ready (snap.)

S: 8.

T: How can you see them in two parts?
S: (Student comes up to the card.) I saw 5 here and 3 here.
T: Say the number sentence.
S: 5 and 3 makes 8.
T: Flip it.
S: 3 and 5 makes 8.
T: Who sees 8 in two different parts?
S: (Come up to the card.) I see 6 here and 2 here.
T: Say the number sentence.
S: 6 and 2 makes 8.
T: Flip it.
S: 2 and 6 makes 8.

Continue with other cards and decompositions of 8.

Count Teen Numbers (4 minutes)

Note: If alternating between counting the Say Ten way and regular way is challenging for some students, consider scaffolding this activity by doing it first with the Rekenrek.

T: Count from 11 to 20 and back to 11 the Say Ten way.
S: Ten 1, ten 2, ten 3, ten 4, ten 5, ten 6, ten 7, ten 8, ten 9, 2 tens, ten 9, ten 8, ten 7, ten 6, ten 5, ten 4, ten 3, ten 2, ten 1.
T: Count from 11 to 20 and back to 11 the regular way.
S: 11, 12, 13, 14, 15, 16, 17, 18, 19, 20, 19, 18, 17, 16, 15, 14, 13, 12, 11.
T: Now, I want you to change the way you count each time. We'll say the first number the regular way. Then, we'll say the next number the Say Ten way. Listen to my example. 11, ten 2, 13, ten 4, 15, ten 6. Now, it's your turn.
S: 11, ten 2, 13, ten 4, 15, ten 6, 17, ten 8, 19, 2 tens.
T: Count back from 20 to 11, starting with the regular way.
S: 20, ten 9, 18, ten 7, 16, ten 5, 14, ten 3, 12, ten 1.

Teen Numbers on the Rekenrek (4 minutes)

Materials: (S) Personal Rekenrek (Lesson 10)

Note: This fluency activity supports the grade-level standard of understanding teen numbers as ten ones and some more ones.

T: Show me the number 12 in two parts on your Rekenrek with one part 10 ones on your top row.
S: (Show 12 on their Rekenreks.)
T: Now, show me 12 again, but this time, with 10 ones that are all red.
T: Now, show me 12 again, but this time, with 10 ones that are all white.

Continue with other teen numbers.

Concept Development (25 minutes)

Materials: (S) 20 linking cubes, personal white board

T: Use your personal white board as a work mat. Partner A, count out 13 cubes on your mat. Partner B, count out 15 cubes on your mat.

T: Now, each of you move your cubes to show the number the Say Ten way. Partner A, tell me your number the Say Ten way.

S: (Partner A only.) Ten 3.

T: Partner B, tell me your number the Say Ten way.

S: (Partner B only.) Ten 5.

T: How can we tell which number is greater? You both have 10 ones. True?

S: Yes.

T: So, let's look at the extra ones. Which number is greater—3 ones or 5 ones?

S: 5 ones!

T: So, which number is greater—ten 3 or ten 5?

S: Ten 5.

T: Let's all say 15 is more than 13.

S: 15 is more than 13.

T: Let's say that the Say Ten way. Ten 5 is more than ten 3.

S: Ten 5 is more than ten 3.

T: Now, Partner A, show me 14 on your mat as 10 ones and some ones. Partner B, show 11 on your mat as 10 ones and some ones.

T: Do you both have 10 ones?

S: Yes.

T: So, let's compare the extra ones. Which part is smaller—4 ones or 1 one?

S: 1 one.

T: Talk to your partner about which number is smaller and which number is larger, as well as how you know.

S: (Students talk.)

T: Now, I want both Partner A and Partner B to show 17 on your mat. Show it as 10 ones and some ones.

T: Do you both have 10 ones?

S: Yes.

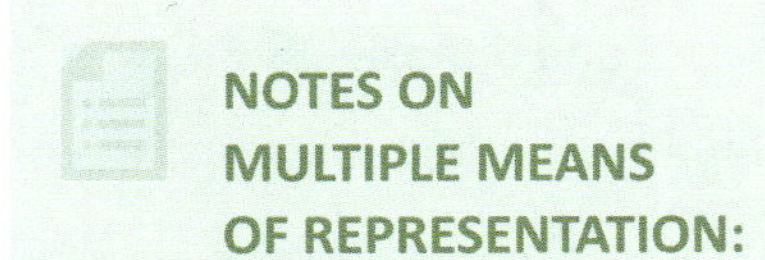

NOTES ON MULTIPLE MEANS OF REPRESENTATION:

Before beginning the lesson, introduce or review key vocabulary for English language learners so that they can keep up with the lesson. Post visuals of key terms such as *greater, smaller, less, more*, and *the same*.

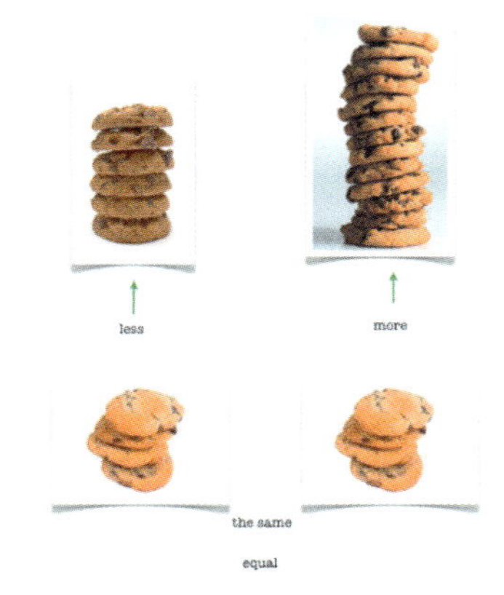

T: How many extra ones do you both have?

S: 7.

T: Is 7 more than 7?

S: No!

T: Is 10 more than 10?

S: No!

T: What should we say about 17 and 17?

S: They're the same! They're equal!

Continue in this manner but without the cubes and personal white boards. Draw two number bonds on the board. Fill one number bond in with 19 decomposed, showing 10 ones as one part. Fill the other number bond with 16 decomposed, showing 10 ones as one part.

T: (Point to 19.) What is the missing part?

S: 9.

T: (Fill in 9.)

T: (Point to 16.) What is the missing part?

S: 6.

T: (Fill in 6.)

T: Compare the extra ones. Which number is more?

S: 19.

T: We are using what we know about comparing the numbers less than 10 to compare numbers that are more than 10.

T: Talk to your partner about that.

S: I know 5 is more than 4, so I know 10 ones and 5 ones is more than 10 ones and 4 ones. → I know that 5 is less than 8, so ten 5 is less than ten 8. → I know that 6 equals 6, so ten 6 equals ten 6. → I know that 10 ones is the same, so it's like both numbers have it. So, it doesn't tell which one is larger or smaller.

Problem Set (7 minutes)

Students should do their personal best to complete the Problem Set within the allotted time.

Note: This work, like many of the lessons in this module, allows students to see the relevance of numbers to 10 as they apply to larger numbers. Students *stand* on the shared structure of the ten in two teen numbers and simply compare the ones to see which number is greater. This bridges to Grade 1 content (**1.NBT.3**).

Student Debrief (7 minutes)

Lesson Objective: Decompose teen numbers as 10 ones and some ones; compare *some ones* to compare the teen numbers.

The Student Debrief is intended to invite reflection and active processing of the total lesson experience.

Invite students to review their solutions for the Problem Set. They should check work by comparing answers with a partner before going over answers as a class. Look for misconceptions or misunderstandings that can be addressed in the Student Debrief. Guide students in a conversation to debrief the Problem Set and process the lesson.

Any combination of the questions below may be used to lead the discussion.

- What was today's lesson about?
- How do you know 11 is less than 15?
- Read each comparison from the Problem Set the Say Ten way and then the regular way. For example, "Ten 3 is more than ten 2. 13 is more than 12. Ten 1 is less than ten 4. 11 is less than 14."
- What do you think I wanted you to learn from the lesson?

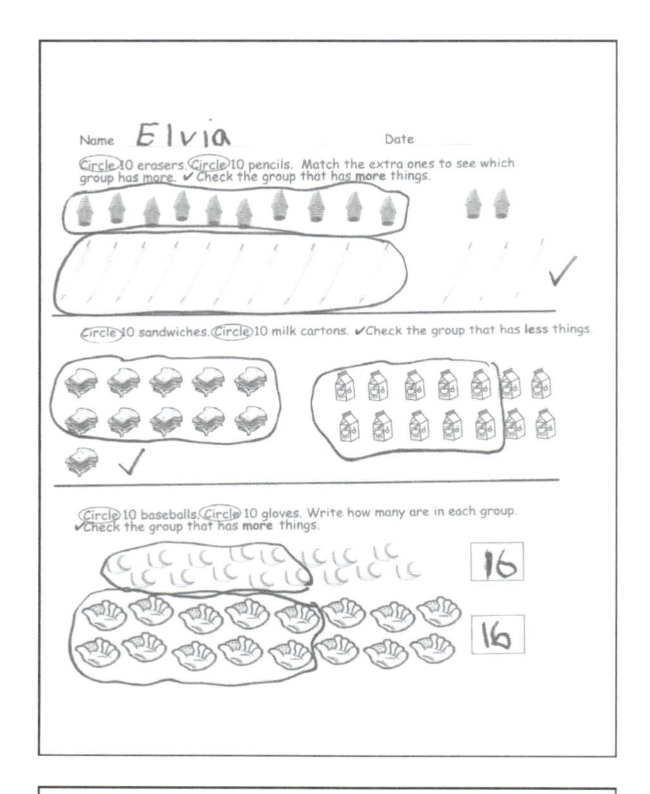

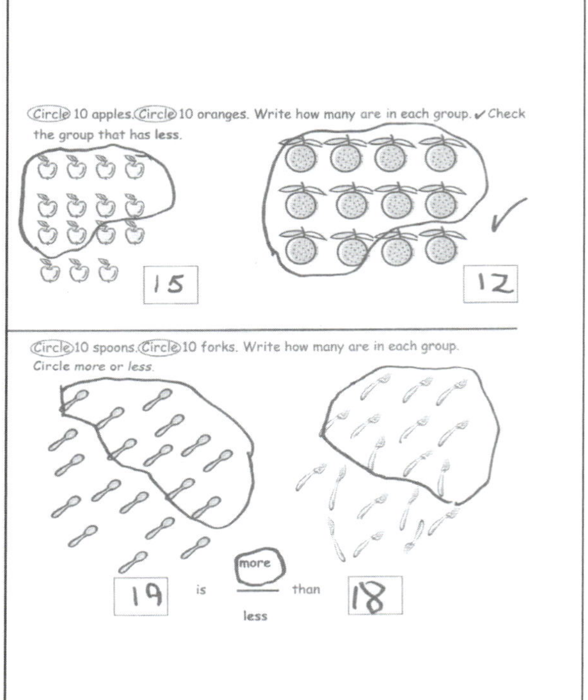

Exit Ticket (3 minutes)

After the Student Debrief, instruct students to complete the Exit Ticket. A review of their work will help with assessing students' understanding of the concepts that were presented in today's lesson and planning more effectively for future lessons. The questions may be read aloud to the students.

Name ______________________________ Date ______________

(Circle) 10 erasers. (Circle) 10 pencils. Match the extra ones to see which group has more. ✓ Check the group that has *more* things.

(Circle) 10 sandwiches. (Circle) 10 milk cartons. ✓ Check the group that has *less* things.

(Circle) 10 baseballs. (Circle) 10 gloves. Write how many are in each group. ✓ Check the group that has *more* things.

Circle 10 apples. Circle 10 oranges. Write how many are in each group.
✓ Check the group that has *less*.

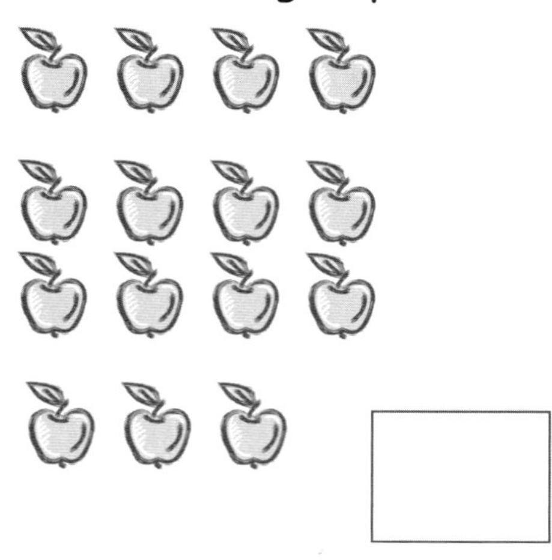

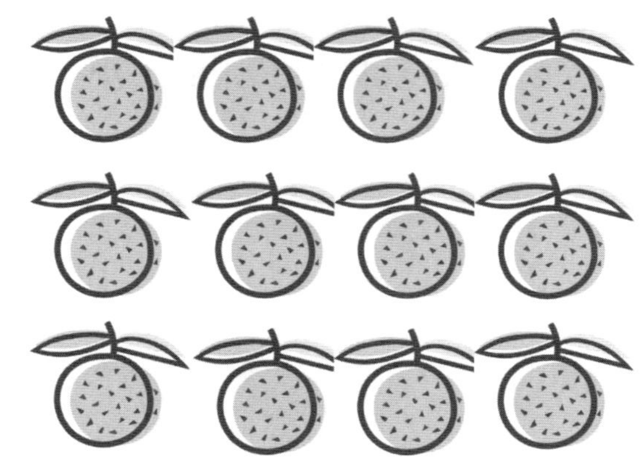

Circle 10 spoons. Circle 10 forks. Write how many are in each group.
Circle *more* or *less*.

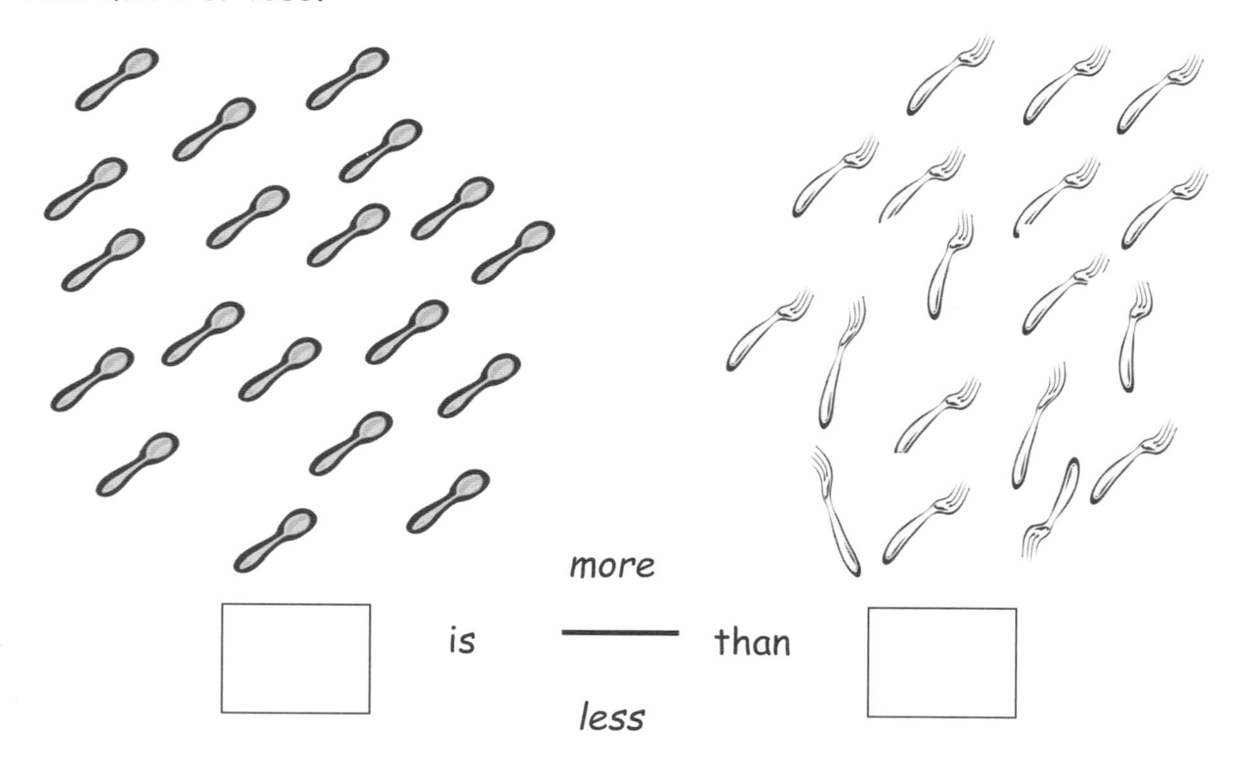

Name ________________________ Date ____________

Count and write the number.
Circle *more* or *less*.

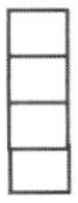

		more	
1	is	(less)	than 4

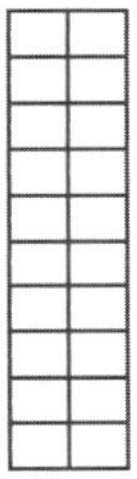

more
_____ is less than _____

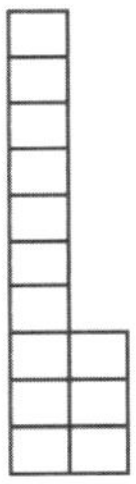

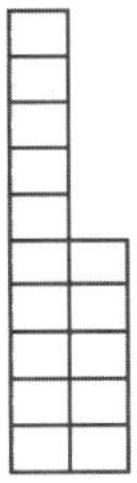

more
_____ is less than _____

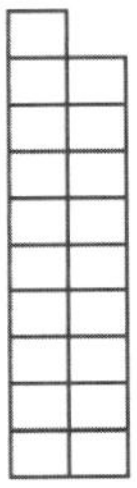

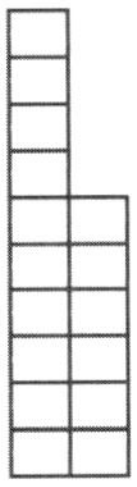

more
_____ is less than _____

Name ______________________________ Date ______________

Fill in the number bond.
Check the group with *more*.

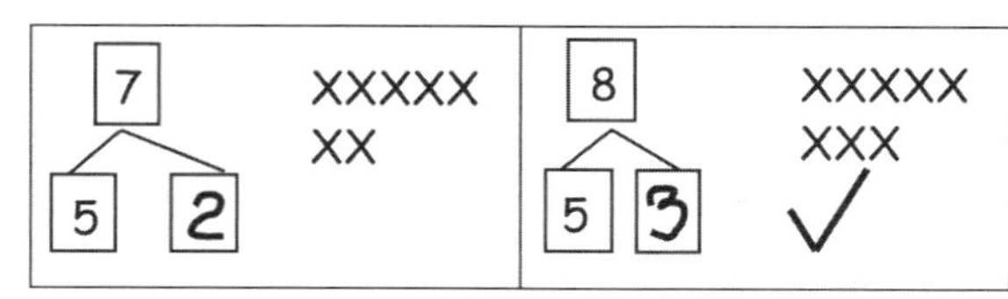

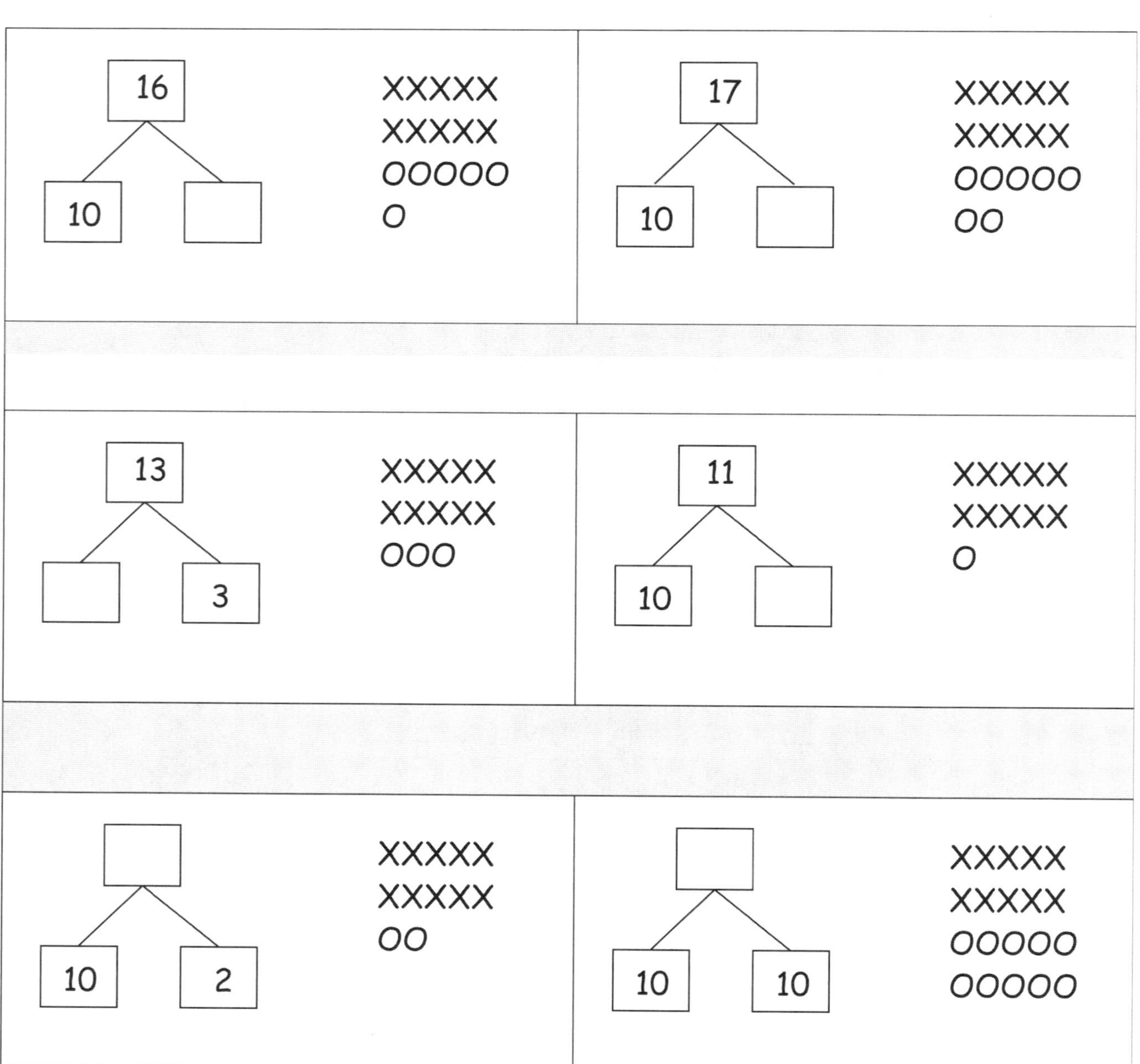

EUREKA MATH

Lesson 23

Objective: Reason about and represent situations, decomposing teen numbers into 10 ones and some ones and composing 10 ones and some ones into a teen number.

Suggested Lesson Structure

Fluency Practice	(12 minutes)
Concept Development	(30 minutes)
Student Debrief	(8 minutes)
Total Time	**(50 minutes)**

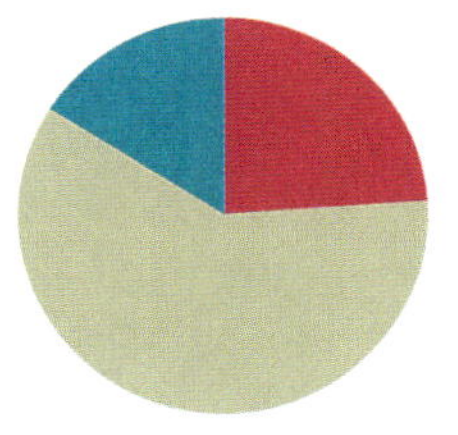

Fluency Practice (12 minutes)

- Number Bonds of Eight **K.NBT.1** (4 minutes)
- Matching Dot and Number Cards **K.NBT.1** (8 minutes)

Number Bonds of Eight (4 minutes)

Materials: (T) Dot cards of 8 (Lesson 6 Fluency Template)

Note: This fluency activity gives students an opportunity to develop increased familiarity with compositions of eight and review number bonds.

Show a dot card, and indicate 7 and 1 as parts.

T: Say the larger part. (Give students time to count).
S: 7.
T: Say the smaller part.
S: 1.
T: What is the total number of dots? (Give time to count.)
S: 8.
T: Say the number sentence.
S: 7 and 1 makes 8.
T: Flip it.
S: 1 and 7 makes 8.

Continue with cards illustrating the number bonds of 5 and 3, 4 and 4, 6 and 2, and 8 and 0.

Matching Dot and Number Cards (8 minutes)

Materials: (S) Teen numeral and dot cards (Lesson 14 Template) (per pair; pictured below)

Note: This activity connects the pictorial representations of teen numbers with the abstract numerals and reinforces teen numbers as 10 ones and some additional ones.

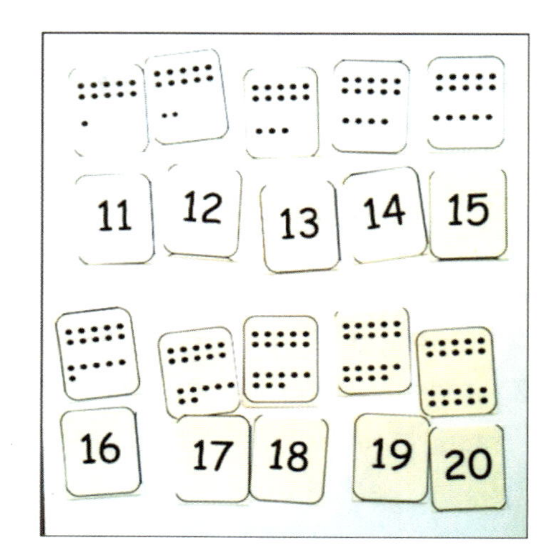

T: Put your number cards in order from smallest to greatest.

T: Match each number card to a dot card.

T: Talk to your partner. What do you notice about your dot cards and your number cards?

S: They all have ten dots. → They all have ones that show the ten. → They all have an extra dot that tells how many extra ones weren't part of the ten ones. → All the dot cards have two parts, and the numbers have two numbers. → Yeah, one of the numbers is one of the parts of the dots.

Concept Development (30 minutes)

Materials: (T) 12 pieces of red construction paper (S) Picture and word problem (Template), number bond (Lesson 7 Template) within a personal white board

Note: The following problems are solved using counting and students' knowledge of decomposing and composing teen numbers. Although addition sentences are included in students' solutions, in this instance, they are another record of the decomposition or the composition of the total that the student counted to find rather than a means of solving the problem. Note that the problems do not ask "How many?" or "How many in all?"

T: (Show 12 pieces of red construction paper in one line, perhaps taped to the board.) Count with me.

S: 1, 2, 3, 4, 5, 6, 7, 8, 9, 10, 11, 12.

T: Draw and show the 12 papers as 10 ones and some ones.

S: Should we draw a number bond?

T: You can draw a picture and make a number bond.

S: Can we write a number sentence?

T: That is another good way to show what twelve is made of.

T: (After working.) Share with your partner how you showed 10 red papers and some more papers.

T: What parts did you break 12 into?

S: 10 and 2.

T: What number sentence did you use to show that?

S: 12 = 10 + 2.

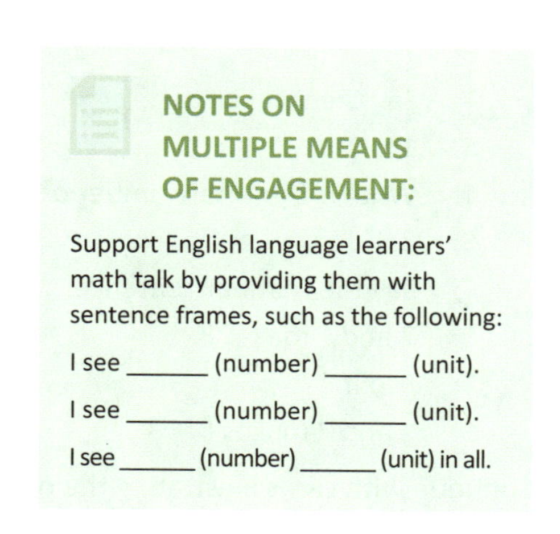

NOTES ON MULTIPLE MEANS OF ENGAGEMENT:

Support English language learners' math talk by providing them with sentence frames, such as the following:

I see ______ (number) ______ (unit).

I see ______ (number) ______ (unit).

I see ______ (number) ______ (unit) in all.

T: Yes, 12 is 10 ones and 2 ones.

T: (Referring back to the red papers on the board.) What can I do with my papers to show that we made two parts?

S: You could put space between the 10 ones and 2 ones to see the parts more easily.

T: Okay, I'll do that. Yes, now we can see that 12 is 10 and 2.

T: Let's do a different problem at a farm. (Pass out the picture and word problem.) Look at the picture with your partner. Talk about what you see.

S: (After talking.) There are 10 geese and 3 pigs.

T: It's easy to see the parts, so let's put them together to find how many animals there are.

T: Work with your partner to show ways to put those parts together.

T: (Pause while students work.) What are some of the ways you put the two parts together?

S: We showed a number bond. → We showed an addition sentence. → We got our Hide Zero cards.

T: When you put the parts together, what was the total of your bond or number sentence?

S: 13.

T: What number sentence did you use to show that?

S: 10 + 3 = 13.

T: Yes, that is how I think of it when I'm putting parts together. When I'm taking them apart, I say it this way: 13 = 10 + 3. Talk to your partner about why you think I do that.

S: One way starts with the big number. → When we put the ducks and the pigs together, we started with the parts. → Like with the animals, we could see the parts really easily, so we wrote those first: 10 + 3 = 13. → It's different with the red papers. → Yeah, like with the red papers, we counted all the papers first and then separated them: 12 = 10 + 2. → Yeah, it was hard to see the groups because the papers were all the same color and in one line.

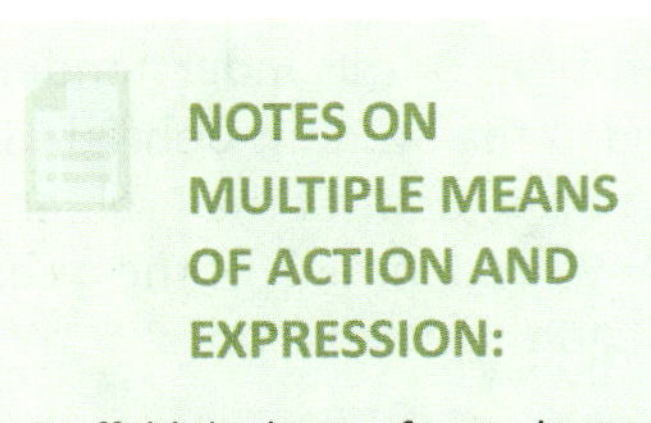

NOTES ON MULTIPLE MEANS OF ACTION AND EXPRESSION:

Scaffold the lesson for students working below grade level by asking them to model with red and blue cubes before expecting them to model with a drawing.

T: I showed the papers like this: 12 = 10 + 2. And I showed the animals like this: 10 + 3 = 13. Talk to your partner about why.

S: The papers were all one color, so we had to find the 10 hiding. So, we started with counting all the papers. → Yeah, with the animals, I counted the pigs first and then the geese.

T: So, with the animals, you thought about the parts first, and with the papers, you thought about the total first?

S: Yeah.

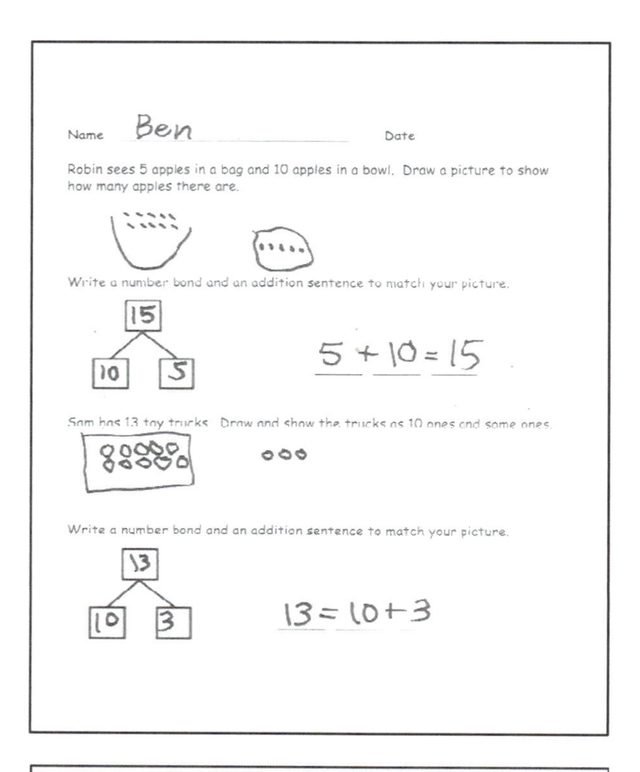

Problem Set (7 minutes)

Students should do their personal best to complete the Problem Set within the allotted time.

Read the stories to them as they work. Because this Problem Set requires reading, it is a good idea to group students by performance level so the situations can be told to students in their small groups.

Student Debrief (8 minutes)

Lesson Objective: Reason about and represent situations, decomposing teen numbers into 10 ones and some ones and composing 10 ones and some ones into a teen number.

The Student Debrief is intended to invite reflection and active processing of the total lesson experience.

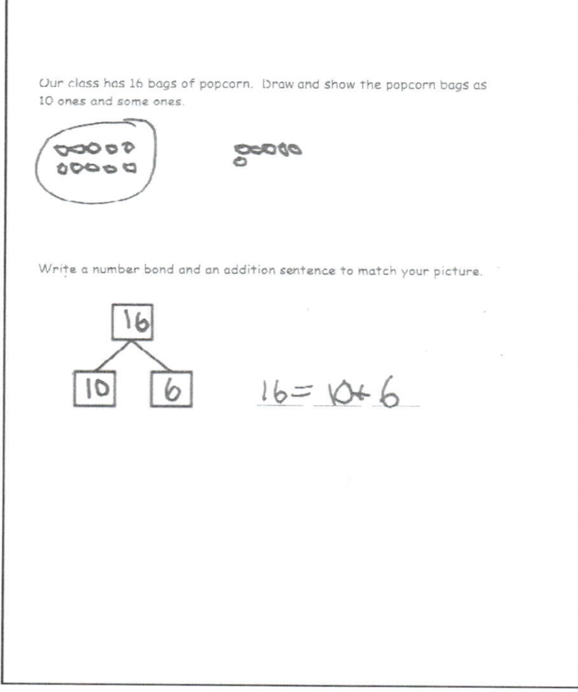

Invite students to review their solutions for the Problem Set. They should check work by comparing answers with a partner before going over answers as a class. Look for misconceptions or misunderstandings that can be addressed in the Student Debrief. Guide students in a conversation to debrief the Problem Set and process the lesson. Any combination of the questions below may be used to lead the discussion.

- Did you start by drawing the parts first or the total first in the story of Robin's apples? The toy trucks? The popcorn bags?
- Explain how your drawing relates to the number bond you wrote.
- Explain how the number sentence relates to the number bond and situation.
- Show how you wrote the number sentence for each situation and whether you started the sentence with the parts or the total. How did you choose your number sentence? Share your thinking.

MP.2

EUREKA MATH

Exit Ticket (3 minutes)

After the Student Debrief, instruct students to complete the Exit Ticket. A review of their work will help with assessing students' understanding of the concepts that were presented in today's lesson and planning more effectively for future lessons. The questions may be read aloud to the students.

Name ______________________________ Date ____________

Robin sees 5 apples in a bag and 10 apples in a bowl. Draw a picture to show how many apples there are.

Write a number bond and an addition sentence to match your picture.

Sam has 13 toy trucks. Draw and show the trucks as 10 ones and some ones.

Write a number bond and an addition sentence to match your picture.

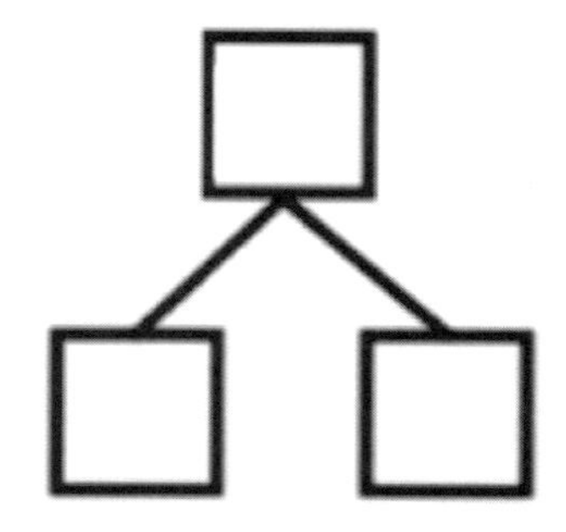

_____ _____ _____

Our class has 16 bags of popcorn. Draw and show the popcorn bags as 10 ones and some ones.

Write a number bond and an addition sentence to match your picture.

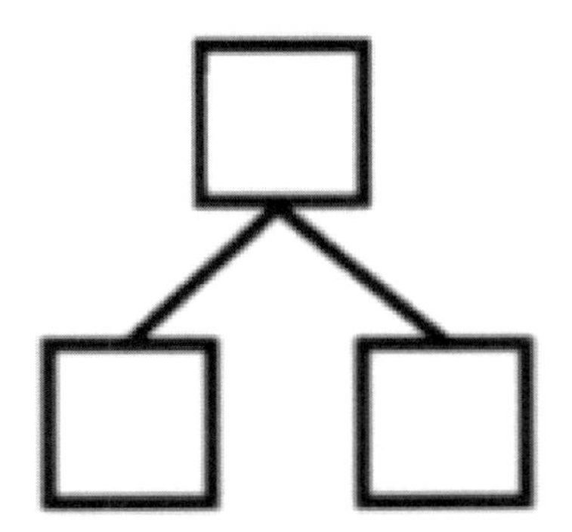

_____ _____ _____

Name ______________________________ Date ______________

There are 12 balls. Draw and show the balls as 10 ones and some ones.

Write a number bond to match your picture.

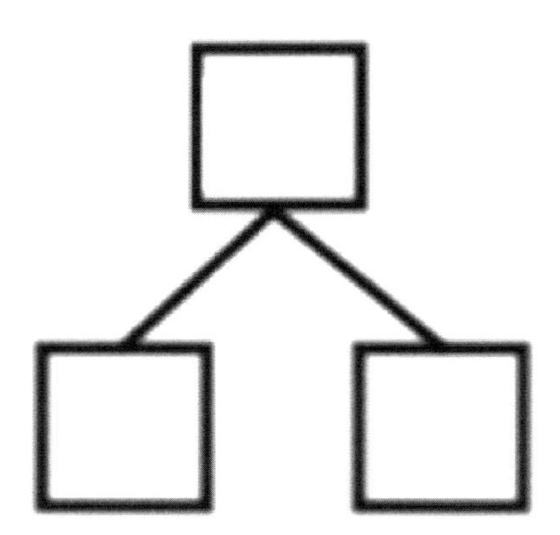

Write an addition sentence to match your number bond.

__________ __________ __________

Name ________________________________ Date ______________

Bob bought 7 sprinkle donuts and 10 chocolate donuts. Draw and show all of Bob's donuts.

Write an addition sentence to match your drawing.

__________ __________ __________

Fill in the number bond to match your sentence.

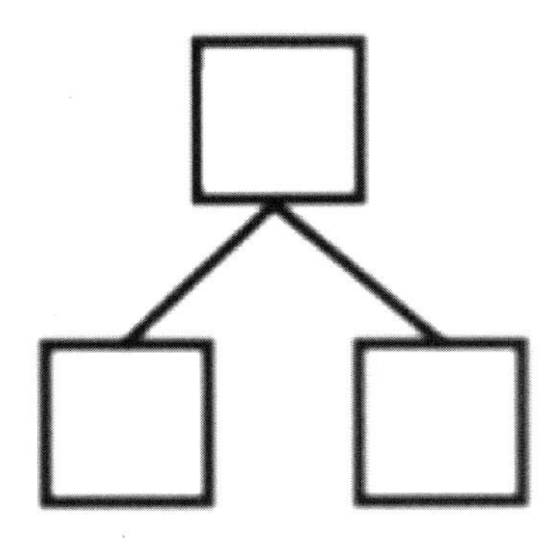

Fran has 17 baseball cards. Show Fran's baseball cards as 10 ones and some ones.

Write an addition sentence and a number bond that tell about the baseball cards.

_______ _______ _______

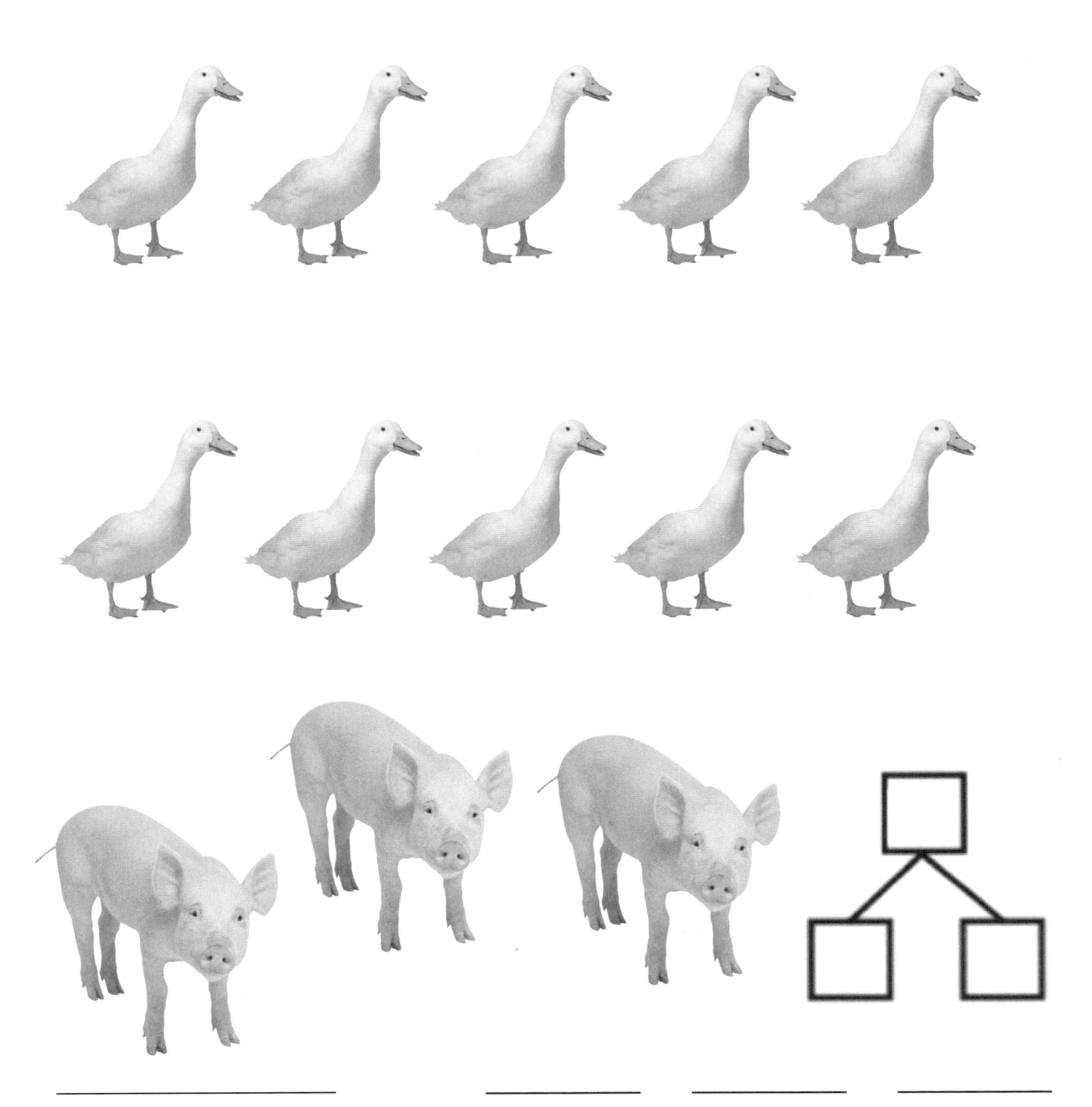

picture and word problem

Lesson 24

Objective: Culminating Task—Represent teen number decompositions in various ways.

Suggested Lesson Structure

■	Fluency Practice	(10 minutes)
■	Concept Development	(35 minutes)
■	Student Debrief	(5 minutes)
	Total Time	**(50 minutes)**

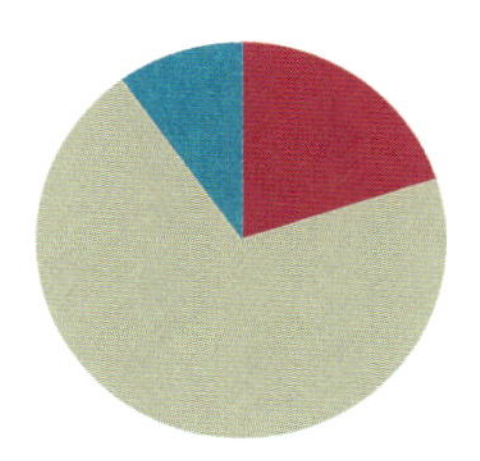

Fluency Practice (10 minutes)

- Help the Frog Catch the Fly **K.CC.4C** (4 minutes)
- Number Bond Hopping Card Game **K.CC.1** (6 minutes)

Help the Frog Catch the Fly (4 minutes)

Materials: (T) Pictorial growth chart 10–20 (Fluency Template 1), frog puppet (popsicle stick with a frog picture)

Note: This activity playfully reinforces the understanding that each successive number refers to a quantity that is 1 larger.

T: (Project the pictorial growth chart 10–20 on the board (Fluency Template 1).) Hold a frog puppet (popsicle stick with a frog picture) on the 10. What number is Froggy on now?

S: 10.

T: Can you help Froggy get the fly?

S: Yes.

T: Tell Froggy what number is 1 more.

S: 1 more is 11.

T: (Make the frog puppet jump to the next stair.) It's working! What number is he on now?

S: 11.

T: Tell him 1 more.

S: 11. 1 more is 12.

T: (Frog jumps.)

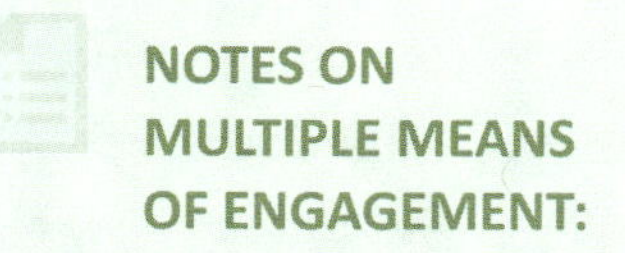

NOTES ON MULTIPLE MEANS OF ENGAGEMENT:

Provide English language learners with individual pictorial growth charts on the sheets they have in front of them. Students can use their fingers to trace the frog's path to the next step. They can then see and say the number the frog is on.

Continue to 20. (Variations: 1 more/2 more. Froggy wants to go back home—1 less/2 less. Consider adding a kinesthetic component—students stand taller or crouch down to reflect the number.)

Number Bond Hopping Card Game (6 minutes)

Materials: (S) Teen numeral and dot cards (Fluency Template 2), Rabbit and Froggy's matching race (Fluency Template 3)

Note: Introducing this game during fluency prepares students to play it again at home.

Complete directions for this game are located in the Homework component of this lesson.

Concept Development (35 minutes)

Materials: (S) 10 bags each with a different teen number of objects inside. Materials for each station: 2-hand cards (Lesson 16 Template), Hide Zero cards: 1 Hide Zero 10 card (Lesson 6 Template 2) and 5-group cards 1–9 (Lesson 1 Fluency Template 2), personal Rekenrek (Lesson 10), 20 centimeter cubes, 20 sticks, 20 beans, 1 small paper plate, 20 linking cubes, blank paper, number bond (Lesson 7 Template)

- Introduction (3 minutes)
- Creating exhibits (32 minutes)

Setup

Unbeknownst to students, Station 1 has a bag with 11 cubes, Station 2 has a bag with 12 cubes, and Station 10 has a bag with up to 20 cubes. Pair students who are generally performing at the same level. Put students performing at higher levels at the stations with 16–20 cubes. Direct each pair of students to one of the stations.

T: Open your mystery bag, and count how many objects are inside. Show this number in different ways using the materials available to you at your station.

T: You are going to create an exhibit showing your number in as many ways as you can.

T: The ways you must show your number include:

- A number bond
- Hide Zero cards
- Rekenrek
- Addition sentence
- Linking cubes

T: Once you have finished the *have to's*, show the number in other ways, too. You will have 20 minutes. At your table are different materials to help you. You do not have to use them all. You may also use paper and pencil.

This culminating lesson is a part of the Kindergarten assessment system. While circulating, use a recording sheet to document what each student does. What representations does the student choose? What skills are obvious? Which materials does he avoid? Which does he gravitate toward immediately? What words is the student using when talking about his teen number? Take a picture of students' work for their portfolio.

T: (After 20 minutes.) Now, we are going to take a tour to see your friends' creations. When I give the signal, move to the next station.

T: Think about what you are seeing at each station. Point to the different ways your friends have shown their number. Talk about each one. What makes it special? (Students spend a little less than one minute at each station.)

Student Debrief (5 minutes)

Lesson Objective: Culminating Task—Represent teen number decompositions in various ways.

The Student Debrief is intended to invite reflection and active processing of the total lesson experience. The following is a suggested list of questions to invite reflection and active processing of the total lesson experience. Use those that resonate when considering what will best support students' ability to articulate the focus of the lesson.

- What are some different ways you saw the teen number represented?

S: Number bonds. → Piles of 10 ones and some more ones. → In circles. → In arrays. → In rows. → With hand cards. → With linking cubes in one long line. → In towers. → In addition sentences. → In story problems. → In pictures. → With Hide Zero cards. → On our Rekenrek.

- Which of these different ways do you feel helps you understand your teen numbers the most? Why?
- How is a number bond different from and the same as an addition sentence?
- How is a pile of 10 sticks and some more sticks different and the same as the number shown with Hide Zero cards?
- What did you notice as you went around the room? How did the exhibits vary?

Close the experience by letting students know that, by understanding their teen numbers, they understand all the numbers better as they move on to Grade 1.

Exit Ticket (3 minutes)

Rather than having an Exit Ticket for this lesson, the teacher is encouraged to record observations as students work with their partners as described in the closing of the Concept Development section of this lesson.

Rabbit and Froggy's Matching Race

Directions: Play Rabbit and Froggy's Matching Race with a friend, relative, or parent to help your animal reach its food first! The first animal to reach the food wins.

- Put your Teen number and Dot cards face down in rows with Teen numbers in one row and Dot cards in another row.
- Flip to find 2 cards that match.
  Place cards back in the same place if they don't match.
  Continue until you find a match. → [dot card] 13
- Write a number bond to match. → 10, 3, 13 → Hop 1 space if you get it right!
- Write a number sentence. → 13 = 10 + 3 → Hop 1 space again if you get it right!

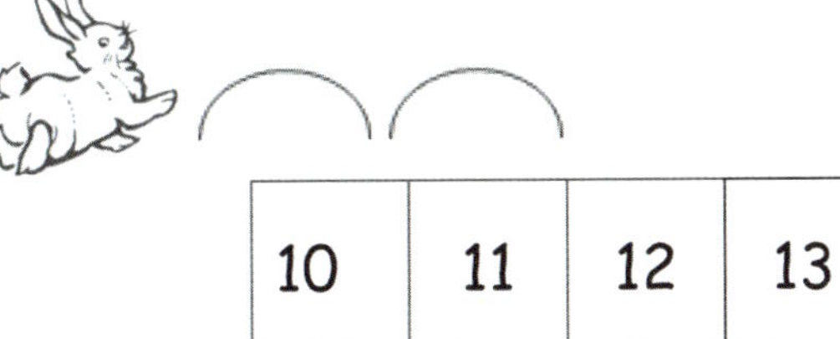

10	11	12	13	14	15	16	17	18	19	20

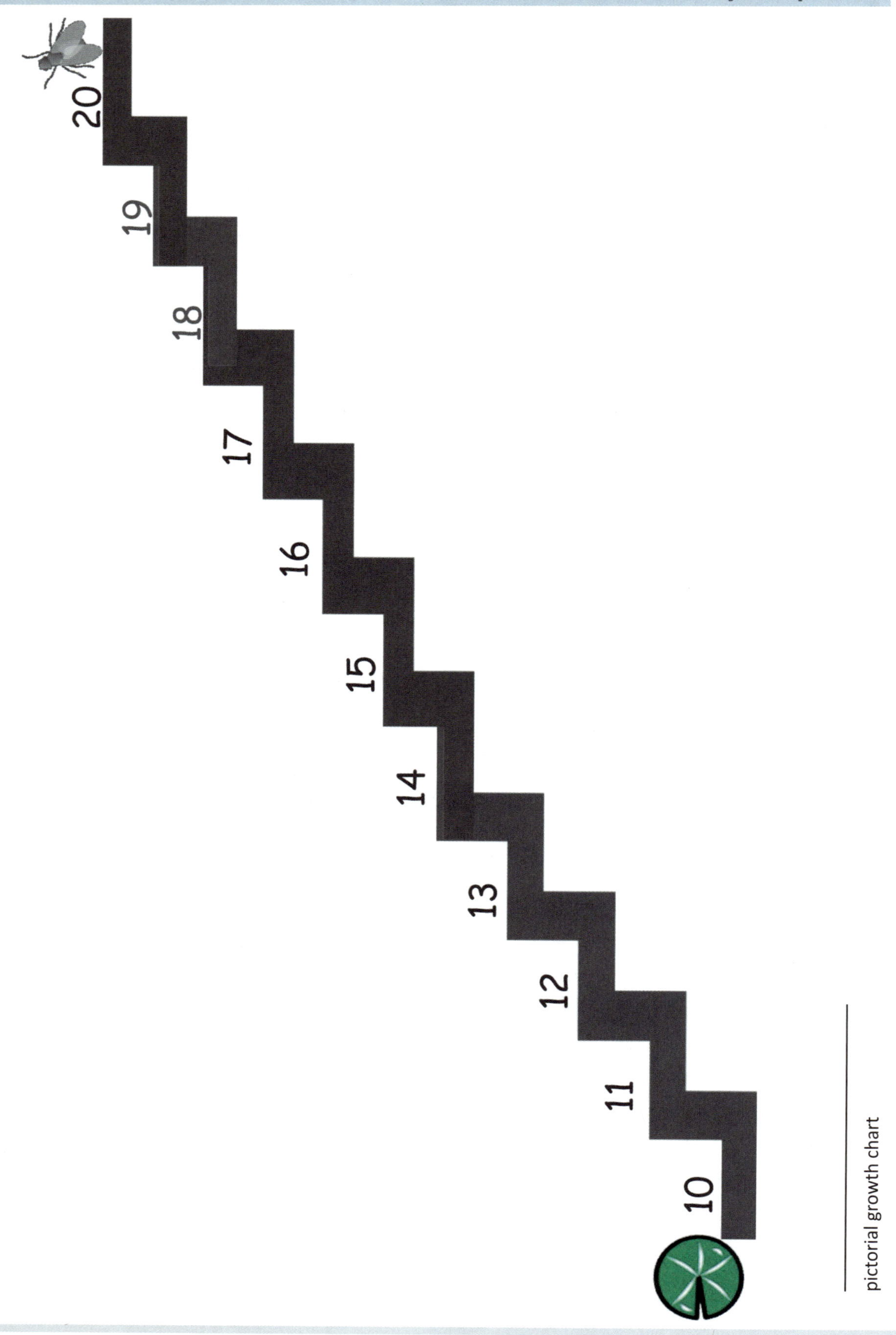
20
19
18
17
16
15
14
13
12
11
10
pictorial growth chart

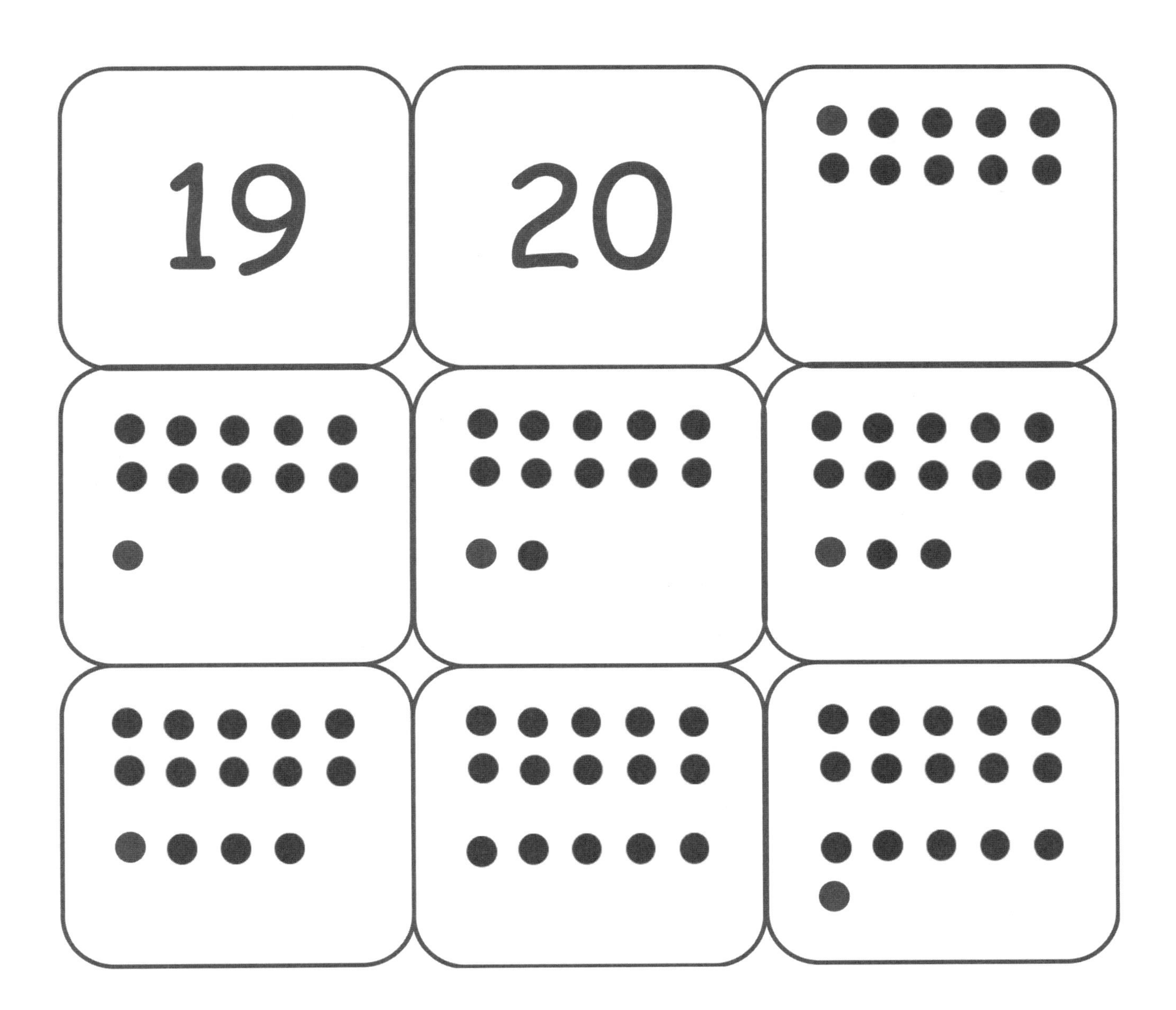

teen number and dot cards

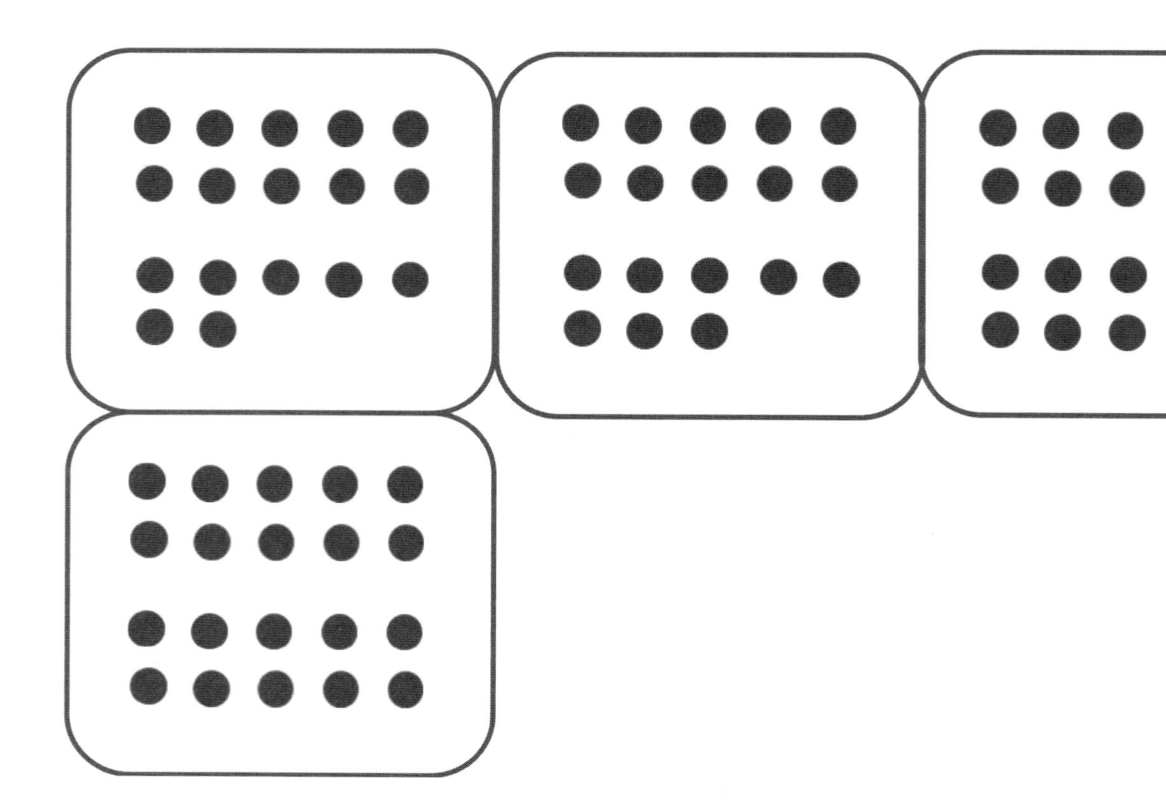

teen number and dot cards

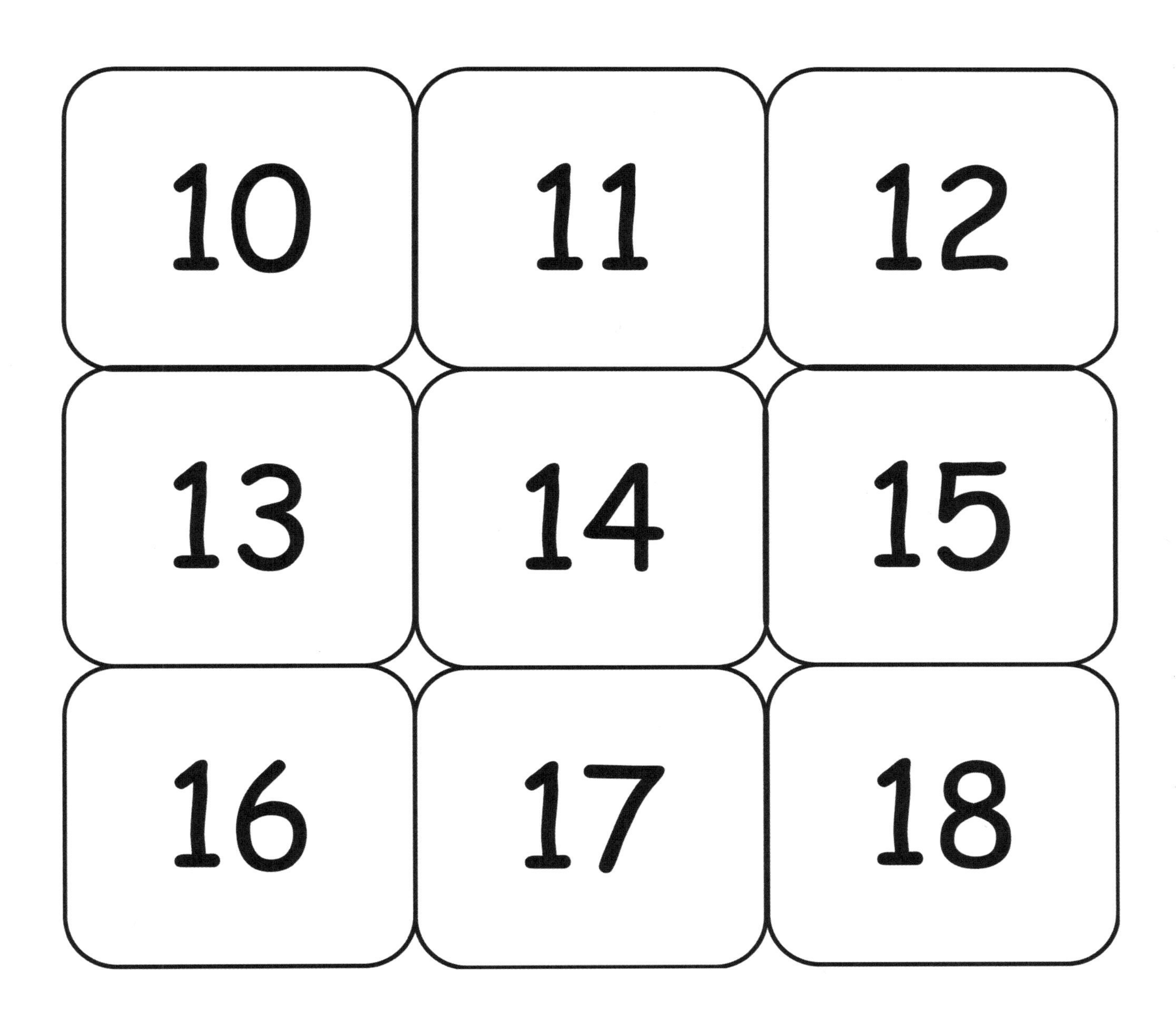

teen number and dot cards

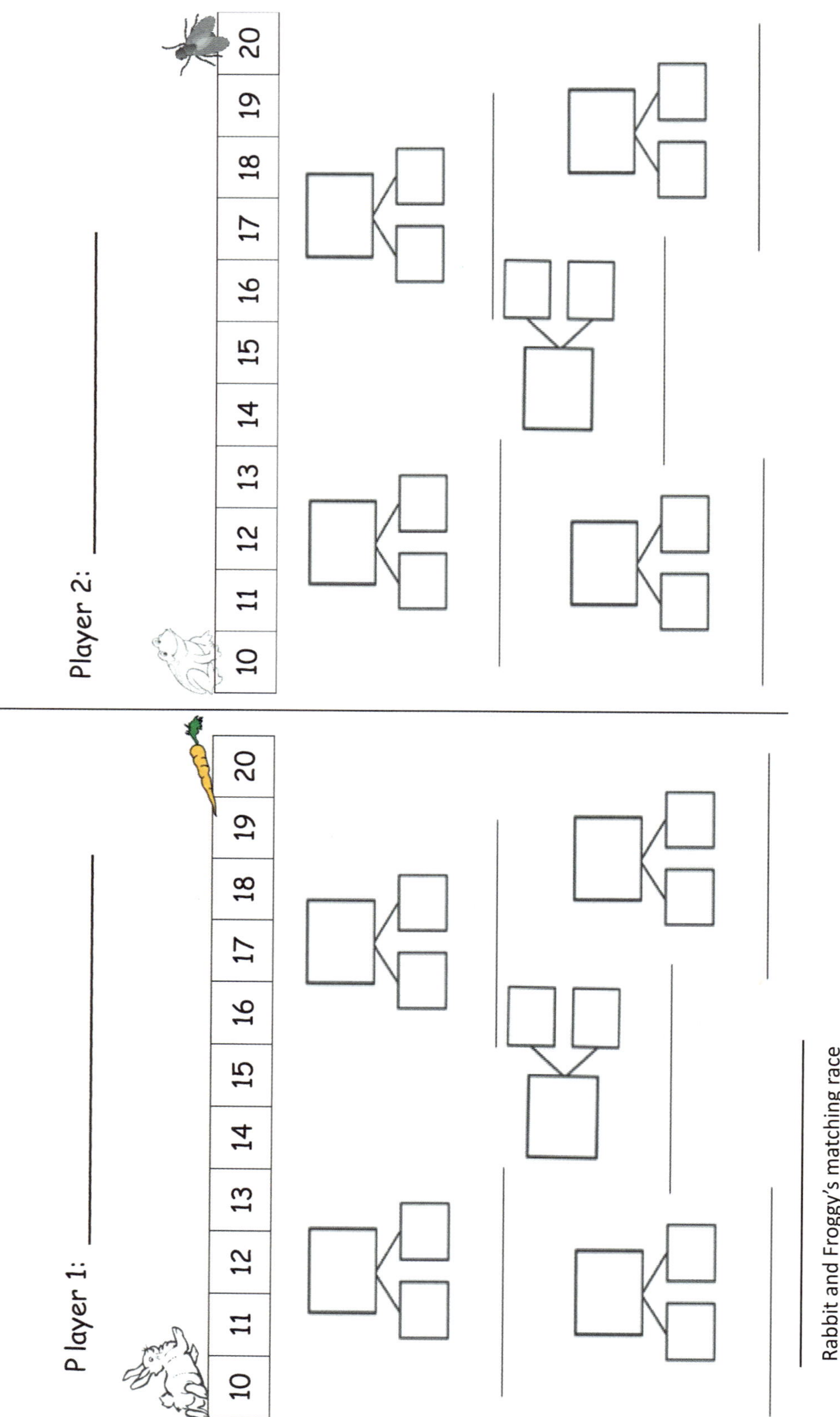
Player 1:
10
11
12
13
14
15
16
17
18
19
20
Player 2:
10
11
12
13
14
15
16
17
18
19
20
Rabbit and Froggy's matching race

Student Name ______________________________

	Date 1	Date 2	Date 3
Topic D			
Topic E			

Topic D: Extend the Say Ten and Regular Count Sequence to 100

Rubric Score ____________ Time Elapsed ____________

Materials: (T) 10 small 10-frame cards (Lesson 15 Template 2)

Set out the 10-frame cards.

T: (Set out two 10-frame cards.) How many dots are on these cards? Touch and count each dot the regular way. Whisper while you count so I can hear you.

T: Please count the dots from 11 to 20 the Say Ten way.

T: Please count by 10s to 100 the Say Ten way.

T: Please count by 10s to 100 the regular way.

T: Start at 28. Count up by 1s and stop at 32 the regular way. (If the student is unable to do this, try 8 through 12, then 18 through 22.)

What did the student do?	What did the student say?

Topic E: Represent and Apply Compositions and Decompositions of Teen Numbers

Rubric Score ____________ Time Elapsed ____________

Materials: (S) 17 centimeter cubes, number bond (Lesson 7 Template) within a personal white board, eraser

T: (Set out 17 cubes.) How many cubes are there? (Note the arrangement in which the student counts. If the student does *not* arrange cubes into a straight line or array, do so for the student.)

T: Separate 10 cubes into a group.

T: Write 17 as a number bond on your personal white board using 10 ones as one of the parts. (Be sure to have students write the numerals.)

T: (Write 17 = ______ + ______.) Make an addition sentence to match your number bond.

T: How are your number bond and your addition sentence the same?

What did the student do?	What did the student say?

End-of-Module Assessment Task Standards Addressed	Topics D–E

Know number names and the count sequence.

K.CC.1 Count to 100 by ones and by tens.

K.CC.2 Count forward beginning from a given number within the known sequence (instead of having to begin at 1).

K.CC.3 Write numbers from 0 to 20. Represent a number of objects with a written numeral 0–20 (with 0 representing a count of no objects).

Count to tell the number of objects.

K.CC.4 Understand the relationship between numbers and quantities; connect counting to cardinality.

b. Understand that the last number name said tells the number of objects counted. The number of objects is the same regardless of their arrangement or the order in which they were counted.

c. Understand that each successive number name refers to a quantity that is one larger.

K.CC.5 Count to answer "how many?" questions about as many as 20 things arranged in a line, a rectangular array, or a circle, or as many as 10 things in a scattered configuration; given a number from 1–20, count out that many objects.

Work with numbers 11–19 to gain foundations for place value.

K.NBT.1 Compose and decompose numbers from 11 to 19 into ten ones and some further ones, e.g., by using objects or drawings, and record each composition or decomposition by a drawing or equation (e.g., 18 = 10 + 8); understand that these numbers are composed of ten ones and one, two, three, four, five, six, seven, eight, or nine ones.

Evaluating Student Learning Outcomes

A Progression Toward Mastery is provided to describe steps that illuminate the gradually increasing understandings that students develop on their way to proficiency. In this chart, this progress is presented from left (Step 1) to right (Step 4). The learning goal for students is to achieve Step 4 mastery. These steps are meant to help teachers and students identify and celebrate what the students CAN do now and what they need to work on next.

A Progression Toward Mastery				
Assessment Task Item and Standards Assessed	**STEP 1** **Little evidence of reasoning without a correct answer.** **(1 point)**	**STEP 2** **Evidence of some reasoning without a correct answer.** **(2 points)**	**STEP 3** **Evidence of some reasoning with a correct answer or evidence of solid reasoning with an incorrect answer.** **(3 points)**	**STEP 4** **Evidence of solid reasoning with a correct answer.** **(4 points)**
Topic D **K.CC.1** **K.CC.2**	Student shows little evidence of counting ability or understanding.	Student shows evidence of beginning to understand counting by 10s and 1s but skips or repeats numbers often, resulting in an inaccurate count.	Student is unable to perform one of the tasks.	Student correctly: ▪ Counts up by 10s using the Say Ten and regular ways. ▪ Counts the dots from 11 to 20 the Say Ten way. ▪ Counts from 28 to 32 the regular way. ▪ Counts a number between 11 and 20 the regular way.
Topic E **K.CC.5** **K.NBT.1**	Student shows little evidence of understanding organized counting, teen numbers, number bonds, or addition sentences.	Student shows a beginning understanding of counting into an array or line, representing teen numbers as number bonds or addition sentences, but answers inaccurately.	Student correctly counts 17 cubes into an array or line and writes the number bond for 17 but cannot write an accurate equation. OR The student writes an accurate equation for 17 but cannot write the number bond or count into an array or line.	Student correctly: ▪ Counts 17 cubes into an array or line. ▪ Separates 10 cubes and correctly writes 17 as the whole and 10 and 7 as the parts of 17. ▪ Writes an accurate addition sentence and reasonably connects both representations.

Class Record Sheet of Rubric Scores: Module 5			
Student Names:	**Topic D:** Extend the Say Ten and Regular Count Sequence to 100	**Topic E:** Represent and Apply Compositions and Decompositions of Teen Numbers	**Next Steps:**

Answer Key

Eureka Math®
Grade K
Module 5

Special thanks go to the Gordon A. Cain Center and to the Department of Mathematics at Louisiana State University for their support in the development of *Eureka Math*.

Mathematics Curriculum

Answer Key

GRADE K • MODULE 5

Numbers 10–20 and Counting to 100

Lesson 1

Problem Set

Footballs

Apples

Sticks

Soccer balls

4

Exit Ticket

Lightning bolts

1

Homework

Triangles – not circled

Circles – circled

Hearts – circled

Diamonds – circled

Triangles – not circled

Faces – 10 circled, 2 not circled

Suns – circled

Squares – circled

Lightning bolts – not circled

Cylinders – circled

Half-moons – not circled

Triangles – not circled

Circles – circled

Rectangular prisms – circled

Trapezoids – not circled

Hearts – not circled

Ovals – circled

Triangles – not circled

Hearts – circled

Triangles – circled

Lesson 2

Problem Set

10 rubber ducks checked; 3

10 gifts checked; 10, 6

10 pigs checked; 2

10 glasses checked; 10, 1

10 small circles and 2 small circles drawn in any configuration

10 ones and 4 ones drawn using lines, circles, or objects of choice

Exit Ticket

10 ones and 3 ones

10 ones and 5 ones

10 ones and 10 ones

10 ones and 2 ones

Homework

2 more circles drawn

3 more half-moons drawn

1 more heart drawn

5 more faces drawn

Lesson 3

Problem Set

10 ice cream cones circled; 5

10 peppers circled; 10, 3

10 apples circled; 10, 2

2 groups of 10 pushpins circled; 10, 10

13 things drawn, 10 circled

18 things drawn, 10 circled

Exit Ticket

10 hearts circled; 3

16 objects drawn, 10 circled

Homework

10 ducks circled; 2

10 diamonds circled; 10, 8

10 faces circled; 4

10 watering pails circled; 10, 1

Lesson 4

Problem Set

10 circles drawn; 3 circles drawn

10 circles drawn; 7 circles drawn

10 circles drawn; 2 circles drawn

10 circles drawn; 9 circles drawn

Exit Ticket

1

10, 5

5

10, 7

Homework

Pictures matched to numbers

Fluency Template 2

10 triangles circled

10 circles circled

10 hearts circled

10 diamonds circled

10 triangles circled

10 faces circled

10 suns circled

10 squares circled

10 lightning bolts circled

10 cylinders circled

10 half-moons circled

10 triangles circled

10 circles circled

10 rectangular prisms circled

10 trapezoids circled

10 hearts circled

10 ovals circled

10 triangles circled

10 hearts circled

10 triangles circled

Lesson 5

Problem Set

10 umbrellas circled; 3 checked; ten three

10 kittens circled; 4 checked; ten four

2 groups of 10 pineapples circled; two ten

10 bananas circled; 7 checked; ten seven

10 hot dogs circled; 1 checked; ten one

Exit Ticket

3; 5

10, 7; 10, 8; 10, 9

Homework

10, 1; 10, 3

10, 4; 10, 6

10, 5; 10, 7

10, 0; 10, 2

10, 8; 10, 10

Fluency Template 2

10 triangles circled

10 circles circled

10 hearts circled

10 diamonds circled

10 triangles circled

10 faces circled

10 suns circled

10 squares circled

10 lightning bolts circled

10 cylinders circled

10 half-moons circled

10 triangles circled

10 circles circled

10 rectangular prisms circled

10 trapezoids circled

10 hearts circled

10 ovals circled

10 triangles circled

10 hearts circled

10 triangles circled

Lesson 6

Problem Set

10 dots, 5 dots; 15

10 dots, 8 dots; 18

10 dots, 6 dots; 16

Exit Ticket

10 dots and 4 dots drawn; 14

14 objects drawn and 10 circled

Homework

10 dots, 2 dots; 12

10 dots, 7 dots; 17

10 dots, 9 dots; 19

10 dots, 4 dots; 14

Lesson 7

Problem Set

0

11, 10, 1

12, 10, 2

10, 3

14, 10, 4

15, 10, 5

10, 6

17, 10, 7

18, 10, 8

19, 10, 9

10, 10

10 smiley faces circled; 13; 10, 3

Exit Ticket

1

10, 4

17; 10, 7

Homework

8

7

10, 6

10, 5

14; 10, 4

13; 10, 3

10, 2

10, 1

10; 10, 0

Lesson 8

Problem Set

10 dots drawn the 5-group way and 1 more

10 dots drawn the 5-group way and 8 more

10 dots drawn the 5-group way and 5 more

10 dots drawn the 5-group way and 4 more

10 dots drawn the 5-group way and 2 more

10 dots drawn the 5-group way and 7 more

20 dots drawn the 5-group way

10 dots drawn the 5-group way and 3 more

Exit Ticket

10 dots drawn the 5-group way and 6 more

10 cubes and 2 cubes colored in

Homework

10 dots drawn the 5-group way and 5 more

10 dots drawn the 5-group way and 3 more

10 dots drawn the 5-group way and 7 more

10 dots drawn the 5-group way and 1 more

10 dots drawn the 5-group way and 2 more

10 dots drawn the 5-group way and 6 more

20 dots drawn the 5-group way

10 dots drawn the 5-group way and 4 more

Lesson 9

Problem Set

10 circles drawn; 2 circles drawn

10 circles drawn; 7 circles drawn

10 circles drawn; 6 circles drawn

10 circles drawn; 3 circles drawn

2 sets of ten drawn and circled

10 and 1 drawn; 1 set of 10 circled

Answers will vary.

Exit Ticket

10 circles drawn; 5 drawn

10 circles drawn; 9 drawn

18 circles drawn; 10 circled

14 circles drawn; 10 circled

Homework

16 objects drawn; 10 circled

20 objects drawn; 2 sets of 10 circled

19 objects drawn; 10 circled

14 objects drawn; 10 circled

12 objects drawn; 10 circled

Lesson 10

Problem Set

Left-hand fingernails colored red

Right-hand fingernails colored black

Corresponding beads below colored to match the hands

Numbers 1 to 10 written beneath the beads

Exit Ticket

Two rows of 5 red beads and 5 yellow beads drawn

20

Hands with fingernails drawn; 10

Homework

Number bond showing that 10 and 3 make 13; fingernails and beads colored to match

Number bond showing that 10 and 4 make 14; fingernails and beads colored to match

Number bond showing that 10 and 1 make 11; fingernails and beads colored to match

Number bond showing that 10 and 2 make 12; fingernails and beads colored to match

Number bond showing that 10 and 6 make 16; fingernails and beads colored to match

Number bond showing that 10 and 7 make 17; fingernails and beads colored to match

Lesson 11

Problem Set

12 written; squares colored to equal 1 more than 11

13 written; squares colored to equal 1 more than 12

Squares colored to equal 1 more than 13

15 written

16 written; squares colored to equal 1 more than 15

17 written; squares colored to equal 1 more than 16

18 written; squares colored to equal 1 more than 17

Squares colored to equal 1 more than 18

20 written

Exit Ticket

Lines drawn to correct numbers; tower completed

Homework

11 written

10 o's and 2 x's drawn to make 12

13 written

10 o's and 4 x's drawn to make 14

15 written

10 o's and 6 x's drawn to make 16

10 o's and 7 x's drawn to make 17

10 o's and 8 x's drawn to make 18

19 written

10 o's and 10 x's drawn to make 20

Lesson 12

Problem Set

Squares colored to equal 1 less than 20

18 written; squares colored to equal 1 less than 19

17 written; squares colored to equal 1 less than 18

16 written; squares colored to equal 1 less than 17

Squares colored to equal 1 less than 15

13 written; squares colored to equal 1 less than 14

12 written; squares colored to equal 1 less than 13

10 written

Exit Ticket

10

13, 11, 10

11, 10, 9

Homework

19 written

10 o's and 8 x's drawn to make 18

17 written

10 o's and 6 x's drawn to make 16

15 written

10 o's and 4 x's drawn to make 14

10 o's and 3 x's drawn to make 13

10 o's and 2 x's drawn to make 12

11 written

10 o's drawn to make 10

Lesson 13

Problem Set

12, 14, 16, 18, 20

11, 15, 16, 19, 20

16

16

15 circles drawn in rows

12 squares drawn in rows

Exit Ticket

12

Second set of blocks circled

Answers may vary.

Homework

Dots drawn to show 10 and 5

Dots drawn to show 10 and 7

Dots drawn to show 10 and 2

Dots drawn to show 10 and 9

Lesson 14

Problem Set

14

12

15

18

3 more circles drawn

4 more triangles drawn

Answers will vary.

Exit Ticket

12

4 more dots drawn

Homework

12

10

9 more dots drawn

10 dots and 8 dots drawn; 18

Answers will vary.

Lesson 15

Problem Set

30, 40, 60, 70, 80, 90, 100

80, 70, 50, 40, 30, 10

2, 4, 5, 6, 7 tens, 8 tens

Exit Ticket

20, 30, 40, 50, 30, 20, 10

Down then up: 9, 8, 6, 5, 3, 2

Homework

90, 80, 70, 60, 50, 30, 20

10 tens, 90, 8, 7, 60, 50, 5, 30, 3, 20, 2, 10, 1

Lesson 16

Problem Set

21, 23, 25, 27, 28, 29

41, 42, 43, 45, 47

93, 94, 95, 96, 97, 98

65, 66, 67; 65, 64, 63

Exit Ticket

50, 51, 54, 55, 56, 57

32, 33, 34, 33, 31

Homework

72, 73, 74, 76, 77, 78

10, 11, 13, 15, 17, 18, and 19

85, 86, 87, 88, 89; 88, 87, 86, 85, 84

31, 32, 33, 34, 35; 34, 33, 32, 31, 30

97, 98, 99, 98, 96

Lesson 17

Problem Set

15, 21, 28, 40

18, 19, 20, 22

27, 29, 30, 31, 32

33, 34, 35, 37, 38, 39, 40

9, 20, 30

Exit Ticket

29 crossed out; 26 written

43 crossed out; 34 written

Second 29 crossed out; 30 written

44 crossed out; 40 written

Homework

3 more dots drawn to make 23

20 more dots drawn to make 27

10 more dots drawn to make 34

8 more stars drawn to make 38

10 more raindrops drawn to make 40

Lesson 18

Problem Set

Last dot in each row colored green

First dot in each row outlined with blue square

Fifth dot in each row outlined with red triangle

Exit Ticket

Last dot in each row colored purple

Homework

Circle 28 colored green; circle 34 colored red

Circle 45 colored yellow; circle 52 colored blue

Circle 83 colored purple; circle 77 colored red

Last number in each row colored black

Lesson 19

Problem Set

21, line to 11

38, 18 circled, line drawn to 18

25, 15 circled, line drawn to 15

32, 12 circled, line drawn to 12

37, 17 circled, line drawn to 17

Exit Ticket

25, 15

12 circled; 32, 12

Homework

37, 1 more drawn, 38

11, 1 more drawn, 12

43, 1 more drawn, 44

25, 1 more drawn, 26

40, 1 more drawn, 41

36, 1 more drawn, 37

Lesson 20

Problem Set

10, 5; 10, 5

17; 10, 7

18; 18

16, 6; 16

14, 10, 4; 14

12, 10, 2; 10, 2

10, 1; 11, 10, 1

Exit Ticket

10 and 2 circled

1 and 10 circled

4 and 10 circled

10 and 8 circled

10 and 0 circled

10 and 10 circled

Homework

5 stars drawn; 15 = 10 + 5; 10 + 5 = 15

10 stars drawn; 17 = 10 + 7; 10 + 7 = 17

9 stars drawn; 19 = 10 + 9; 10 + 9 = 19

10 stars and 4 stars drawn; 14 = 10 + 4; 10 + 4 = 14

10 stars drawn; 20 = 10 + 10; 10 + 10 = 20

Lesson 21

Problem Set

10, 2; 2; 2

13, 3; 3; 3

15; 10; 10

10, 7; 10; 10

18, 8; 10; 8

10, 6; 10; 10

19, 10, 9; 10, 9

Exit Ticket

10, 7; 10; 7

10; 10, 13; 3

Homework

5; 5; 5 cubes drawn

10; 10; 10 cubes drawn

16; 10; 10 cubes drawn

10; 10; 10 cubes drawn

Lesson 22

Problem Set

10 erasers; 10 pencils circled; pencils checked

10 sandwiches; 10 cartons of milk circled; sandwiches checked

10 baseballs; 16; 10 gloves circled; 16

10 apples circled, 15; 10 oranges circled, 12; oranges checked

10 spoons circled, 19; 10 forks circled; 18; *more* circled

Exit Ticket

12, less, 20

13, less, 15

19, more, 16

Homework

6; 7; second group checked

10; 1; first group checked

12; 20; second group checked

Lesson 23

Problem Set

Drawing of 5 apples in a bag and 10 apples in a bowl

15, 10, 5; 5 + 10 = 15

Drawing of 13 toy trucks

13, 10, 3; 13 = 10 + 3

Drawing of 16 bags of popcorn

16, 10, 6; 16 = 10 + 6

Exit Ticket

12 balls drawn

12, 10, 2; 10 + 2 = 12

Homework

17 donuts drawn; 17 = 10 + 7; 17, 10, 7

17 baseball cards drawn; 10 + 7 = 17; 17, 10, 7

Lesson 24

Culminating Activities

Answers will vary.